Springer Texts in Business and Economics

For further volumes:
http://www.springer.com/series/10099

Harald Gleissner • J. Christian Femerling

Logistics

Basics – Exercises – Case Studies

Springer

Harald Gleissner
Berlin School of Economics and Law
Berlin
Germany

J. Christian Femerling
Investa Holding GmbH
Eschborn
Germany

Translation from the German language edition: Logistik by Harald Gleißner and J. Christian Femerling Copyright © 2012 Springer Gabler Springer Gabler is a part of Springer Science + Business Media All Rights Reserved

ISSN 2192-4333 ISSN 2192-4341 (electronic)
ISBN 978-3-319-01768-6 ISBN 978-3-319-01769-3 (eBook)
DOI 10.1007/978-3-319-01769-3
Springer Cham Heidelberg New York Dordrecht London

Library of Congress Control Number: 2013956625

Printed on acid-free paper

Springer is part of Springer Science+Business Media (www.springer.com)

Preface

Globalization and Logistics are closely connected, just as Globalization and Academic Lecturing are. This poses new challenges in studying Logistics and makes textbooks in English an important tool in college and university curricula that deal with transportation, logistics, and supply chain management.

The idea of offering a textbook in two languages was conceived after the success of the first volume of *Logistik: Grundlagen – Übungen – Fallbeispiele*, published in 2008. At that time, it was an early textbook primarily intended for students enrolled in bachelor degree programs in Logistics, and was extremely well received among academics and industry experts. Based on their own teaching experience, the authors later decided to prepare an English edition of the book to support international academic environments.

In 2012, Gleissner and Femerling published a second edited volume in German, expanded by two additional chapters, "Logistics Infrastructure" and "Logistics and Finance," which are also included in this English edition. In comparison to the German edition, the bibliography of this edition contains additional references concerning relevant US and British literature.

The goal of the book is to present basic logistics concepts and principles in a format that is useful for students as well as practitioners dealing with logistic issues for the first time.

The chapters therefore will help readers by providing:

- Learning objectives at the beginning of each chapter
- In-depth case studies
- Corporate applications in the industry
- End-of-chapter questions
- Chapter-specific references
- Chapter-specific recommendations for further reading

For application and further information, readers are advised to consult *Gleissner and Möller: Case Studies in Logistics* from 2011.

We owe considerable thanks to all companies involved for their cooperation in providing insights into their business in the form of examples and case studies. We also wish to thank our publisher for encouraging us to pen the English edition of this book. Our special thanks for finishing the book in English go to Patrick Speckamp

(London, UK), Ralph G. Kauffman (Houston, Texas, US), Vin Nguyen Truc (Perth, Australia), Claudia Wittmann (Berlin, Germany), and last but not least Anja Foerster and Stefan Baier (Berlin, Germany) for much-appreciated editing work. It was a pleasure working with all of them.

Berlin, Germany
May 2013

Harald A. Gleissner und J. Christian Femerling

Abbreviations

AG	Aktiengesellschaft = Public Limited Company (PLC)
AGV	Automated Guided Vehicle Systems
ANSI	American National Standards Institute
APS	Advanced Planning and Scheduling Systems
ARA	Antwerp-Rotterdam-Amsterdam Range
B2B	Business to Business
B2C	Business to Customer
BASF	Badische Anilin- & Soda-Fabrik
BMG	Baugruppen- und Modulfertigung GmbH
BMW	Bayrische Motorenwerke
BSC	Balanced Score Card
BSHG	Bosch und Siemens Hausgeräte GmbH
bt	Billion Ton
CD	Cross Docking
CEN	Comité Européen Normalisation
CEP	Courier, Express, Parcel
CIF	Cost, Insurance, Freight
CMI	Co Managed Inventory
CPFR	Collaborative Planning Forecasting Replenishment
CRM	Customer Relationship Management
CRP	Continuous Replenishment Programs
CW	Central Warehouse
DB	Deutsche Bahn
DCF	Discounted Cash Flow
DCM	Demand Chain Management
DDP	Delivered Duty Paid
DIN e. V.	German Institute for Standardization
DIY	Do It Yourself
DOS	Days of Stock
DP	Desktop Purchasing
DPWN	Deutsche Post World Net
DTD	Document Type Definition
DW	Distribution Warehouse
E-Commerce	Electronic Commerce

E-Procurement	Electronic Procurement
E-Purchasing	Electronic Purchasing
E-Sales	Electronic Sales
EAN	European Article Number
ECR	Efficient Consumer Response
EDC	European Distribution Centers
EDI	Electronic Data Interchange
EDIFACT	Electronic Data Interchange for Administration, Commerce and Transport
EDP	Electronic Data Processing
ELA	European Logistics Association
EPC	Electronic Product Code
ERMTS	European Railway Transport Management System
ERP	Enterprise Resource Planning System
EU	European Union
EWI	European Economic Institute
EXW	EX Works
FeFo	First EXPIRED-FIRST OUT
FiFo	First In-First Out
FOB	Free On Board
FRG	Federal Republic of Germany
ft	Feet
FUL	Full Truck Load transport
GDP	Gross Domestic Product
GLN	Global Location Number
Glonass	Globalnaja Nawigazionnaja Sputnikowaja Sistema
GmbH	Gesellschaft mit beschränkter Haftung = Limited Company (Ltd)
GPS	Global Positioning System
GRAI	Global Returnable Asset Identifier
GSM	Global System for Mobile Communication
GSM-R	Global System for Mobile Communication Railways
GTIN	Global Trade Item Number
ha	Hektare
HGB	Handelsgesetzbuch= German Commercial Code
HP	Hewlett Packard
HR	Human Resource
IATA	International Air Transport Association
INCOTERMS	International Commercial Terms
ICAO	International Civil Aviation Organization
ID	Identification
IP	Internet Protocol
ISDN	Integrated Service Digital Network
ISO	International Standards Organization
IT	Information Technology

JIS	Just in Sequence
JIT	Just in Time
Kg	Kilogram
Km	Kilometer
KPI	Key Performance Indicator
L x W x H	Length × Wide × High
LANs	Local Area Networks
LiFo	Last In-First Out
LPG	Liquid Petroleum Gas
LTL	Less Than Truck Load Transport
m	Meter
MFS	Material Flow System
MRP I	Material Requirements Planning
MRP II	Manufacturing Resource Planning
NVOCC	Non-Vessel-Operating Common Carriers
OBU	On-Board (Computer) Unit
OCR	Optical Character Reading
ODETTE	Organization for Data Exchange by Tele Transmission in Europe
OEM	Original Equipment Manufacturer
p.a.	per anno
PC	Personal Computer
PLC	Programmable Logic Controller
PLU	Product-Lookup
pm	Past
PPS	Production Planning Systems
PU	Polyurethane
PW	Production Warehouse
QR	Quick Response
R&D	Research & Development
RDT	Remote Data Transmission
RFID	Radio Frequency Identification
ROI	Return on Investment
RoRo	Roll-On/Roll-Off
RW	Regional Warehouse
SCE	Supply Chain Execution
SCD	Supply Chain Design
SCM	Supply Chain Management
SCP	Supply Chain Planning
SEDAS	Standardregelungen einheitlicher Datenaustauschsysteme – Standard Regulations of Uniform Data Exchange Systems
SPC	Special Purpose Company
SSCC	Serial Shipping Container Code
t	Ton

TEU	Twenty Foot Equivalent Unit
TFL	Thiel Fashion Lifestyle
Tkm	Ton kilometer
TÜV	Technischer Überwachungsverein
TV	Television
UMTS	Universal Mobile Telecommunication System
UPC	Universal Product Code
US	United States
USA	United States of America
VDA	German Association of the Automotive Industry
VMI	Vendor Managed Inventory
VPN	Virtual Private Network
VW	Volkswagen
WANs	Wide Area Networks
WC3	World Wide Web Consortium for Standardization of www-Technologies
WMS	Warehouse Management System
WOTIF	Work On Time In Full
WTO	World Trade Organization
XML	Extensible Markup Language
XSL	Extensible Style Sheet Language

Contents

List of Figures

List of Tables

1 Introduction

The structure of this book is oriented towards the componential functions of logistics. The logistical task spectrum is characterized by a great degree of variety and dissimilarity. This is due firstly to logistics' roles, functioning within both overall economic as well as business processes, in all value-added steps from the processing of raw materials to the end-user. The persons concerned with these value-added steps are suppliers, manufacturers, producers, and service providers but even wholesalers and retailers whose goal is to satisfy the demands and wishes of their customers. Secondly, complexity arises due to dynamics that are simultaneously a result of business activities and promoted by logistics. In doing so, logistics finds itself constantly seeking a balance within the competing fields of performance, cost, and quality.

Figure 1.1 positions the individual topics addressed in this book into the broader context of the functions of logistics.

Chapter 2 lays the groundwork for the range of logistical concepts. The ensuing Chap. 3 puts this basic knowledge into a topical context, thereby illustrating the broad range of application of logistics currently used in the economy. Both chapters are a prerequisite for the study of logistics; therefore, they are placed above the tripartite model of performance, cost, and quality.

As logistics needs its own specific infrastructure, Chap. 4 presents logistical infrastructure facilities. In addition to the familiar traffic infra-and traffic suprastructure, this chapter also discusses other logistically significant facilities; namely, logistics real estate and structures of information and communication networks.

Chapter 5 about transport services and logistics services, Chap. 6 about warehousing, transshipment and picking, as well as Chap. 7 about inventory and provisioning management give a detailed account of the logistical core functions and therefore assume a central role in this book. They constitute the most essential body of knowledge for any future logistics expert.

Chapter 8 deals with logistics network planning. In particular, basic decision-making processes and possible structures of warehousing networks and transportation networks are addressed.

H. Gleissner and J.C. Femerling, *Logistics*, Springer Texts in Business and Economics, DOI 10.1007/978-3-319-01769-3_1, © Springer International Publishing Switzerland 2013

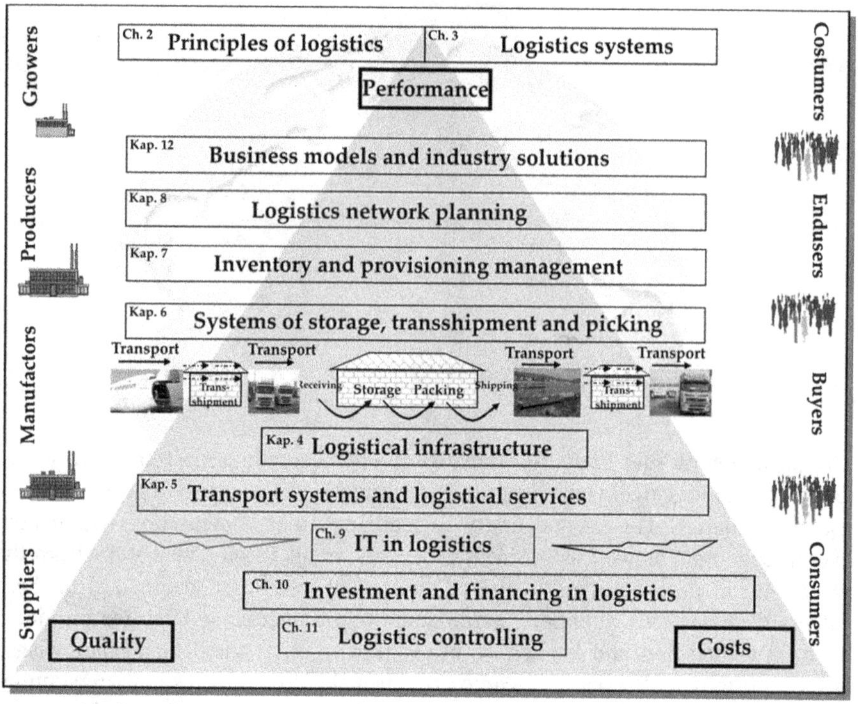

Fig. 1.1 Logistics and its branches

Logistics systems have become inconceivable without the professional use of data processing. Thus, Chap. 9 presents the most important and functionally necessary supporting instruments of information technology, identification technology, and communication technology.

Chapter 10 introduces the reader to the topic of financing operations for stock, logistics real estate, and movable properties (equipment). This financial approach leads to an extended view on the supply chains including the monetary flows and generating the financial supply chain.

Controlling (used in the sense of managerial accounting) plays a pivotal role in the planning and monitoring of logistics systems. Chapter 11 will discuss both the basic approaches to cost and performance accounting in logistics and the functionalities of logistics controlling. Thus, the topics of Chaps. 9 and 11 provide a frame for the core functions outlined in Chaps. 5, 6, and 7. All systems, facilities, and processes mentioned in the previous chapters assume the expenditure of capital.

In order to provide a context to the practice of logistics companies, Chap. 12 presents the typical business models and industry solutions. Thus, the basic knowledge is brought into relation with applied logistics systems. At the same time, it illustrates the multi-faceted range of logistics solutions in practice.

2 The Principles of Logistics

Learning Objectives

As the field of logistics has undergone rapid developments both in scientific research and commercial practice, a multitude of new contents, point of views, and terminologies exists. The key aspects are explained in this chapter. The focus is on the development which has seen the transition of logistics from being perceived in a purely functional way to the modern view of management-oriented, holistic supply chain management. Subsequently, the individual levels of logistics, the involved agents, institutions and their key activities will be discussed. In this way, the readers are introduced to logistic actions and decisions, enabling them to identify with logistical issues by applying the most current understanding of logistics.

Keywords

- Basic definitions of logistics
- Significance of logistics in the context of general business administration
- Classification and differentiation of logistics in the economic process
- The logistical system of performance with its components and key activities
- Agents and key elements of logistics in the national economic context
- Distinction between the terms procurement logistics, production logistics, and distribution logistics, and their basic decision fields
- Delivery policy as the basis for cost and price calculation

H. Gleissner and J.C. Femerling, *Logistics*, Springer Texts in Business and Economics,
DOI 10.1007/978-3-319-01769-3_2, © Springer International Publishing Switzerland 2013

2.1 Definitions and Significance of Logistics

In a world which increasingly relies on the division of labor and fragmentation of work flows, goods and commodities need to be transported from their place of origin to their place of consumption or place of use. The span time between the production and the utilization of goods has to be bridged. This is also true for both the people and the information involved in the fragmented work flow and which are available in different locations and at different times within the system. The tasks and activities associated with this within the economic process were first systematically subsumed under the term *logistics* in mid-twentieth century. Initially, the term emerged in a military context since military systems are characterized by their tendency for concised classification.

The European Committee for Standardization CEN (Comité Européen Normalisation) defines logistics as[1]:

- "... the planning, execution and control
- Of the movement and placement of people and/or goods and
- Of the supporting activities related to such movement and placement, within an system organized to achieve specific objectives."

Here, the term *system* is to be understood as a dynamic unit of interconnected elements and subsystems, connected with each other in specific relations. These elements and subsystems form the system structure and – by virtue of their interaction with each other – result in the system behavior. The defining characteristic of a system is its orderliness.[2]

The purpose of logistics is to plan, organize, coordinate, and implement the bridging of the dimensions of time and space within a system. Logistics is thus one of the most important functions of the economy. This ranges from the procurement of raw materials to their processing and to the delivery of the goods to the end-user.

The central tasks of logistics can be graphically described by the so-called *seven R's*, as depicted in Fig. 2.1. There are not any clear-cut definitions as to what constitutes *right*. The *right* thing to do always individually results from the particular elements involved in the logistical task, i.e. from the products, the quantity, the spatial relations, the customer requirements etc.

In the course of the development of modern business studies, the notion of logistics in the context of the production and distribution of goods has changed significantly since its emergence in the mid-1950s. In 1955, the article *Note on the Formulation of the Theory on Logistics* by Morgenstern was published, in which logistical approaches were transferred from military contexts to the field of business studies for the first time.[3] As time went on, logistics developed from serving a simple, unidimensional operational support function to being a networked,

[1] European Logistics Association (ELA) (2005), p. 54.

[2] Cf. Greschner (1996), p. 9 and Jirasek (1977), p. 17 et seq.

[3] Cf. Morgenstern (1955), p. 129 et seq.

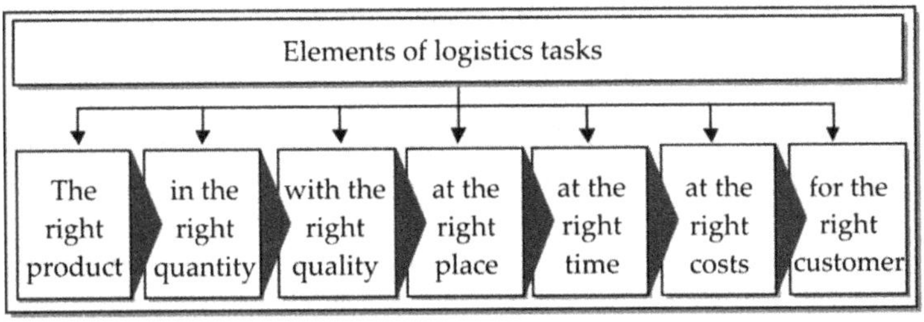

Fig. 2.1 Elements of logistics tasks
(Cf. Klaus 2002, p. 11; Pfohl 1972, p. 28 et seq.; Jünemann 1989, p. 18)

multidimensional process that spans all value-added steps and aims to optimize holistic and complex operational and economic activities. This change of logistical understanding now comprises the entire sequence of activities depicted in Fig. 2.1,

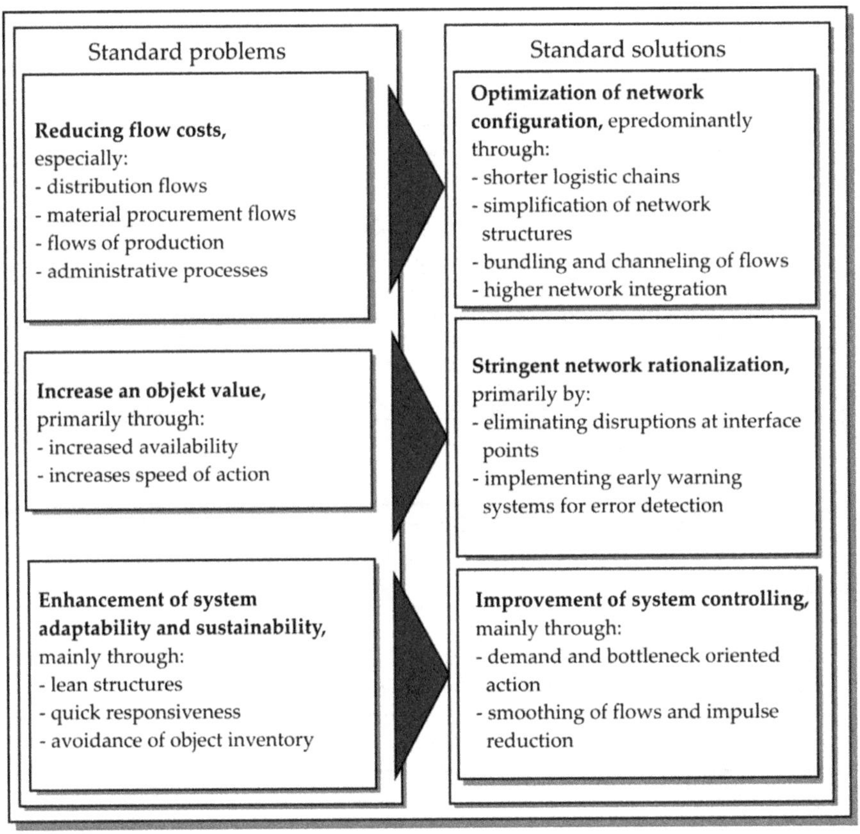

Fig. 2.2 Logistical problems and solutions (Cf. Klaus 2002, p. 30)

which includes the actual transport of goods from their origin to the consumer or end-user. Klaus offers his scientific theory of the *third meaning of logistics* (*Die dritte Bedeutung der Logistik*), which discusses this development and the modern logistical understanding of Flow Management – an understanding of logistics that has become necessary in today's globalized economy which is based on the division of labor. Flow Management conceptualizes the management of continually moving flow systems and defines logistics as an economic phenomenon comprising activities and processes in chains and networks in order to transform goods and

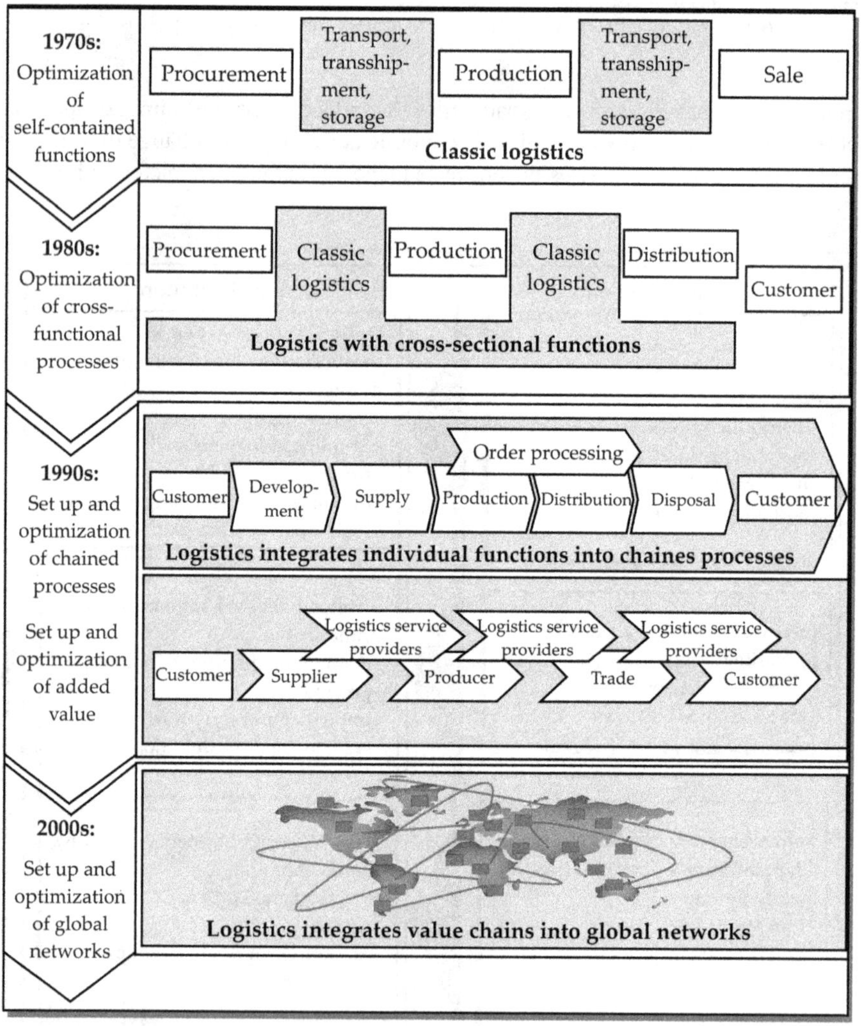

Fig. 2.3 Changes in the understanding of logistics (Cf. Baumgarten and Walter 2000, p. 2)

the information pertinent to them.[4] According to this practice-oriented explanation, logistics is characterized by the following operational goal clusters:

- Cost reduction of the logistics process in particular and of the total process in general
- Increase of the adaptability of the logistics system to changes in demand and environment
- Value enhancement of the logistical objects (goods) through or during the logistics process

By assigning properties to logistics that increase the value of the goods, many courses of action open up for logistics in a macroeconomic context. Thus, modern business management understands logistics as a performance component which is relevant for success. The identified goal clusters can also be seen as standard problems of the logistical task accomplishment. The following possible solutions to these problems can be offered (see Fig. 2.2).

The change in the understanding of logistics over time is illustrated in Fig. 2.3. It proceeds from the original, rather simplistic view of *transporting, handling, warehousing* with its emphasis on the physical aspect of the logistical task and moves on to the current, management-oriented view of integrated value chains.

2.2 Logistical System of Services

Based on the preceding chapter's definitions of logistics, the services that logistics renders will be more specifically defined in the following.

The logistical system of services features a wide range of system components, which are illustrated in Fig. 2.4. The *core services* of logistics, such as order processing, storage and transport, are directly supported by the logistical information services and the additional services necessary for the logistics process. In practice, the element of transshipment is rarely dealt with separately since it is usually closely connected to storage and transport. The core element of *order processing* is somewhat virtual in nature but constitutes an indispensable element for the initiation and monitoring of logistics processes. It is the order processing which gives logistics its dynamic character. Order processing comprises the processing and monitoring of the order data from the time of order until the arrival of the goods at the customer's site and of the return of the shipment documents. In many cases this includes the initiation of invoicing processes, e.g. by simultaneously delivering the invoice together with the goods.

The core element of *storage* denotes the warehousing of goods, which have mostly been produced for an anonymous market to offset the quantitative and chronological divergence of production (supply) and sales (demand). It has to be noted at this point that the functions of picking and packing, which are closely

[4] Cf. Klaus (2002), p. 26 et seq.

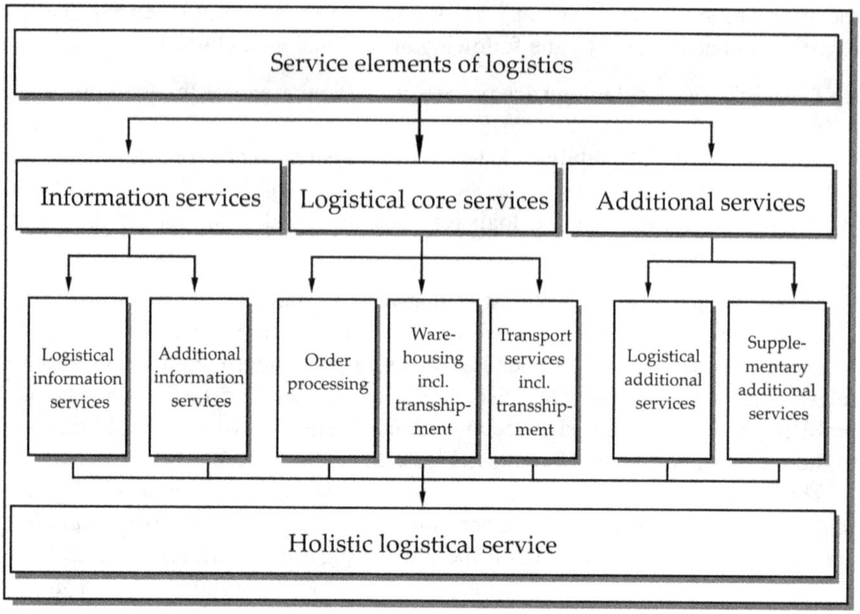

Fig. 2.4 System of logistical services (Cf. Isermann 1994, p. 25)

connected to the core element of storage, fall into the category of additional services. Picking means the compilation of sellable articles according to individual customer orders to form package and shipping units.

Finally, the core function of *transport* bridges the spatial distance from the site of production to the storage location to the delivery location at the customer's site. Thus, it offsets the spatial divergence of supply and demand. Again it has to be pointed out that both storage and transport include the function of transshipment.

Logistical information services produce and utilize data about the goods being transported within the logistics system. This data includes information about the forwarder, the recipient, the type of goods, order number, quantity, weight, volume etc. The logistical information is the key to planning, monitoring, and supervising in the provision of the actual logistics service. A data-driven logistical activity closely connected to the core service of storage is the inventory management (see Chap. 7). The volume of inventory normally has a direct but often underestimated influence on the financial situation of the company, resulting from the cost of capital commitment incurred by the volume of inventory (see Sect. 10.2).

Additional services of logistics include, for instance, picking, palletizing of goods for economical shipping (easier loading and transport), additional packaging for protection, identification and easier handling of goods and commodities, repackaging of goods from bulk packaging into smaller package units etc.

Supplementary information services and additional services can usually be incorporated into the logistical core process without expanding further organizational effort. In this way, substantial additional value can occasionally be created for the contractor of the logistical service. Additional information services may, for example, refer to additional information about the goods such as prices, quantity discounts, and cash discounts. They may also refer to obvious facts regarding the customer's commercial situation or, for instance, regarding the number of residents, size of the backyard or location of the private property.

Information of this kind can later be analyzed for marketing purposes. Additional services also comprise promotional or advertising packaging of goods. Additional services can be rendered at any given location, i.e. not only at the forwarder's site but also, as it is often the case in the automobile sector, directly at the contractor's site or at a third-party site in transit of the goods between origin and destination.

The ensuing case studies aim to illustrate the logistical system of services.

Case Study 2.1: Motor Scooter Assembly

The forwarding agency *Wackler* has been commissioned with the logistic operations of a DIY superstore chain. *Wackler* provides its client with additional services. They import motor scooters from a Chinese manufacturer for the DIY superstore chain. The content of the shipment is communicated to Germany via data transmission upon dispatch at the port in Hong Kong [*logistical information service*]. The motor scooters are disassembled into their components, then shipped to the port of Rotterdam and subsequently taken by truck to the *Wackler* location in Göppingen [*logistical core service*]. First, the boxes containing the components are stored there [*logistical core service*] before they are successively assembled by a specially formed and trained team of *Wackler* employees [*additional service*]. Having been technically inspected and certified by the TÜV (German Association for Technical Inspection), the motor scooters are packed for transport [*additional service*] and sent to the individual DIY stores [*logistical core service*].

Case Study 2.2: Module Assembly

BMG (a module assembly company) is a fully owned subsidiary of the logistics provider Schnellecke. BMG has been entrusted by Volkswagen Braunschweig with the assembly of the modules *Subframe* and *Cornermodule*, pertaining to the car models Golf and Passat, for Volkswagen Sachsen Ltd. in Zwickau [*additional service*]. The service provider is also responsible for the disposition and storage of the primary material [*logistical core service*] and for the delivery of the assembled modules to the assembly line [*logistical core service*]. The assembly and delivery for each particular production sequence is carried out according to the logistics planning of Volkswagen Sachsen Ltd.

Case Study 2.3: Price Tags and Textile Finishing

Apart from logistical core services (administrative transaction, storage, handling, transport), the logistics service provider Thiel Fashion Lifestyle (TFL) offers a broad spectrum of additional fashion services in the field of textile procurement, production and distribution logistics. This includes, for instance, all types of picking, price tagging, labeling, and textile finishing.

Upon pickup at the production site (e.g. in Middle Eastern Europe), TFL forwards the finished goods of its clients Hugo Boss and Gery Weber to its logistics center in Macedonia [*logistical core service*]. The client conveys all information relating to the order to TFL beforehand via remote data transmission (RDT). TFL then logistically processes this information [*logistical information service*]. Among the additional services is the finishing of the hanging garments. Specially trained staff members check the quality of the textiles. Subsequently, the garments are finished using a tunnel finishing system. First, the hanging garments are dampened, then stretched and finally dried. After the textiles have been finished and are ready for sale, they are packed into special hanging cartons, which is to ensure crease-free and undamaged transport of the goods [*additional service*].

Another additional service offered by TFL is the labeling of the goods. Besides price tags, TFL additionally fits the garments with anti-theft tags. Depending on the customer requirements, the order-related picking is followed either by the storage of the goods, the transport to the customer's central warehouse, or the transport to the point of sale by means of TFL's comprehensive European network [*logistical core service*].

2.3 Agents and Elements of Logistics

The business performance within logistics networks is determined by various agents, processes, and not at least by the different kinds of goods. The agents, available means of labor, and different kinds of goods constitute a frame of action within which the logistical processes are affected.

The now all-encompassing process of the transformation of goods from the source to the drain can also be described in national economic terms. Referring to the different aggregate levels in economics, a distinction has to be drawn between macrologistics, micrologistics, and – deriving from these two – metalogistics (see Fig. 2.5).

Macrologistics on the one hand describes the traffic system and, on the other hand, the suitable infrastructure for the logistical core functions of warehousing and transshipment. Thus, macrologistics represents the macroeconomic perspective.

Micrologistics describes microeconomic systems with logistical functions which in most cases are characterized by the respective sector. From a shipping agent's viewpoint, industrial logistics is different from retail logistics in that they each cater their logistical performance to varying time standards and to the respective differences in the shipper's and recipient's supply and demand behavior. The logistical service is a secondary service for the consignor, i.e. it is not the company's main purpose.

In service logistics, the logistical operation is a primary service, i.e. the actual object of the company. Depending on the kind of services provided by the respective companies, we can distinguish between carriers, logistics service providers in the narrow sense (forwarders, shipping companies etc.), and infrastructure

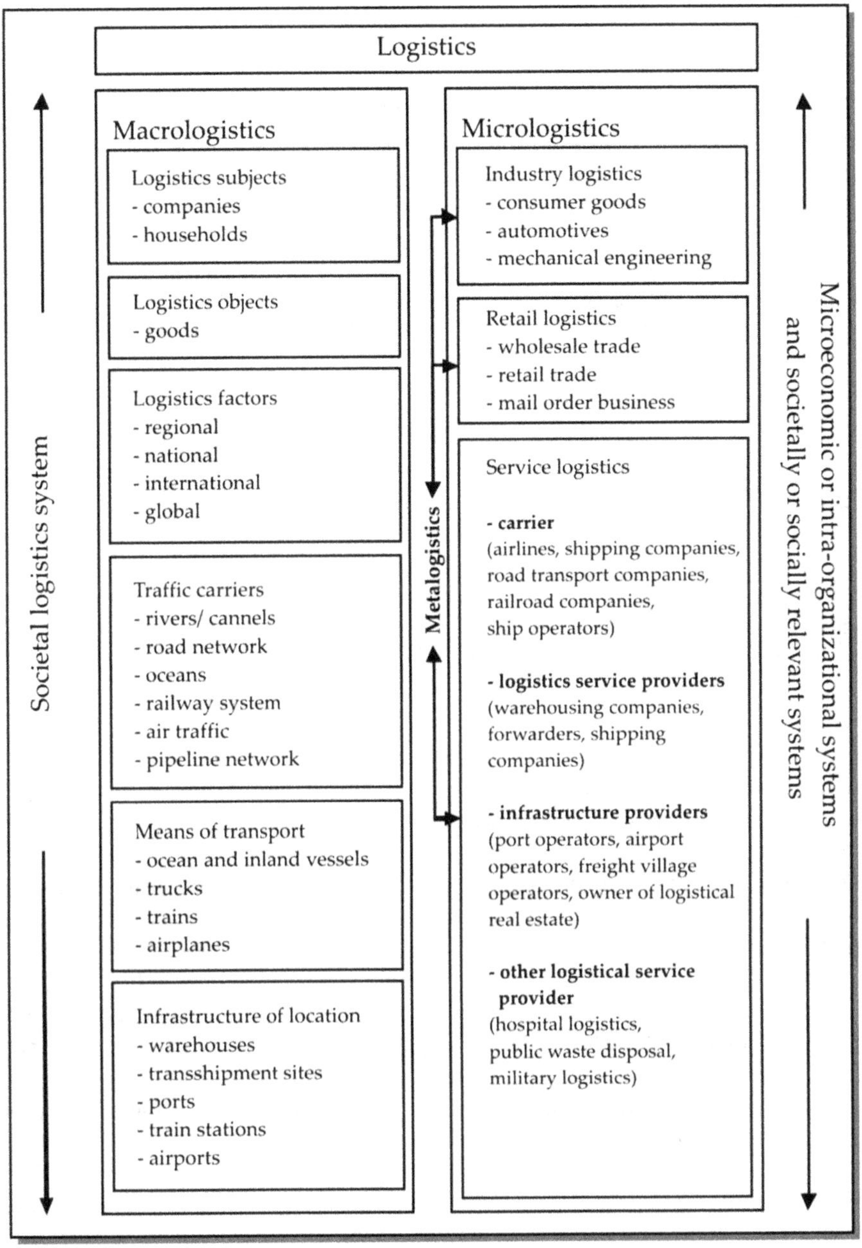

Fig. 2.5 Elements of logistics (Cf. Pfohl 1990, p. 14 and Ihde 2001, p. 134 et seq)

providers (warehousing service providers). Furthermore, societal or social systems such as hospital logistics or public waste disposal can be subsumed under the term micrologistics.

Metalogistics describes every kind of institutional or functional cooperation between independent organizations within micrologistics in order to enable logistical performance. These cooperations may be established, for instance, between industrial and commercial companies to facilitate the joint organization of warehousing and the supply of final customers, between logistics service providers to build up a nation-wide distribution network covering all regions, or between different traffic carriers (railway, inland vessel, road, air traffic).[5]

Thus, logistical processes can always be categorized according to individual elements pertinent to the systematization components of macrologistics, metalogistics, and micrologistics.

2.4 Distinction between Procurement Logistics, Production Logistics and Distribution Logistics

Following the functional distinctions of logistics and the transport and sector-specific categorization of logistics, we shall now consider the functional distinctions of logistics systems according to the phases of the flow of goods. This ranges from the supply of goods to the disposal or recycling of waste. The range of these functions, with reference to conventional producers, is depicted in Fig. 2.6.

As it is the case in many areas of operational organization, the distinction between the tasks and their definitions as well as the assignment of functions (order processing, warehousing, transport) to certain forms of organization (inventory management, procurement logistics, production logistics, distribution logistics, or logistics) is very difficult and only possible in highly general terms. In operational practice, the individual functions are diversely assigned within the organizational structures of the companies.

The basic task areas of business enterprise are shown in detail in Fig. 2.7.

Procurement logistics is concerned with the organization and the physical processes involved in the transport and supply of the input factors for the corporate process. This applies both to industrial companies and trading companies.

The procurement process aims to ensure the economic supply of the materials or commodities which are to be processed. In practice, the purchase with its market-oriented and contracting tasks is either subsumed under the term procurement (see Fig. 2.7) or is given its own business function. Manufacturing companies often refer to this phase in the flow of goods as materials management (see Fig. 2.6). However,

[5] Cf. Pfohl (2010), p. 14 et seq.

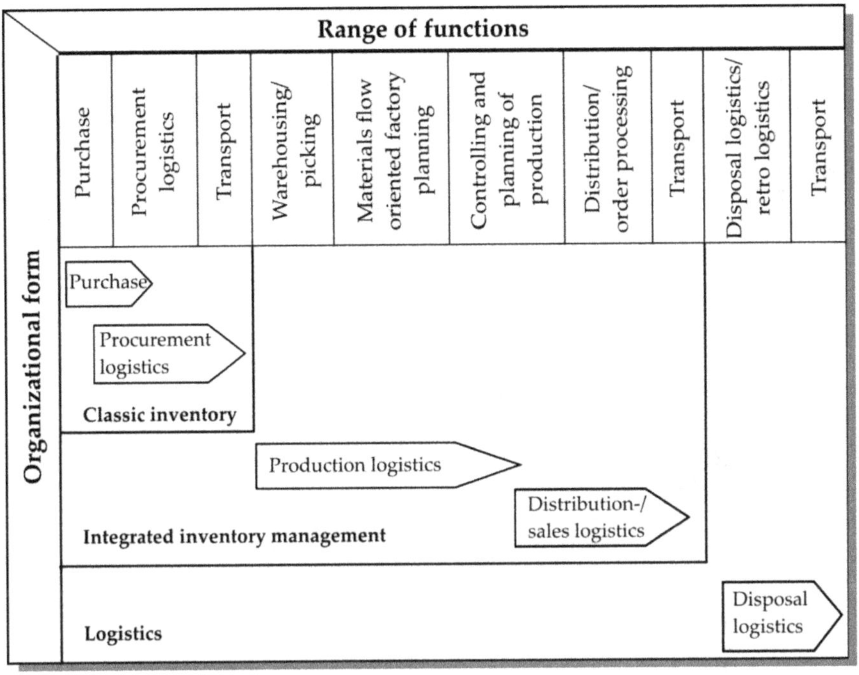

Fig. 2.6 Functional distinctions of logistics (Cf. Schulte 2009, p. 2)

materials management may also include production and distribution. Due to this ambiguity, we shall avoid the term materials management.

Production logistics is mainly associated with manufacturing companies. It deals with all tasks pertinent to the planning and controlling of those internal processes that relate to the materials flow, storage, and internal transport. Additionally, production logistics plays an important part in planning and integrating both the preceding procurement logistical processes and the subsequent distribution logistical processes. If there is a high degree of integration as a result of a flow or network-oriented understanding of logistics, production logistics may be integrated into the logistics network as an immediate link. Thus, IT systems are able to receive direct input from inventory control systems for the planning and controlling of production programs (see Sect. 9.3).[6] Production logistics is less characterized by major physical flows of goods over long distances rather than by intelligent planning of the processes and the provision of goods within a smaller context. This is why it can also be understood as a special field of production.

Distribution logistics is primarily concerned with the coordination of all processes that serve to deliver the goods to the recipient or to the point of sale for consumption by the end user. Distribution logistics describes the interaction of

[6] Cf. Vahrenkamp (2008), p. 88 et seq.

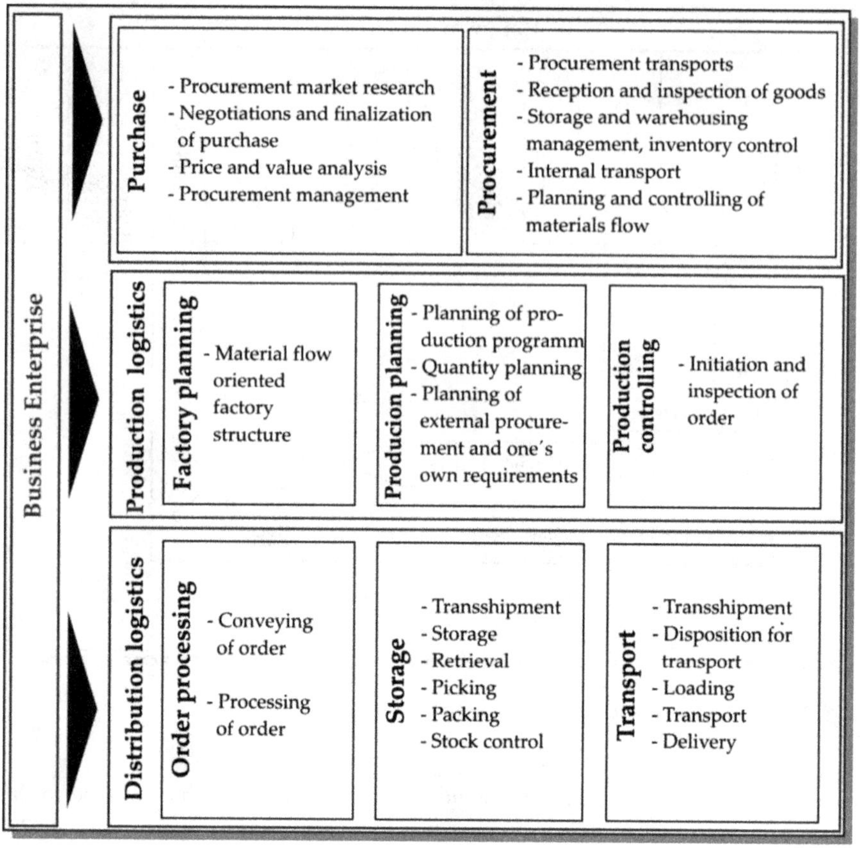

Fig. 2.7 Differentiation of business enterprise (Cf. Schulte 2009, p. 267)

transport and storage processes within logistics systems for the distribution of a company's goods. Since the distribution process includes all elements of the logistical performance spectrum in a variety of ways (see Fig. 2.4), it will be given special emphasis in the following.

2.5 Service Level

The elements of logistical business performance were outlined in Sect. 2.2. In order to determine the extent and quality of business performance, we need to define a service level. An operationalized service level allows for the measurability of the service provided and also serves as a basis for the cost and price calculation. The individual determinants of the service level are illustrated in Fig. 2.8.

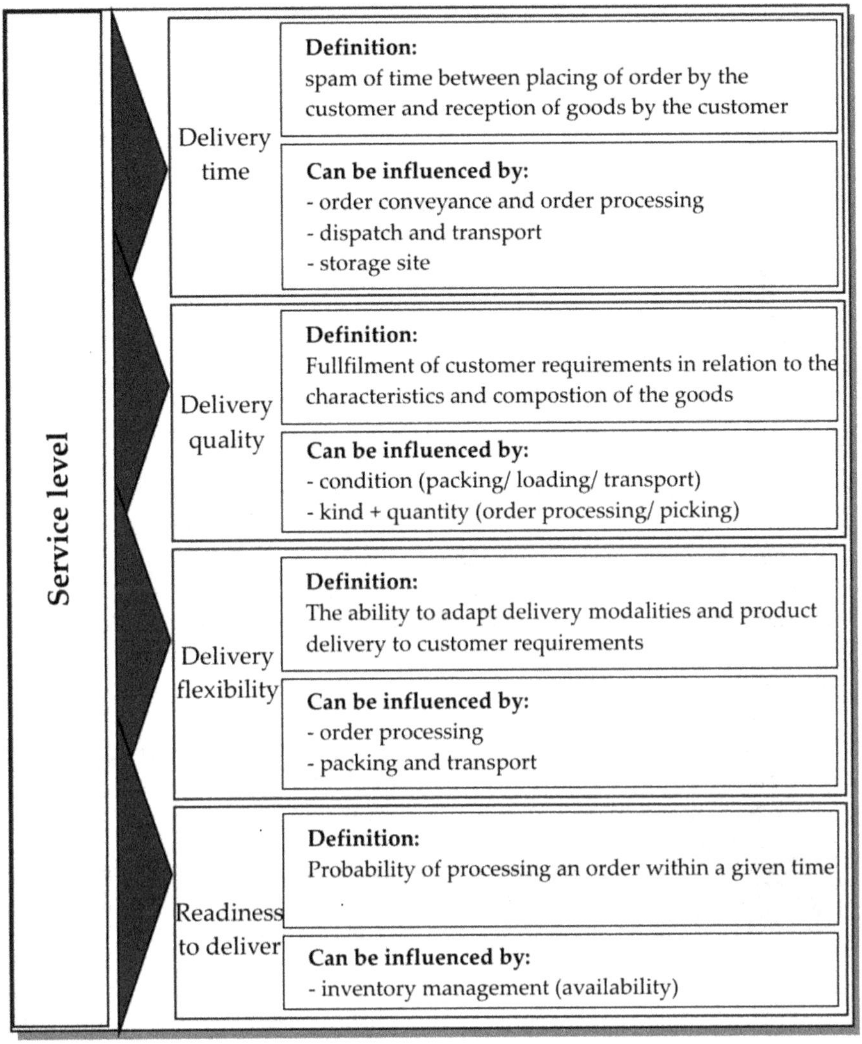

Fig. 2.8 Determinants of the logistical service level (Cf. Pfohl 1972, p. 177 et seq)

The service level is not an arbitrary variable. It should be defined as part of a logistical service offer for a specific group of goods and clients.

The service level and the features of the delivery service are not solely determined by customer requirements. Factors such as service levels of competitors, capabilities of the company including earning power (contribution margin) of the objects (goods and commodities), and the extent of the value to be added by logistics should also be taken into consideration. A further distinction has to be drawn between regular service and emergency service.

In many cases, however, the service level is determined by the goods themselves. A prime example of this is food logistics, where commodities need to be moved quickly within the logistics networks so that they reach the final customer fresh.

Case Study 2.4: Service Level Components in Mail Order Business

Many mail order companies offer their customers a diversified delivery service. The regular delivery time is 2–3 working days. This case study shows one typical delivery service of a full range assortment mail order company offered on the German market.

Other additional offers tailored to customer requirements affect the delivery time and flexibility. In their catalog and on their webpage, the example company labeled their individual service offers with the symbols next to the text on this page. As regards the delivery time they offer a *24 h delivery service*.

From Monday to Thursday, they guarantee next day delivery if the customer orders by 12 noon. If the order is placed by 12 noon on a Friday, delivery will be effected the following Monday. Orders which are placed between Friday 12 noon and Sunday inclusively are delivered on Tuesday. This applies to articles that are delivered in postal parcels. Similar conditions apply for bulky articles, such as fridges or TV sets, which are delivered by forwarding agencies.

Furthermore, the mail order company offers an *evening delivery service* in many ZIP-code areas, which guarantees the delivery of major articles by a forwarder from Monday to Friday between 5 pm and 9 pm if the customer orders by phone or online [delivery time and flexibility].

For many articles the customer may state a *preferred delivery day*. Depending on the article, delivery is then effected from the 3rd or 5th day onwards after placing the order online or on the phone.

A preferred delivery day between Monday and Friday may be stipulated for articles which are sent by mail. It is not possible, however, to request a certain time for the delivery [delivery time and flexibility].

Neckermann also offers a disposal and assembly service. If a new device or furniture has been ordered, the customer may request the disposal of the old device or furniture by Neckermann upon delivery [flexible delivery].

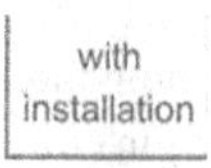

The delivery of tumble dryers, washing machines, dishwashers, fridges, and freezers includes unpacking, removal of lock fasteners, and installation. Furniture may be assembled on request.

By means of checks during picking and before the packing of customer parcels and by using padding material, a high standard of delivery quality is ensured. Articles which are delivered by forwarders receive special repackaging to protect them from damage [delivery features]. In this case, delivery to the customer is ensured by a pre-announcement via telephone (advising). The availability of the articles [readiness to deliver] is maximized by issuing test catalogs prior to the publication of a new catalog, timely observation of customers' ordering behavior during a particular catalog season, and evaluating the ordering behavior of previous catalog seasons.

Review Questions

1. What kind of problems does logistics deal with?
2. Describe the development of the understanding of logistics up to Klaus' *third meaning of logistics.*
3. Describe the *7-r-approach* of the modern understanding of logistics.
4. Distinguish between macrologistics and micrologistics.
5. Give a concrete example for each element of metalogistics and explain its connection to macrologistics or micrologistics, respectively.
6. Dinstinguish between procurement logistics, production logistics, and distribution logistics.
7. What are *agents* in logistics? Give examples.
8. What may influence each element of the service level?
9. How can the right service level for a product or company be found?
10. What is the difference between inventory management and procurement logistics?

Additional Literature

Beckmann, K. (2007). *Logistik.* Rinteln: Merkur-Verl.

Bloomberg, D. J., Lemay, S., & Hanna, J. B. (2002). *Logistics.* Upper Saddle River: Prentice-Hall.

Coyle, J. J., Langley, C. J., Gibson, C. H., & Novack, R. (2013). *Supply chain management: A logistics perspective* (9th ed.). Mason: South-Western College Publishing.

Eichler, B. (2003). *Beschaffungsmarketing und -logistik.* Herne: NWB-Verlag.

Gudehus, T. (2005). *Logistik-Grundlagen, Strategien, Anwendungen* (3rd ed.). Berlin: Springer.

Ihde, G. B. (2001). *Transport, Verkehr, Logistik* (3rd ed.). München: Vahlen.

Klaus, P. (2002). *Die dritte Bedeutung der Logistik – Beiträge zur Evolution logistischen Denkens.* Hamburg: Deutscher Verkehrs-Verlag.

Lambert, D. M., & Burduroglu, R. (2000). Measuring and selling the value of logistics. *International Journal of Logistics Management, 11*(1), 27–50.

Murphy, P. R., & Wood, D. F. (2010). *Contemporary logistics* (10th ed.). Upper Saddle River: Prentice-Hall.

Pfohl, H. C. (2004). *Logistikmanagement* (2nd ed.). Berlin: Springer.

Rümenapp, T. (2002). *Strategische Konfigurationen von Logistikunternehmen, Ansätze zur konsistenten Ausrichtung in den Dimensionen Strategie, Struktur und Umwelt.* Wiesbaden: Univ.-Verl.

Rutner, J. M., & Langley, C. J. (2000). Logistics value: Definition, process, and measurement. *International Journal of Logistics Management, 11*(2), 73–82.

Schulte, C. (2009). *Logistik – Wege zur Optimierung der Supply Chain* (5th ed.). München: Vahlen.

Vahrenkamp, R. (2007). *Logistik-management und strategien* (6th ed.). München: Oldenbourg.

3 Logistics Systems

Learning Objectives

Logistics is characterized by thinking in processes, value chains, and networks. It includes parameters such as performance, quality, value, and cost. At the end of this chapter the reader will have gained an understanding of the basic components and organization variables of the logistics systems.

Apart from the technical and economic aspects, which increasingly have an impact on the efficiency of the logistics systems, it is important to discuss the fundamentals and functions of a company's internal and external logistics structures, and how the relationship with their agents is managed. In this way the reader will obtain information on competencies that goes beyond the scope of basic logistics knowledge.

Keywords

- Logistics and corporate strategy
- Basic structures of the logistics systems
- Logistical chains, networks, and processes
- Demand and supply chain
- Agents and elements of logistics
- Organization of logistics
- Relationship management in supply chains

H. Gleissner and J.C. Femerling, *Logistics*, Springer Texts in Business and Economics,
DOI 10.1007/978-3-319-01769-3_3,

3.1 Logistics and Corporate Strategy

Logistics is both a competitive tool and a means of rationalization. Thus, logistical services may generate separate opportunities to develop strategic competitive advantages. On the other hand, the efficient organization of the logistics systems allows for the development of rationalization potential, which, in turn, may give a company a sustainable competitive advantage.

The two basic approaches of competitive strategies are *cost leadership* and *differentiation*. Moreover, another strategy called *focus scope* is frequently employed. The focus scope strategy is the niche strategy dependent on a company's strategic orientation (cost leadership or differentiation) and the scope of the targeted market.[1]

These basic strategies differ from one another according to what kind of competitive advantage is desired and according to the scope of the targeted competition field. The strategies of cost leadership and differentiation target the entire market of a particular business area. The focus scope strategy, however, is aimed at individual market segments.

Comprehensive cost leadership pursues the goal of consistent cost orientation across all business functions. Cost leadership is primarily based on a large market share, on the basis of which economies of scale in the procurement, production and distribution stages are to be implemented. It is the organization of logistics systems which substantially supports the cost leadership strategy. Thus, the network structures in procurement and distribution logistics determine the warehousing and transport costs (see Chap. 8). Further cost estimation factors include the application of transport, warehouse, picking, and material transport technologies. The degree of automation of the applied technologies significantly influences cost structure and cost effect (see Sect. 11.5).

Besides the physical logistics processes, the organizational design of logistics influences the cost structure. For example, this applies to order processing, inventory planning, and the disposition of materials (see Chap. 7). The costs of the associated processes, in turn, are influenced by the IT, information and communication systems used (see Chap. 9). Allocating logistics capacities and processes to logistics service providers is a strategy that has increasingly been pursued by companies. This form of logistical outsourcing has primarily been utilized in distribution logistics, with service providers running distribution centers and managing the distribution of goods (see Sects. 4.2.6 and 10.2.3). On the one hand, this outsourcing process aims to make the costs more variable. On the other hand, it is expected to reduce the costs through the realization of economies of both scale and scope by the service provider (see Sect. 8.5).

Companies pursuing a differentiation strategy try to achieve a unique selling point for a customer's highly-valued demands. By reaching a unique selling point for certain products or services, it is often possible for a company to establish higher prices. Assuming that the prices achievable in the market will not be compensated by the (additional) costs of differentiation, a company will enhance its profit prospects.

---

[1] Cf. Porter (1985), p. 11.

In addition, differentiation generally enables more customer loyalty, which helps prevent new market entries and substitution products. Among the forms of differentiation we can distinguish between the criteria of quality, design, image, and service. Thus, service differentiation is aimed at providing services which offer the consumer or customer additional benefits apart from the actual product. Differentiation strategy is best implemented in/within the logistics service. Examples are incredibly fast or reliable delivery services, or additional disposal services.[2]

3.2 Supply Chains and Networks

The modern understanding of logistics, which is based on the third level of logistical development (see Chap. 2), calls for process-oriented management of all logistical elements in the business cycle. As opposed to the individual examination of these elements, a highly complex picture of the logistics processes thus emerges which involves extensive division of labor. In contrast to the management of rather straightforward processes (e.g. the transport of goods between a supplier and a customer), this complexity requires much more monitoring and managerial effort.

The terms *supply chain management (SCM)* and *demand chain management (DCM)* conceptualize these managerial tasks. Here, the words supply or demand only describe where the logistics process is initiated, i.e. on the supply side or on the demand side. Since in most cases logistics makes commodities available on the market according to the supply volume and most companies offer their commodities on an anonymous market, the term SCM is far more frequently used.[3] Ultimately, the terms SCM and DCM are a linguistic advancement of the terms *controlling* or *management* of logistics systems across several steps in the value chain. Supply chain management is enhanced by the comprehension of financial and monetary flows, the coherent internal organizational units and external participants (see Sect. 10.1.3).

In order to accurately portray the complexity mentioned above, the term supply chain – describing a string of logistical elements – has in many cases been

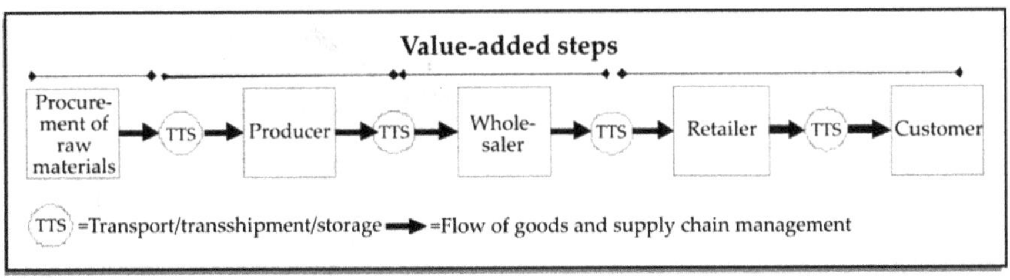

Fig. 3.1 Model of a logistics chain (supply chain)

[2] Cf. Herter (1999), p. 81 et seq.

[3] Cf. Marbacher (2001), p. 18 et seq.

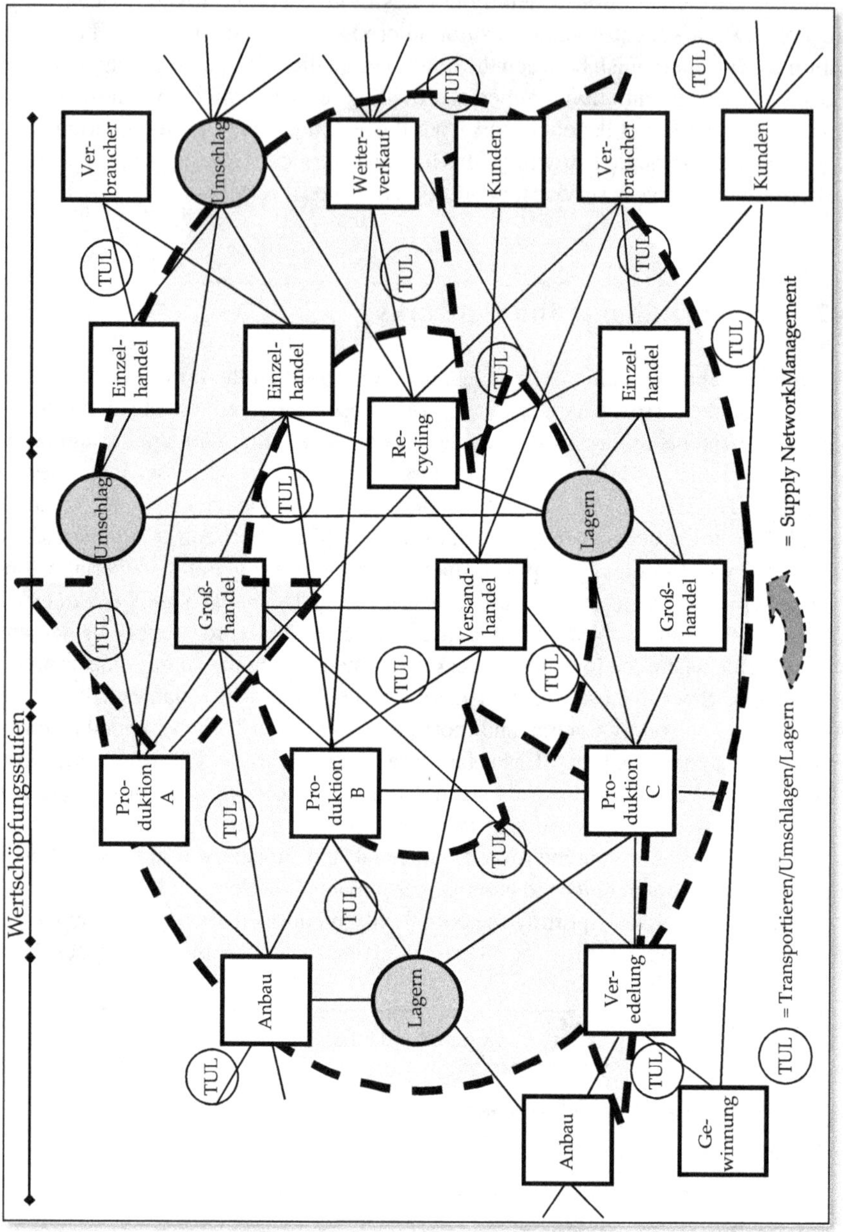

Fig. 3.2 Model of a complex logistics network

superseded by the term supply network. This term comprises all elements, agents, and processes involved in both the physical and virtual (informational) flow of goods from the source to the drain. While Fig. 3.1 depicts a relatively simplistic supply chain, Fig. 3.2 shows a schematic representation of a complex network across several value-added steps.

In operational practice, all forms of relationship networks are encountered, ranging from less complex ones to highly complex ones. Included in this are simple, isolated and self-contained networks with only one element, point to point connections with two agents, and multi-level supply chains or highly branched logistics networks. The main types of agents forming part of supply chains are[4]:

- Raw material producers (growing, extraction etc.)
- Manufacturers, producers (suppliers, industry, processors etc.)
- Wholesalers, distribution provider, importers, exporters
- Retailers
- Users, consumers

It is the objective of logistics and logistics management (supply chain or network management) to influence the physical and informational transactions of the logistics processes in a manner that meets the specific requirements and efficiently combines the elements and methods consuming as few resources as possible and at the lowest cost. Such efficient logistics processes may entail a significant competitive advantage for the companies involved. This competitive advantage is comprised of

- *Cost advantage* – through low factor consumption, low information cost and high productivity
- *Value advantage* – through the increase in value of the products during the logistics process by means of timely or fast availability, additional treatment of the goods, or additional services associated with the goods.

Case Study 3.1: Detergent Supply Chain

The supply chain of a detergent manufacturer starts with the purchase of detergent by a customer at a retail shop. The removal of the detergent from the shelf is registered during the payment process at the checkout and supply is automatically requested. As soon as the need for supply has reached a certain level the relevant wholesaler is requested to supply the amount of detergent needed and delivery to the retailer is effected. Meanwhile, the dispatching process at the wholesaler initiates a supply request for detergent to the detergent manufacturer. The manufacturer obtains the intermediate products required for the detergent production from a variety of suppliers which, in turn, are supplied by other suppliers. The packaging, for example, is produced by a supplier which obtains its cardboard in an upstream stage of the supply chain from a paper manufacturer, which, in turn, orders wood from a forestry (see Fig. 3.3).

[4] Cf. Chopra and Meindl (2009), p. 22.

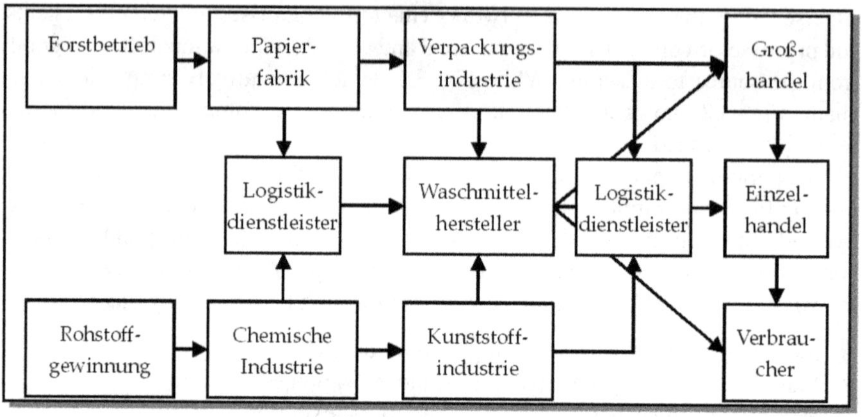

Fig. 3.3 Model of a detergent supply chain

The management of the supply chain requires an efficient IT network to control the flow of goods across all the stages (see Chap. 9). The retailer provides the customer with a ready-to-sell version of the detergent. In all the upstream stages of the supply chain, sales data is needed for internal planning and delivery is carried out on order. In the same way, time frames for delivery are stipulated in those stages, in which, for instance, discount prices may be attained upon which the retailer may implement a sales campaign. Moreover, special cases such as direct delivery by the manufacturer to the wholesaler (e.g. a laundry shop) always need to be taken into account in such supply chains.[5]

3.3 Management of Logistics Systems

3.3.1 Basic Management Systems

The basic logistical management strategies derive from marketing strategies. The *push principle* conceptualizes the supply of goods on the market without specific demand for those goods. This means that action for the initiation of a logistics chain is taken by the manufacturer and thus at the beginning of the logistics chain. The push principle is the traditional strategy to supply goods to the market. Large production lots and fully utilized transport units result in cost advantages. This, though, must be set against the corresponding inventory costs and the sales risk.

This principle is used for low-value goods in the distribution of consumer goods, especially in the event of a sales campaign where products are pushed onto the market at low prices.

[5] Cf. Chopra and Meindl (2009), p. 21.

The *pull principle,* on the other hand, initiates the logistics chain at its end. This means that the logistics process is activated when the end user (consumer) requires specific goods and thus creates demand. This principle is normally applied for high-quality investment goods. Due to the generally difficult sales situation and the competition-induced need to reduce costs in the supply chain, the pull principle is also increasingly used for consumer goods and low-price goods. This principle entails an interesting cost advantage as a result of reduced inventory costs and sales risks. It is, however, also associated with relatively long delivery times and increased costs for small (single) shipments.

In order to offer customers better availability of goods that pass through the logistics channel according to the pull principle, elaborate forecasting methods are increasingly being used. These are based on detailed analyses of previous customer behavior and on intelligent logistical management techniques. Thus, the time between the placing of the order by the customer and delivery can sometimes be shortened significantly and the disadvantages associated with long delivery times mentioned above may be offset (see Sect. 8.3).

Another approach in managing the supply chain processes is measuring the cycle time of the logistics process. A distinction can be drawn between the following cycles[6]:

- Procurement cycle of the suppliers
- Manufacturing cycle of the manufacturers
- Replenishment cycle in trade
- Customer order cycle of the consumers

This classification is modeled according to the functional distinction of logistics (see Sect. 2.4) and can be broken down into the time spent on the informational processing of the order (acceptance of order or purchase by customer at point of sale), on the physical processing of the order (picking, packing), and on the delivery (transport) or reception by the customer.[7]

Apart from these basic management techniques, there is a multitude of other management systems. One of them is the so-called order to cash cycle system which focuses on shorter time spans between order placement and payment by the customer. This perspective is above all in the interest of the industry sector (sometimes also in the trade sector if it is part of the supply chain) since the time until receipt of the sales revenues is shortened and cost effective capital commitment is therefore reduced (see Sect. 10.1.3). In this way, efficient logistics processes and a networked supply chain management may substantially expedite the order to cash cycle.

[6] Cf. Chopra and Meindl (2009), p. 27 et seq.

[7] Cf. Delfmann and Reihlen (2003), p. 9.

3.3.2 System Leadership

The core question around system leadership is to determine which one of the agents at different value-added steps within a logistics chain or network assumes leadership in managing these steps. Joint control by several agents normally yields dissatisfying solutions due to shared responsibility. If several value-added steps within a supply chain are linked to each other, the question of system leadership appears. This question for instance is discussed and negotiated between

- The raw material supplier, the component manufacturer, or the end product manufacturer; here, the end product manufacturer often dominates
- The end product manufacturer and the trade sector; depending on the market power of the parties involved, various constellations between the industry sector and the trade sector may result here
- The wholesalers and the retailers if several levels of trade are involved. In these instances many small retailers are confronted with few powerful wholesalers, which also assume leadership
- The shipping agents (industry and trading sector) and the logistics service providers entrusted with the logistics processes

System leadership entails supremacy of control, which enables the system leader to generate logistical synergy potential, such as cost advantages. Barring distinctly cooperative companies, system leadership is usually wielded by industrial and commercial companies which prevail due to their market power. However, it is also possible to pass system leadership on to third parties, for example to logistics service providers. If several service providers are involved in the supply chain, it is possible to designate one of them as system leader (Lead Logistics Provider). The car industry offers many examples that illustrate this issue since car manufacturers often assume leadership over their suppliers or service providers with concepts such as Just in Time (JIT) or Just in Sequence (JIS) (see Sect. 7.3.3). Consumer goods manufacturers and trading companies have taken similar courses of action by implementing instruments for Efficient Consumer Response (ECR) (see Sect. 7.5).

3.4 Organizational Variables in Logistics Systems

3.4.1 Organizational and Operational Structures of Logistics

Logistical task management and supply chain management require a corresponding organization of these functions within a company. On the one hand, it is crucial to set up a structure for the allocation of tasks and competencies within the respective areas of responsibility. On the other hand, the functions and processes need to be structured. A suitable way to view the corporate structure is to distinguish between

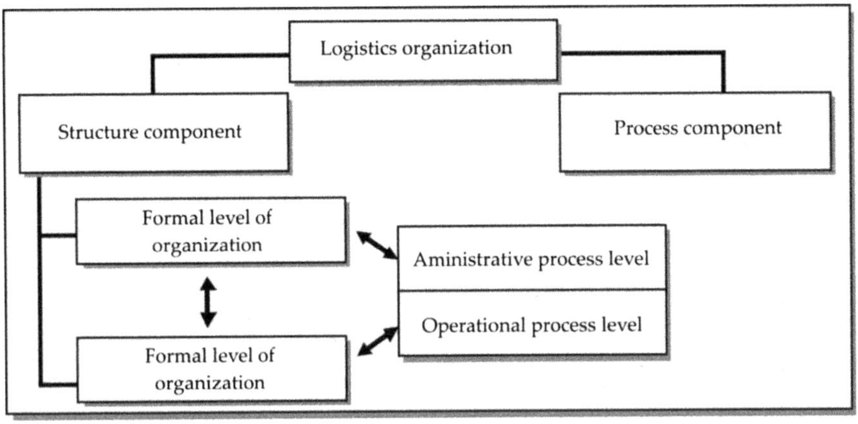

Fig. 3.4 Determining elements of logistics organization structures (Cf. Klaas 2002, p. 131)

organizational structures and operational structures. As is highly characteristic of logistics and, above all, of supply chain management, giving priority to either organizational-oriented aspects or operational-oriented aspects heavily depends on whether the company features more flow-oriented or process-oriented structures. This type of logistical organization comprises both structural components and process components (see Fig. 3.4).[8]

The structural components comprise both the formal levels of organization and the physical levels of the value-added structure, which enables the provision of services in a supply chain. Included in the value-added structure are all personal (the 'know-how' and staff), material (inventory), technical (warehouse equipment), and geographical (locations and buildings) factors and institutions in terms of their kind, number, capacity, and spatial distribution within a supply chain. The operational processes of the logistics organization include basic processes such as transport, transshipment, storage, packing, signing etc. The administrative processes comprise tasks such as planning, disposition, order processing, and control. Figure 3.5 shows the relationship between structural components and process components. It illustrates the supply chains of individual companies as well as cross-company supply chains (see Chap. 4).

The aim of the organizational structure is to arrange and (sub-) divide the company into positions, departments, and divisions and to coordinate them. A position is the smallest organizational unit within a company. Tasks, responsibilities and materials are given to these positions. Positions with managerial or leadership functions are additionally given authorities and competencies. If several positions,

[8] Cf. Klaas (2005), p. 12 et seq.

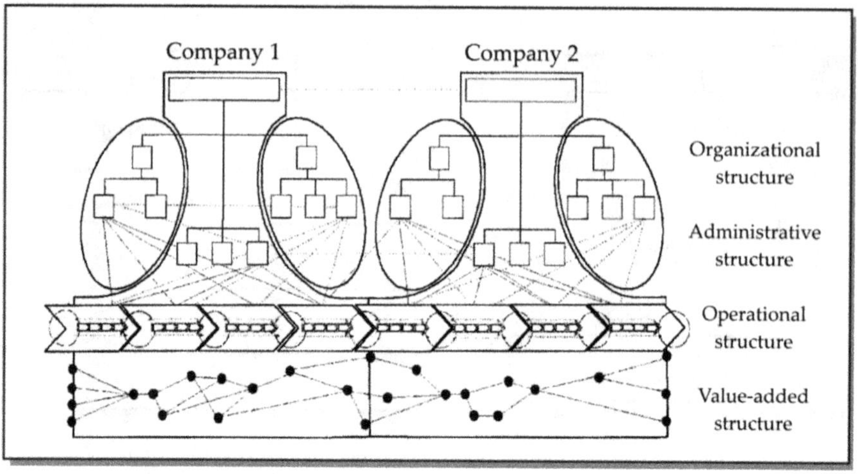

Fig. 3.5 Organization and supply chain (Cf. Klaas 2002, p. 148)

such as dispatcher, warehouse manager and customs manager, are subsumed under one organizational unit, a new department – as for instance a transport planning department or export department – may be set up. Several departments, in turn, form a division. In this case this would be the shipping division. By means of job descriptions (responsibilities) and organizational directions, clear regulations are issued to assign rights and duties to the holder of the position and to stipulate authorities and reporting lines. Job descriptions outline the tasks and responsibilities of a position, its superordinate, co-ordinate, or subordinate status, and its relation to other departments. Organizational structures are visualized in organograms.

In practice, summaries and documentations of job descriptions as well as organograms are compiled in organizational handbooks. Such handbooks are an integral in designing the operational structure, whose main objectives include the appropriate sub-division of the overall process into specific operations, the determination of the ideal operation sequence and supporting the flow of information, forms and documents.

The basic principles of organizational design, namely the division of labor (specialization) and coordination, also apply to logistical organization structures. Thus, the overall logistics task needs to be sub-divided into different smaller tasks – such as disposition of goods, order processing, or transport planning – and subsequently brought together again. The kind of specialization is decisive for the organizational design. Specialization may be performance-oriented, i.e. functional positions or departments are set up. This is what is referred to as functional organization structure. On the other hand, it is also possible to specialize in certain objects if subtasks of a company focus on these objects. These objects may be products, product groups, markets, regions, or customer segments. This is referred to as divisional organization structure.

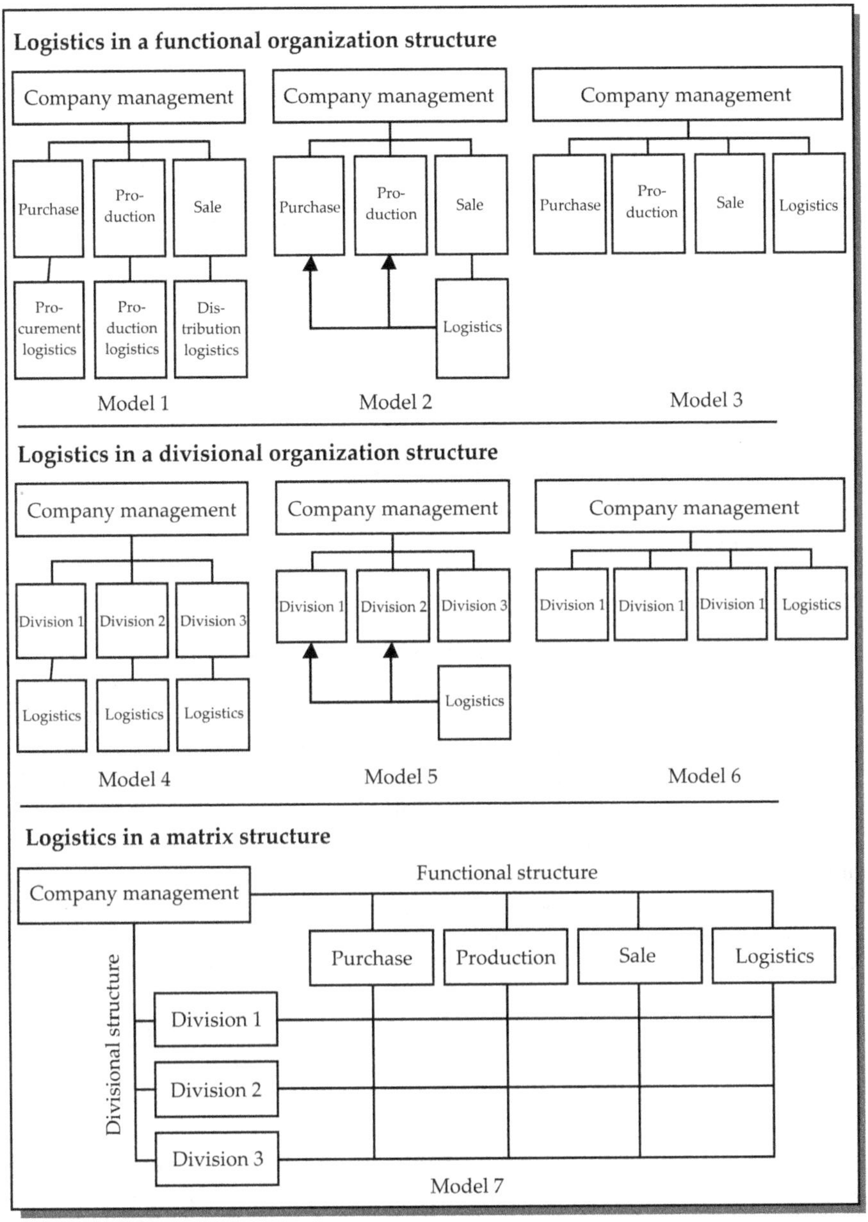

Fig. 3.6 Logistical organization models (Cf. Schulte 2009, p. 559)

Regarding logistics organization, Fig. 3.6 shows logistical organization forms which can be distinguished as follows[9]:

- Functional Organization Structure
  - Model 1: Logistical sub-functions, such as procurement logistics, production logistics and distribution logistics, are generated to match the respective functional areas of purchase, production, and marketing.
  - Model 2: The cross-functional coordination is effected by setting up a logistical functional area.
  - Model 3: An independent central department is established for logistics which operates on the same level as purchase, production, and marketing.
- Divisional Organization Structure
  - Model 4: Each division operates its own logistics department.
  - Model 5: The logistical processes are coordinated by an independent logistics department.
  - Model 6: An independent central department carries out the logistical tasks for the respective division.
- Matrix Structure
  - Model 7: Logistics is an independent functional area responsible for all logistical operations.

Apart from these organizational designs, there are a number of other logistical organization structures which can be employed by small to medium-sized businesses or by businesses with a small volume of logistics tasks. By establishing a central staff unit, it is thus possible to coordinate the flow of goods and materials centrally without having to set up a new department. In these cases, however, the management responsibilities are given to other functional areas such as production or sales. That is why staff units may often only have a say by virtue of their methodical and professional competencies as for instance budgeting or simulation calculation, cost-benefit analysis, or key figures development. The same is true for expert panels (conferences, committees, meetings) and teams that assume advisory functions and/or perform preparatory work for decision-making. Teams are especially suitable for collaborative and temporary tasks and projects.

Along with the development of logistics (see Chap. 1), there has been a change in the company's logistical organization structure.[10] This change is characterized by a shift away from rigid, hierarchy-oriented organizational structures towards more flexible, process-oriented operational structures. This organizational form is also referred to as process chain management.[11] In the case of a process-oriented organizational design, the criterion of process efficiency is essential. Process

[9] Cf. Schulte (2009), p. 558 et seq.

[10] Cf. Pfohl (2010), p. 238 et seq.

[11] Cf. Ihde (2001), p. 343.

efficiency is to be understood as the orientation of the internal performance processes towards the goals of the company as a whole.[12]

The main characteristics of process orientation are[13]:

- The process-specific allocation of resources
- The assignment of processes to process owners
- A marked customer orientation

A clear resource allocation reduces the level of resource interdependency between individual processes, which, for example, leads to faster cycle times. The appointment of process owners with corresponding discretionary, control, and coordination competencies foregoes the setup of additional managing, controlling, and coordinating authorities. Customer orientation allows customer requirements to be considered across all departments and not only those with immediate concern, for instance, sales or distribution. There has been an ongoing appreciation of logistical functions in operational practice in line with these developments. It is reflected both in the hierarchy layout of logistics areas (department management, corporate management) and in the functional areas themselves (logistics centers, site logistics, plant logistics).

However, knowledge of logistics concepts, in many cases, has been circulating in companies and individual departments so that sometimes it does not appear necessary to further anchor professional expertise in solid knowledge, at least with regards to logistics. It seems more important to optimize cross-company alignment between the agents in the supply chains.

3.4.2 Inter-Organizational Optimization Through Supply Chain Management

In the preceding chapters, we discussed concepts of logistics and supply chain management which were mostly based on material-flow techniques or information techniques or featured solutions based on special methods and models. Apart from these concepts, it is also the ways of collaboration between the agents in supply chains, and thus their relationships among each other, which are becoming more and more important. This has to do with the fact that supply chains are becoming increasingly complex and globalized and also with the growing number of companies involved and the resulting higher information technological interconnectedness.

Numerous changes reinforce these tendencies. Thus, traditional customer-supplier relationships are evermore developing into innovation-oriented partnerships. The choice of suppliers and service providers is in many cases no longer based on a comparison of offers but on the level of existing development competence. Additionally, pricing processes and rebate policies are undergoing changes which make

[12] Cf. Frese (2005), p. 317.

[13] Cf. Dehler (2001), p. 141.

for a replacement of annual price rounds without disclosure of the calculations by continuous price and cost reductions based on the disclosure of target costs.

Owing to the cross-company responsibility and management of the flows of goods, supply chain management also needs to undergo organizational changes. This gives rise to logistical organization designs based on networks which subsume all cross-company relationship structures comprising several partner companies within a service network that trade with each other cooperatively.[14] A prerequisite for cooperation is the will to cooperate and the genuine purpose of overall optimization. To ensure this right from the beginning, cooperations should be encouraged at senior level by the respective corporate management as there often are initial misconceptions and opposition to be dispelled.

Ideally, the qualities of the agents complement each other in supply chains based on the division of labor and thus lead to improved productivity and increased revenues and profits, which is also referred to as *win-win-situation*.[15] However, this improvement potential is not generated automatically. Moreover, there are a number of problems which require appropriate relationship management. It seems interesting that the information and communication technological components do not constitute a problem in the realization of supply chain management concepts but that problems arise in the formation and maintenance of partnerships.

One of the main reasons for this is the fact that companies often lack familiarity with and experience in managing business relations. This is especially true for handling the apparent paradox of building up sustainable but yet highly flexible partnerships in a fast changing environment. Problems also frequently arise with regard to choosing the right business partners. Furthermore, operationalizing the cost and benefits optimization, which is being achieved through relationship management, often poses problems. Thus, the goals of relationship management are[16]:

- The results for the partner (successful relationship, fair relationship)
- The activities of the partner (specific investments, specific stakes)
- One's own goals (successful relationship, fair relationship)

Obviously, in operational practice there are numerous hindrances to the realization of these goals. In particular, choosing the right partner is of paramount importance since long-term cooperation might be desired. Business relationships may thus be characterized as dominant, confrontational, cooperative or amicable. Partnerships between agents in a business relationship should feature the following characteristics:

- High technological standard
- Existing development and innovation activities
- Distinct skills in single functional areas

[14] Cf. Stölzle (1999b), p. 587.

[15] Cf. Scheer and Borowski (1999), p. 9.

[16] Cf. Stölzle (1999a), p. 224 et seq.

It is essential to build up reputation and trust. This can be achieved by agreeing on exclusivity, risk-taking, fair distribution of decision-making rights between the partners in a non-centralized way, and by demonstrating reliability.[17] Based on this, differing interests can be expressed and mutual obligations stipulated. However, this may lead to mutual dependency between the partners and loss of control over one's own resources. Trust, and thus the disposition to enter into long-term partnerships, can only be brought about if the partners are under the subjective impression that investing in the business relationship is proportionate to the benefits resulting from this relationship. This can be measured in monetary dimensions, such as increased volume of turnover or profits, but also in non-monetary dimensions.

The risk of opportunistic behavior after the disclosure and exchange of information, such as sales figures, cost data or profit margins, may pose another problem. If this information is leaked to the dominating partner, they might use it to improve their own market position by demanding additional services or discounts.

An example of this would be the market position of the trade sector compared to that of the manufacturer. Since the trade sector is closer to the consumer and because it has the manufacturer's products in its product range, the manufacturer's success is in part heavily dependent on the trade sector. Another example is the provision of sales data at the point of sale. If the cooperation between trader and manufacturer is characterized by mutual trust, this data can be provided without any problems and rationalization potential may be realized. This, however, is still rarely the case and often the manufacturer is only given aggregated central warehouse data which is far less detailed and does not allow for or at least complicates the supply of subsidiaries.

3.4.3 Intra-Organizational Behavior and Changed Staff Requirements

The deliberations outlined above about the inter-organizational prerequisites of logistics and supply chain management cannot conceal the fact that an improvement of inter-organizational coordination still requires coordinating the intra-organizational processes. This means that inter-organizational and intra-organizational coordination are mutually dependent and only jointly enable further optimization. The mentality of department and sector optimization is to be superseded by a mentality of thinking in holistic, customer-oriented processes. Goal conflicts between the organizational units should be avoided.[18] In this context it is important to bear in mind the mutual dependencies which, above all, occur between the areas of marketing, sales, and logistics. These reciprocities, also referred to as trade-offs, are established by the

[17] Cf. Stölzle (1999a), p. 229 et seq.

[18] Cf. Pfohl (2010), p. 233.

actions or decisions of one organizational unit, thereby modifying the decision area of another organizational unit in a goal-relevant manner.[19]

These trade-offs result from possible goal conflicts between the logistics department and other business functions. Due to market conditions, the marketing and sales departments may demand comprehensive 24/7-availability of a broad and deep product range, often without considering the associated inventory costs and the cost effects of an extensive product range. Thus, marketing goals may (initially) play a counterpart to a desired logistical efficiency.[20] Suitable cost and benefit calculations as well as controlling instruments therefore need to make goal conflicts transparent internally (see Chap. 10). Examples of cross-company solutions are Vendor Managed Inventory (VMI), Collaborative Planning Forecasting Replenishment (CPFR) etc. (see Sect. 7.5).

In order to develop and realize the right solutions for inter-organizational and intra-organizational processes it is important that the people involved are not only professionally but also methodically competent and that they display a high level of soft skills. On the other hand, they need to have the necessary authority to be able to make and enforce decisions. This entails increased requirements for staff members in the overall optimization of supply chain logistics. Purchase decisions, for example, may thus have repercussions on the logistics level which the purchaser needs to take into account and has to answer for. It is not least due to these changing requirements that opposition is encountered. Advanced vocational training courses can help alleviate these negative effects.

Cooperation concepts of ECR turn former product managers of producers into sales-oriented commodity group managers, whose autonomous management of their commodity group takes the specific interests of their partners into account. One should not underestimate the influence of the employees within growing organizational structures that can impact the implementation of such changes. The setup of supply chains usually calls for dispensing with familiar working models and hierarchical models and makes it necessary to embrace change. This, however, also means that both openness and trust must be prevalent among the employees so that the right solutions can be found and fair distribution of the jointly achieved benefits can be effected.

---

[19] Cf. Frese (2005), p. 242 et seq.

[20] Cf. Zentes (2004), p. 256.

Review Questions

1. Give examples of organizational logistics structures.
2. What are logistics systems?
3. What are the basic principles of controlling logistics systems?
4. What is the significance of system leadership and how can it be influenced?
5. Describe different organizational and operational logistics structures within companies.
6. What is the difference between process-oriented organization and conven tional forms of operational organization?
7. What is a process owner?
8. What kind of competencies do employees in logistics organizations need to have?
9. What are the main challenges of logistical relationship management?
10. Draft a job description for a transport planning role.

Additional Literature

Bowersox, D. J., Mentzer, J. T., & Speh, T. W. (1995). Logistics leverage. *Journal of Business Strategies, 12*, 36–49.

Chopra, S., & Meindl, P. (2009). *Supply chain management*. Upper Saddle River: Prentice-Hall.

Christopher, M. (2005). *Logistics and supply chain management*. Harlow: Pearson Education Limited.

Gleißner, H. (2000). *Logistikkooperationen zwischen Industrie und Handel*. Göttingen: Cuvillier.

Göpfert, I. (2005). *Logistik Führungskonzeption*. München: Vahlen.

Klaas, T. (2002). *Logistik-Organisation*. Wiesbaden: DU-Verlag.

Laurent, M. (1996). *Vertikale Kooperationen zwischen Industrie und Handel*. Frankfurt: Deutscher Fachverlag.

Pfohl, H.-C. (2010). *Logistiksysteme: Betriebswirtschaftliche Grundlagen* (8th ed.). Berlin: Springer.

Rogers, D., & Tibben-Lembke, R. S. (1999). *Going backwards: reverse logistics trends and practices*. Reverse Logistics Executive Council: University of Nevada, Reno.

Schulte, C. (2009). *Logistik-Wege zur Optimierung der Supply Chain* (5th ed.). München: Vahlen.

Sebastian, H.-J., & Grünert, T. (Ed.) (2001). *Logistikmanagement: Supply chain management und E-Business*. Stuttgart: Teubner Verlag.

Spengler, Th., Voß, St., & Kopfer, H. (Ed.) (2004). *Logistikmanagement: Prozesse, Systeme, Ausbildung*. Heidelberg: Springer.

Stock, J. R., & Lambert, D. M. (2001). *Strategic logistics management* (4th ed.). Homewood: McGraw-Hill Irwin.

Zhao, M., Droge, C., & Stank, T. O. (2001). The effects of logistics capabilities on firm performance. *Journal of Business Logistics, 22*(2), 91–100.

4 Logistical Infrastructure

Learning Objectives

Logistical infrastructure is the backbone of logistics systems. Apart from the transport infrastructure, the suprastructure including the logistics locations and real estate as well as the telecommunication infrastructure are becoming increasingly important for logistics. It is the aim of this chapter to illustrate the logistical functions of the various infrastructural facilities. To this end, the numerous complexities and interdependencies of the individual infrastructures will be identified and amalgamated into the logistical infrastructure.

Keywords

- Transport infrastructure
- Infrastructure, superstructure
- Individual properties, logistics parks
- Terminals
- Transshipment facilities
- Freight villages
- Satellite navigation
- Disaster Recovery
- Business Continuity

H. Gleissner and J.C. Femerling, *Logistics*, Springer Texts in Business and Economics, DOI 10.1007/978-3-319-01769-3_4,

4.1 Basic Terminology, Types and Functions

A capable *logistical infrastructure* is prerequisite for modern logistics systems in which efficient logistical processes are to be carried out. In general terms, infrastructure means the entirety of sustainable facilities and supply channels that can be used by private households and companies. We distinguish between *economic infrastructure* and *social infrastructure*. Economic infrastructure comprises transport infrastructure, information infrastructure, and (tele)-communication infrastructure.[1] Social infrastructure includes institutions for education, healthcare, culture and security within a state and society. In contrast to this macroeconomic view, the microeconomic dimension of logistical infrastructure refers to the definition of the spatial and technical structural properties and to the dimensioning of a logistics system. This, for example, includes warehouses, means of transport, conveyors, storage and picking technology and facilities as well as the corresponding information and communication systems.[2] It is therefore in a comprehensive and holistic sense that the concept of logistical infrastructure covers the micro- and macro-economic infrastructure facilities of transport infrastructure, site infrastructure, building infrastructure, and communication infrastructure (real estate).

Transport infrastructure comprises the infrastructure for passenger transport and freight transport. This includes all physical traffic routes such as railroads for trains and city railroads, roads, tunnels, waterways and pipelines including their control systems, and handling facilities. These are also termed linear or networked traffic infrastructure or transport infrastructure. In contrast, *site infrastructure* includes airports, train stations, inland ports and sea ports, which are also referred to as point-shaped traffic or transport infrastructure. Logistical infrastructure in this sense not only comprises traffic facilities but also facilities for transshipment and storage at the site's terminal. These facilities are also termed *suprastructure* (see Fig. 2.5).

The information exchange necessary for logistics processes is made possible by telecommunication and its corresponding infrastructure. Telecommunication infrastructure includes telephone networks, fibre optics networks and their network nodes, satellites with their associated ground stations, and data centers. This is also referred to as web-based or digital infrastructure. As part of all the infrastructure facilities mentioned so far we can also summarize buildings, structures and properties in different shapes and sizes under the term *real estate infrastructure*.

---

[1] Also Cf. Pfohl (2000), p. 349 et seq., who uses the term macro-logistical infrastructure to cover the infrastructure of goods and information flows, as well as transport policy.

[2] Cf. Arnold et al. (2008), p. 901.

4.2 Transport Infrastructure and Suprastructure

4.2.1 Transport Routes and Transport Networks

Depending on the respective transport mode, transport infrastructure can be broken down into the national and international networks of air transport, rail transport, inland waterway transport, sea transport, and road transport.[3]

The EU features a *network of air routes* which comprises more than 450 airports operated by more than 130 airlines. Hub-and-spoke networks have been set up by airlines in order to operate these air routes in the most efficient manner (see Sect. 8.4.2). Within these networks, the busier routes are bundled together by central airports (hubs). Less busy routes start from the spokes and are then joined to the hubs.

The longest *railroad networks* in Western Europe are located in Germany (38,206 km) and France (30,832 km). The quality of railroad networks within Europe differs as to the degree of electrification and the expansion of high-speed train networks. Differing track gages, e.g. in Spain and Russia, necessitate transshipment in cross-border transport. By 2020, Spain will convert its railroad network to the standard European gage.

Inland waterway networks are made up of navigable waterways that serve as a link between inland ports and/or sea ports. Depth and breadth of waterways are defining for their utilization. The longest networks can be found in Finland (7,884 km), Germany (6,636 km), the Netherlands (6,183 km), and France (5,384 km).

Sea transport networks provide routes for either short sea shipping or international ocean shipping. Short sea shipping is carried out on pre-defined offshore routes with good connections to the Hinterland. Sea routes determine the connections on the high seas.

Extensive *road networks* make a region accessible. Within Europe, Germany (12,363 km), Spain (11,622 km), and France (10,801 km) feature the longest highway network. Germany distinguishes between federal highways (*Bundesfernstraße/Bundesautobahn/Bundesstraße*), state roads (*Landesstraße*), county roads (*Kreisstraße*), and municipal roads (*Gemeindestraße*).

Transport infrastructure is increasingly controlled by the *information and communication infrastructure* of systems that are associated with the respective transport mode. This serves the purpose of improving the transport infrastructure's quality. Examples for inland waterway transport include river information systems (water levels, lock allocation etc.) and automatic indentification systems. In sea transport, automatic identification systems are provided in the form of Internet applications and portals.[4] Numerous telematics systems are available for road traffic transport in order to track vehicles and goods, to monitor rest and driving periods, and to capture data about tolls, traffic congestions and diversions.

---

[3] Cf. Rühl (2008), p. 37 et seq.

[4] Cf. Biebig et al. (2004), p. 444 et seq.

4.2.2 Airports

Air transport is organized according to the following (sub-) systems[5]:

- Air transport services offered by airlines
- Set-up and operation of air traffic control and control systems
- Set-up and operation of airports (airport services)

Airports fulfill important infrastructural functions. Their main purpose is to provide *infrastructure relevant to air transport.* This mainly includes the provision of runway systems, passenger terminals, and connections to ground transport systems (see Fig. 4.1).[6] Airports can be subdivided into landside and airside areas as well as air freight buildings.[7] The terminal is the main *landside* building. In general it consists of two areas separated by border and passport control through which all passengers and flight personnel need to pass preflight. Check-in (ticket collection, seat reservation, luggage check-in, security control) takes place in the front part of the terminal.

Frequently, this part also features shops, restaurants, and exhibition areas (non-aviation area). Waiting areas and accesses to the gates are located behind this part.

The *airside area* of an airport comprises the system of runways which is needed to enable aircraft movements (take-off, landing). Taxiways, maneuvering areas, hangars, and cargo centers also form part of this area.

Depending on the location and distribution area of airports and thus on their position in the hierarchy of central facilities, we can distinguish between *regional* or *national* airports for feeder and distribution services and *international major airports* serving as hubs and transit points. In this context, an airport's significance as a logistics location and its interconnection with the surrounding region are displayed by nearby *airport industrial parks, airport terminals* or *airport logistics centers* for air cargo dispatch.[8] However, rather than necessarily offering air cargo services, logistics service providers in many cases only take advantage of an airport's prominent position within the transport system and of the existence of a highly developed transport infrastructure.

The development of *cargo airports* was brought about by the importance of air cargo and due to the limited service offers of existing airports such as capacity bottlenecks and bans on night flights.[9] In most cases, the hubs of mail and parcel service providers (integrators) are located at these airports (see Sect. 5.5.2). This, in

[5] Cf. Koch (2006), p. 93 et seq.

[6] Cf. Maurer et al. (2003), p. 69.

[7] Cf. Arnold et al. (2008), p. 770.

[8] Cf. Ihde (2001), p. 214; Vahrenkamp (2005), p. 301 et seq.

[9] Cf. Meeder (2003), p. 492.

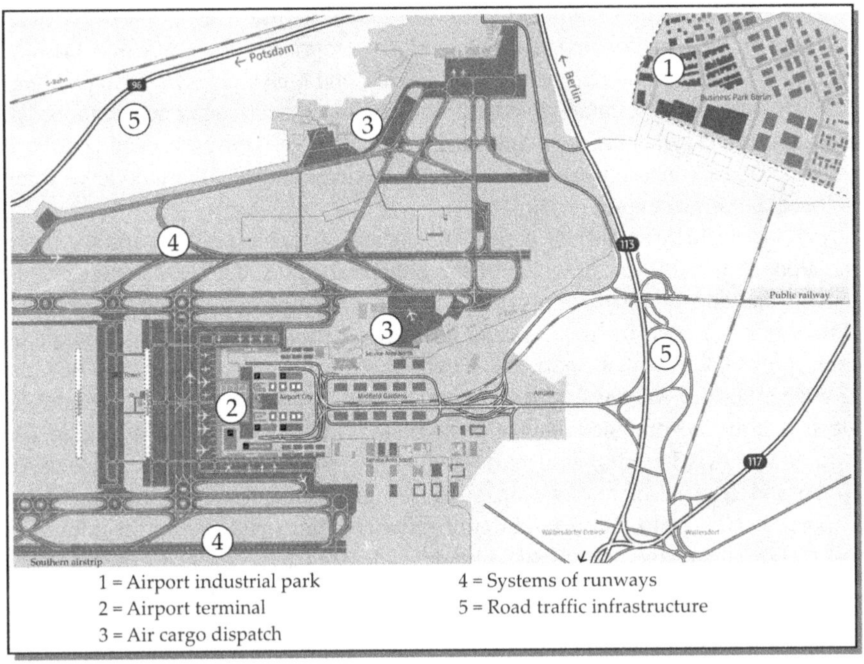

Fig. 4.1 Schematic outline of Berlin's airport 'Willy Brandt' (Berliner Flughäfen 2011, p. 10)

turn, directly affects the set-up of the airport's infrastructure. The service portfolio of an airport and thus that of an airport operator is of direct relevance for logistics. The most relevant factors are:

- Guaranteed 24 h operations in all weather conditions
- Minimal ground times for air cargo
- Existence of suitable cargo facilities with sufficient space for consolidation and deconsolidation of cargo and for intermediate storage
- Capacities for direct transshipment of passengers, luggage, and cargo

4.2.3 Sea Ports and Inland Ports

Sea ports or inland ports are infrastructure facilities consisting of berths for seagoing and/or inland vessels. They serve as an interface between land and sea transport and fulfill the functions of transshipment and storage as well as pre-carriage and on-carriage of goods. *Inland ports* are used for transshipment by inland vessels, coasters, and smaller sea vessels. *Sea ports* are used for feeder transport by sea vessels or inland vessels. They can be located at the sea and along canals and rivers. Inland ports are located along rivers, canals, and at inland lakes.[10]

[10] Cf. Ihde (1991), p. 99.

With regard to a port's infrastructure, we generally distinguish between the infrastructure and the suprastructure. *Infrastructure* comprises immovable port facilities such as docks, shipping channels, track systems, and quays. *Suprastructure*, in turn, includes movable and immovable facilities such as crane systems, warehouses, and industrial trucks. Also included in this superstructure are information and communication systems which illustrate the interdependencies between the individual forms of logistical infrastructure.

Port infrastructure is largely dependent on the development of the ports and on the functions they fulfil. In the wake of the industrialization, the development of the railway and due to globalization, ports have become a *gateway* to the Hinterland.[11] Hinterland is a term for those inland territories behind a port which maintain an exchange of goods with the port.[12] The *infrastructural connections to the Hinterland* through pipelines, railways, waterways, roads, and by air are of great importance for the economic development of the Hinterland.[13] The demand for incoming and outgoing seaward (mass) transports and improved transport connections in the Hinterland lead to increased ship sizes, the construction of deeper approaches to ports as well as to the installation of railway sidings and depots for docks and quays. This also exemplifies the interdependencies between different logistical infrastructures. Major factors in the development of ports are the fact that increasingly bigger ships are built as well as the containerization of transport flows (see Sects. 5.3.4 and 5.4.2). Both these developments call for an increased industrialization of transshipment in ports, which, in turn, poses significant challenges for a port's infrastructure and suprastructure.

4.2.4 Rail Stations and Railroad Systems

Rail stations are generally defined as railroad systems with at least one switch, providing a starting and ending point for trains and allowing them to swerve or turn.[14] *Rail infrastructure* includes all railroad systems, properties, buildings and other facilities necessary to perform or secure passenger or cargo transport by rail. An overview of elements relevant to rail infrastructure is given in Fig. 4.2.

Depots serve to load and unload rail cars and wagons. Facilities for loading and unloading may come in the simple form of transshipment areas between freight cars and road vehicles. These areas can be located beside or in between the rails. There

---

[11] Cf. Nuhn (2005), p. 110.

[12] Cf. Biebig et al. (2004), p. 290 et seq.

[13] Cf. Pfohl (2000), p. 352.

[14] Cf. Berndt (2001) p. 82 et seq.

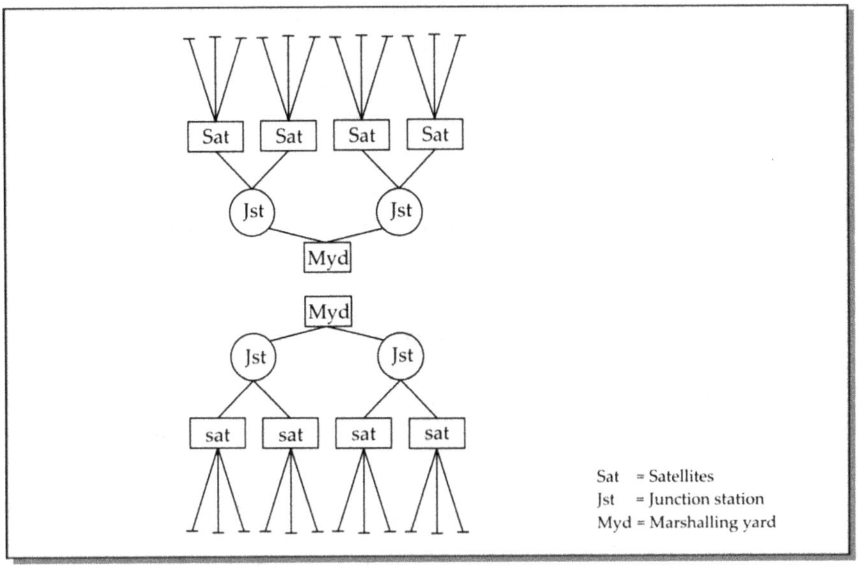

Fig. 4.2 Rail infrastructure (Cf. Berndt 2001, p. 44)

are different ways in which transshipment can be carried out. For example, ramps, cranes, or tipping devices for bulk goods may be utilized.[15]

Train units are formed or decoupled at *marshalling yards*. At these yards, a large number of freight cars can be decoupled from incoming trains within a short time in order to form new freight trains.

Rail sidings also form a part of railroad systems. They provide access to the rail freight network and serve for the loading and unloading of rail cars. In most cases, they are located on the premises of the sending or receiving company. Rail sidings enable companies to transport rail cars directly from the sender to the receiver, thereby avoiding time-consuming and costly transshipment procedures.

4.2.5 Terminals and Transshipment Facilities

Transshipment terminals (transshipment points) are (external) sites at which transshipment activities take place. These activities include the loading and unloading of means of transport, sorting of goods, storage of goods, and release of goods from the warehouse (see Sect. 6.5.2).[16] Transshipment terminals exist for all means of transport and can also be seen as a hub where different transport modes converge.

[15] Cf. Berndt (2001), p. 94 et seq.

[16] Cf. Boysen (2008), p. 1286.

Air cargo terminals are transshipment points in air freight transport between ground-based modes of transport (rail/road) and airplanes.[17] The facilities and equipment for transhipment of air cargo are in the air cargo centers of major airports (see Sect. 4.2.2). The structural layout of a terminal is influenced by the following kinds of transhipment activities[18]:

- Export: receipt of cargo landside and loading onto an airplane
- Import: airside receipt of cargo from an airplane and loading onto a truck
- Transfer or transit: receipt of cargo from an airplane and transfer onto another airplane

Sea port terminals or inland port terminals are furnished with facilities and equipment needed for the loading and unloading of sea vessels and inland vessels. They also feature areas for storage and distribution of goods. Depending on the type of goods being transshipped and their corresponding mode of transport, various *kinds of terminal* can be distinguished[19]:

- Terminals for liquid goods (petroleum products or chemicals) or bulk goods (ore, coal, crop, fertilizers)
- Terminals for general cargo (machines, technical equipment)
- Container terminals
- RoRo Terminals

Handling at these terminals can be carried out using cranes and container gantry cranes, by sucking and pumping the goods into storage areas, tanks, and silos or through direct transshipment. Which one of these handling techniques is utilized depends on the type of goods being transshipped, on the type of on-carriage, and on the available transport connections.

The infrastructural facilities of bulk good terminals may comprise buildings and halls (quay sheds) or open-air storage areas. The loading and unloading of bulk goods – such as coal and ore – is carried out by means of conveyor belts, through free fall in feeding pipes, or by crane. The unloading of a ship at the port of destination can be effected by using the ship's own devices such as conveyor belts or cranes. Liquid goods such as oil are transshipped in special oil terminals. These are equipped with special transhipment devices as, for example, pumps on land for loading and unloading, platforms with loading arms, tanks for intermediate storage, and pipelines.

General cargo with different measurements and properties are loaded and unloaded at *general cargo terminals*. Since this is usually very costly due to the high number of staff members required, powerful transhipment facilities such as gantry cranes and truck cranes are frequently used.

---

[17] Cf. Arnold et al. (2008), p. 766 et seq.

[18] Cf. Mensen (2007), p. 304.

[19] Cf. Brinkmann (2005), p. 137 et seq.

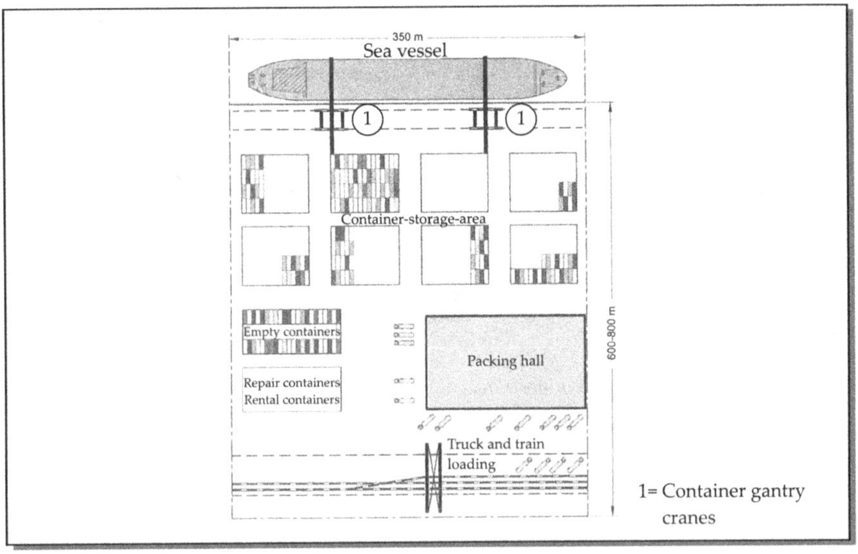

Fig. 4.3 Layout of a container terminal (Brinkmann 2005, p. 239)

A *container terminal*, in contrast, enables the efficient transshipment of goods into containers. The design of a container terminal needs to take into account ship sizes, the types of containers to be dispatched (open-top container, cooling container, hazardous material container, empty containers etc.), container sizes (40, 20, 10 ft.) as well as container capacities (see Sect. 5.4.2) (see Fig. 4.3).

RoRo (Roll-on/Roll-off)-terminals are needed for transport with RoRo ships. This technique was developed for ferry transport, in which ships are loaded through gates and ramps at the stern, side or bow of the ship. Automotive trailers, chassis, large-sized pallets, train cars and containers or vehicles carrying them load the ship by rolling onto it. RoRo-terminals are suitable for smaller ports without landside transshipment facilities but with connections to the road and railroad network.

Terminals in railroad transport are above all transshipment points of combined railroad transport in the form of transshipment terminals and container terminals. They can be small areas for loading and unloading or large transshipment stations. The most important infrastructural elements of these terminals are road infrastructure for road vehicles (driveways for entrance and exit, parking lanes, loading lanes, traffic lanes), railroad infrastructure in the form of siding tracks which connect trains to the main rail network and storage areas.[20]

The following case study illustrates a rail terminal in the form of a railport of *Deutsche Bahn*.

[20] Cf. Berndt (2001), p. 118.

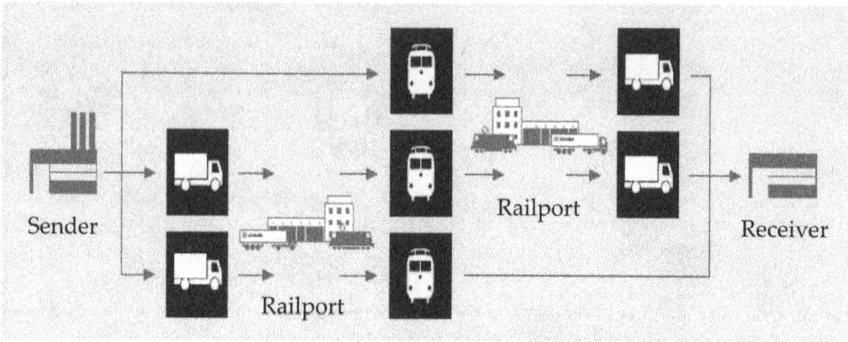

Fig. 4.4 Schematic layout of a door-to-door logistics chain (DB Schenker Rail GmbH 2011)

Case Study 4.1: Railport

A *railport or rail cargo logistics center* is a multifunctional logistics location with direct rail link in order to connect several transport modes. Due to its dominant position within the transport system, this is usually road transport with connections to the highway network and direct access to the distribution network in the catchment area. It is also possible and reasonable to provide connections to inland waterways, airports, and pipeline networks.

Deutsche Bahn AG and its transport and logistics department *DB Schenker* pursue the idea of offering their customers an integrated door-to-door service by making use of these multifunctional and multimodal rail logistics centers throughout Europe (see Fig. 4.4). The development of this concept was triggered by the idea of more deeply integrating the railroad system in order to unlock new logistics potential. This also helps decongest the road infrastructure and thus decreases the ecological impact by using eco-friendly modes of transport. Railports are mostly built in the vicinity of already existing freight depots or cargo terminals.

Railports offer the possibility of transshipping and storing many different types of goods:

- Units of combined cargo traffic (e.g. container, swap bodies, trailers)
- Cranable large-sized goods (e.g. steel products, pipes, wood, machines)
- General cargo (e.g. palletized goods, paper rolls, furniture)
- Bulk goods (e.g. clay, fertilizers, scrap metal, building materials)

However, the services offered at railports go much beyond the mere storage and transshipment of goods. Goods also need to be prepared for on-carriage and distribution throughout the region. This involves services such as picking, inventory management, quality checks, clearing, and delivery or collection of the goods to the region or from the region.

An ideal railport usually features the following components:

- Traffic routes of the different transport modes
- Areas dedicated to the operation of transshipment devices (e.g. gantry cranes, truck cranes, forklift trucks)
- Open areas for handling and storage
- Sheltered or indoor areas to carry out logistics services

Transshipment and storage activities can thus be adapted to requirements of the specific types of goods. Open areas, warehouses, and transshipment halls usually have direct rail and road access.

Depending on the size and number of terminals, the most common transshipment techniques involve cranes (bridge cranes, gantry cranes) and forklift trucks (frontload trucks, sideload trucks, telescopic handlers).

Cross-docking terminals as another type of transshipment terminals in which consignments are directly transshipped between incoming and outgoing trucks without intermediate storage. They are mostly long buildings with numerous gates. They are mainly utilized in consumer goods logistics (retail chains), by mail and parcel service providers (hub-and-spoke systems), by transport companies and carriers (general cargo networks) and in the automobile industry (OEMs, suppliers) (see Sects. 5.5, 7.4, 8.4.3).[21] Different technical devices may be used, depending on industry and function. These include ground conveyors (vehicles), material flow technology, and sorting technology (conveyor belts, chain conveyors, sorting robots) (see case study 6.3).

4.2.6 Logistics Real Estate, Logistics Parks, and Freight Villages

Logistics real estate can be described as a building with appropriate infrastructure and technology built on a suitable property for the purpose of performing transshipment, storage, and all services associated with it (picking, packing, quality checks, (sub) assembly).[22] Therefore, a number of different types and categories of logistics real estate can be distinguished. On the one hand, there are *individual properties* including warehouses, transshipment halls, distribution centers, and logistics service centers.[23] On the other hand, there are *real estate parks* comprising several buildings and sites, which can serve as logistics parks, industrial parks, or freight villages.[24]

Warehouses may be used as central warehouses, regional warehouses, production warehouses, or distribution warehouses. Their purpose is the storage of goods, and they serve as a link between procurement and distribution. The way they are

---

[21] Cf. Boysen and Tiedemann (2010), p. 54.

[22] Cf. Börner-Kleindienst (2006), p. 453; Nehm et al. (2009), p. 24.

[23] Cf. Börner-Kleindienst (2006), p. 454.

[24] Cf. Nehm and Schryver (2007), p. 233 et seq.

equipped is dependent on the range of functions they fulfil and on the logistics processes to be carried out on the property (see Sect. 8.3). In most cases, warehouses offer good delivery capabilities for a speedy transshipment of goods.

Transshipment halls serve to distribute goods that arrive through long-distance and short-distance transport. Since goods are stored for only a short time and because the throughput rate is accordingly high, transshipment halls feature a small height and depth but many gates for usually double-sided delivery. The gates need to be designed in a way which enables the loading and unloading of large trucks, utility vans, as well as station wagons.

Compared to warehouses and transshipment halls, *distribution centers* are significantly larger and have a greater number of loading gates. In addition, they are frequently equipped with (automated) storage and picking systems. The type and quantity of goods and commodities to be handled have a significant bearing on the size and capacities of a distribution center. The spectrum of goods ranges from all kinds of consumer goods (food, necessities, brown goods, white goods) to various industrial goods. Perishable and temperature-sensitive products, for example, require facilities for low-temperature storage which, in turn, raises certain challenges for the property (layout, accesses, insulation). Machines and facilities require an especially stable ground as well as special lifting and transport devices (cranes) (see Sect. 6.6).

Logistics service centers have the purpose of concentrating several suppliers, warehousing companies, and transport companies in one place so that production can take place nearby (e.g. just-in-time delivery). To this end, goods from different sources need to be consolidated and processed in, for instance, production-logistics centers (light industrial).

A *logistics park* is an area where several logistics properties are concentrated in close proximity. The individual properties may differ with regard to their type and size as well as to their tenancy and ownership relations. The main purpose of a logistics park is the realization of synergies through the common use of infrastructural facilities such as gas stations, forklift truck fleets, workshops, offices etc. The administration and maintenance of the infrastructural facilities is at the responsibility of the park owner. Since these parks are usually managed as an integrated whole, it is safe to say that they are cheaper to maintain than an individual property. Logistics parks may also be designed in an industry-specific way and according to required logistical concepts. For this reason, there are numerous concepts for logistics parks, which are also termed supplier parks. These are especially prevalent in the automobile industry (see Sect. 7.4.3).[25]

Due to the high accumulation of logistics real estate, freight villages can be seen as an extended form of logistics parks. Freight parks, however, serve purposes related to transport and traffic policy and to the supply and disposal of goods in specific regions (see Sect. 8.4.4).

[25] Cf. Becker (2006), p. 41.

Table 4.1 Location and object requirements of logistics real estate (Cf. Muncke et al. 2008, p. 202)

Location requirements	Object requirements
Macro location • Active promotion of economy and investment • Relationship between economy and government / administration (investment climate, according to the time required to obtain permits and the amount of regulations) • Low tax burden • Availability of workforce • Good living standards for executive staff *Micro location* • Commercial and logistical use secured by planning and building law. • 24h accessability via the national road network (at a distance of max. 1-3 km) • Proximity to airports (max. 1hr driving time) • Technological infrastructure is available and expandable • Possibility of carrying out compensatory measures • Supply with retail and services close-by	• Property on ground level with good potential for expansion, property completely fenced in • 45-60 % of property built-over or ready to be built on (preferably 5 ha or larger) • 5-10% allocated office space • Single-floor warehouse with at least 10,000 m² • Warehouse height 10-12 m; distance between pillars > 12.5 m • Ground bearing capacity > 5,000 kg/m² • Ceiling sprinkler system; gas heating • Great number of ramps and one gate on ground level per section (at least one ramp for each 1.000 m² of warehouse); double-sided delivery for transhipment warehouses • Maneuvering space and yard space of at least 35m depth; sufficient parking spaces for cars, trucks and semi-trailers • Flexible utilization which ensure multi-functionality and adaptability; possibility to form sub-divisions, high alternative use capacity

Regardless of the individual category and the actual design of logistics real estate, specific requirements for locations and objects are increasingly becoming apparent. These requirements are listed in Table 4.1.

4.3 Information Infrastructure and Communication Infrastructure

4.3.1 Telecommunication Infrastructure

Along with transport infrastructure, an efficient information and communication infrastructure is vital for the development of logistics concepts and for the performance of logistics processes. This infrastructure is to a great extent based on telecommunication infrastructure and can be set up using different networks (landline, mobile telephony, radio network, microwave radio relay), depending on the services offered (GSM, UMTS) and on the data transfer.[26] There is, however, a tendency towards the development of a uniform network infrastructure through which all services can be offered and utilized. This network infrastructure is based on the Internet Protocol (IP) which replaces the circuit-switched networks with a packet-switched network infrastructure.

Telecommunication networks can exhibit different structures (network typology) which utilize various types of hardware and access methods for data connection and transmission. These, in turn, determine the rate of transmission, the data throughput, and data security concepts. Thus, local networks (LANs – Local Area Networks) consist of several computers and external devices (printers, scanner etc.) which are interconnected in one building. Internet access is given via a router. In contrast to LANs, WANs (Wide Area Networks) cover a large geographical area. Commercial WANs are designed for maximum capacity utilization and consist of circuit-switched connections, point-to-point connections, packet-switched connections, and Virtual Private Networks (VPN). In order to support these services and to achieve high transmission speeds, optical transmission media (fiberglass) are used more and more frequently for broadband infrastructure.[27]

All these hardware and software installations are usually hosted in IT rooms, server rooms, or in data centers which display a specific infrastructure. The infrastructure of data centers includes the provision of rooms, energy supply, air conditioning, and object security.[28]

The development of IT and computer technologies does not only substantially influence the kind of services rendered in data centers – it also affects the property and room layout. Data centers are a specific type of real estate which are also termed collocation centers, IT centers, IT hotels, server hotels, telehouses, and so forth.[29]

We can also see an increase in the number of data center parks in which companies rent cages, several rooms, or entire buildings which are then equipped or (re-) constructed according to the customer's specifications. These types of solutions offer the advantage of redundant provisioning of building infrastructure as well as the possibility to provide office space with emergency workplaces. These workplaces can be utilized in emergency situations (disaster recovery, business continuity) in order to continue business without interruption, which is also becoming increasingly important in a logistical context.

[26] Cf. Heiserich et al. (2011), p. 361 et seq.

[27] Regarding this technology Cf. Keller (2011), p. 141 et seq.

[28] Cf. Rittweger and Rossbach (2006), p. 239.

[29] Cf. Seitz (2004), p. 29.

4.3.2 Satellite Systems and Satellite Navigation

World-wide telecommunication is to a great extent based on satellite systems. This technology makes it possible to set up a comprehensive infrastructure that offers services with high data transmission rates. A logistical example of this is the ERMTS (European Railway Transport Management System), which offers several projects and services in the area of railroad information systems, as for example the international GSM-R-network (Global System for Mobile Communication Railways). This network is a platform for commercial railroad radio systems.[30]

Apart from the support of telecommunication services, satellites also offer *satellite navigation* as one of their core functions. This method enables the determination of an object's position. Using suitable technologies and programs, modern satellite navigation makes it possible to determine the coordinates of locations based on their distances to at least three satellites. The construction of the European satellite system *Galileo*, due to be operational by 2013, is of paramount significance for the commercial use of satellite systems. Galileo will be a system of the European Union which adds to the already existing state-owned US satellite system GPS (Global Positioning System) and to the Russian system Glonass (Globalnaja Nawigazionnaja Sputnikowaja Sistema). The system will comprise 30 satellites. Galileo will make it possible to offer different services which vary in regard to accuracy, number of signals, and reliability of service.

Commercial satellite navigation services are especially suitable for logistics and can be used for the navigation of continental transport, telematics platforms, for locating purposes in aviation and shipping, and as research platforms for transport and logistics systems.[31] The additional availability of these applications in comparison to existing systems is mainly due to the system's high accuracy and its world-wide availability. Central to these applications is the localization and tracking of goods which are transported in a multi-modal manner. This requires constant location of the respective carriers and of the goods transported, and all parties involved in the supply chain need to be able to continuously obtain information across all transport modes and independent of their location (location information).[32]

---

[30] Cf. Berndt (2001), p. 122.

[31] Cf. Schenk et al. (2011), p. 573 et seq.

[32] Cf. Clausen and Inninger (2009), p. 43 et seq.; Elsenbach (2006), p. 445.

Review Questions

1. What does logistical infrastructure mean?
2. What is the difference between superstructure and transport infrastructure?
3. What is the significance of information and communication infrastructure for logistics systems?
4. Name the infrastructure facilities of an airport that are relevant for air cargo transport.
5. Name the most important factors that need to be considered when setting up a container terminal at a sea port.
6. Describe the functions of a marshalling yard.
7. Differentiate between different types of logistics real estate.
8. What are the most important location and object requirements for logistics real estate?
9. In what way are logistics parks different from freight villages?
10. What is the significance of telecommunication infrastructure for logistics?

Additional Literature

Böse, J. W. (2011). *Handbook of terminal planning*. New York: Springer.

Bowersox, J. D., Closs, D. J., & Cooper, M. B. (2009). *Supply chain logistics management*. New York: McGraw-Hill.

Coyle, J. J., Bardi, E. J., & Novack, R. A. (2000). *Transportation* (5th ed.). Cincinnati: South-Western College Publishing.

Deutsche Bahn AG. (2007). Transport & logistik: Drehkreuze. In *Mobility networks logistics*. Berlin.

Dornier Consulting GmbH. (2006). *Erfolgsfaktoren und Bewertungsmöglichkeiten von PPP-Projekten für Verkehrsinfrastruktur: Eine internationale und verkehrsträgerübergreifende Analyse*. Berlin.

Ernst, M., & Walpuski, D. (1997). *Telekommunikation und Verkehr*. München: Vahlen.

Hatton, G. (1994). Designing a warehouse or distribution centre. In J. Gattorna (Ed.), *The Gower handbook of logistics and distribution management* (pp. 175–193). Adlershot: Chartered Institute of Purchasing & Supply.

Kazda, A., & Caves, R. E. (2007). *Airport design and operation*. Amsterdam: Elsevier.

Klauenberg, J. (2009). Modellierung und Typisierung von Distributionszentren in Seehäfen. Dissertation Universität Rostock, Hamburg.

Song, D.-W., & Panayides, P. H. (2012). *Maritime logistics: A complete guide to effective shipping and port management*. London/Philadelphia: Kogan Page.

Stock, J. R., & Lambert, D. M. (2000). *Strategic logistics management*. New York: McGraw-Hill.

Vahrenkamp, R. (2005). Logistik: Management und Strategien, 5., vollständig überarbeitete und erweiterte Edition, München: Oldenbourg Verlag.

5 Transport Systems and Logistics Services

Learning Objectives

Transport services and logistics services are an integral part of logistics systems. This chapter aims to give an overview of the existing means of transport, the modes of transport and their services, the role that logistics service providers take, and the development they have undergone. In this way, we will equip the reader with a basic understanding of the concepts and service products within the transport and logistics service economy in order to utilize them for the various forms of co-operation.

Keywords

- Transport development
- Modes of transport and transport services
- Categories of modes of transport
- Transport networks
- Transport chains
- Transport technology
- Freight carriers and forwarders
- CEP service providers and integrators
- Contract logistics

H. Gleissner and J.C. Femerling, *Logistics*, Springer Texts in Business and Economics, DOI 10.1007/978-3-319-01769-3_5,

5.1 Transport Basics

A *transport system* comprises the elements of the goods transported (transport object), the means of transport used (mode of transport, traffic routes), and the transport process as an organized transport sequence (transport chain). The process of service provision results in *transport services* being carried out. Transport services cause movements of people or goods. We can distinguish between internal and external transport services. Internal transport services, for instance, occur within a plant among different departments of a warehouse. External transport services take place between suppliers and customers or between plants and warehouses. Transport processes carried out without actual goods being transported are referred to as empty transport. The criteria for a transport service are a providing location, a destination, and the goods transported. Goods are generally not manipulated during the transport process, i.e., merely the technical process of changing location takes place. The transport service performance is measured in tonne kilometers (tkm).

Companies may use their own vehicles to carry out transport services. This is referred to as private haulage. There has been a tendency, however, towards logistics service providers being commissioned with the transport. The means of transport used for this include air, rail transport, road, and water; all of which are also termed *modes of transport*. The vehicles employed in these modes of transport are trucks, railway carriages, as well as inland and ocean vessels. The totality of both means and modes of transport is the pipeline network.

Transport service is a basic element of *logistics services*. These services comprise all relevant interdependencies between production, transfer, and consumption and are integrated into the value chains of senders and receivers. Apart from the transport component, logistics services also include the subservices of transshipment, storage, sorting services, and auxiliary services. Transshipment consists of the loading and unloading of vehicles. Storage serves to modify the temporal properties of goods and to bridge time. Sorting services, such as sorting and sorting out, alter the quantitative and categorical composition of goods. Auxiliary services include services such as packing, insuring, customs clearance, as well as less significant manipulations of goods and quality checks.

Production and cross-border consumption processes based on the fragmented work flow, and thus the *mobility of goods,* only become possible through transport, traffic, and logistics services in conjunction. The degree of the mobility of goods on a national level and cross-border transports are directly linked to the intensity of the exchange processes and to the level at which division of labor is implemented.

5.2 Significance and Development

The development of the transport and logistics sector largely depends both on the economic development, conveyed in terms of economic growth, and on the further development of the logistic structures and processes within the value adding systems of the sectors of industry, trade, and service. These can be assessed by looking at the supply chain strategies of companies.

One of the main strategies is decreasing the vertical range of manufacture, i.e., the outsourcing of value-adding steps to suppliers and service providers. This strategy continues to become more popular, which results in a significant increase in the Europe-wide and world-wide flows of goods between companies. The increasing division of labour leads to more freight traffic (measured in tons) and larger transport volumes (measured in tons per kilometer). Both of these parameters are experiencing a growth which is disproportionate to the growth of production (GDP). This development is referred to as *volume of goods effect*. It conceptualizes a decrease in the share of typical bulk goods transported while the share of high-quality general cargo transported increases due to changed procurement and production strategies. An example of such strategies would be the modularization of products and the procurement of these modules (modular Sourcing), thereby increasing the value of the goods at each delivery stage. This development is termed *goods structure effect*.

Apart from an increase in the division of labor and growing site diversification, logistic processes are undergoing changes, especially in the transport sector. This is initiated by modern logistic concepts, including, for instance, the concepts of Just-In-Time (JIT), Efficient Consumer Response (ECR) or Vendor Managed Inventory (VMI) (see Chap. 7). These concepts require high-quality transport services, increased adherence to delivery dates, and speedy transport of small transport lots, i.e., the transport of small shipment sizes with high transport frequency. Effects resulting from this are sometimes referred to as *logistics effect*.

The so-called *integration effect* is brought about by opening up the national markets through free trade areas and WTO treaties in conjunction with an expansion of market and procurement areas, and by establishing world-wide production networks. Increased globalization amplifies these effects. Correspondingly, Global Sourcing, Global Production, and Global Distribution are strategies that are very common among companies. A striking indicator of globalization is, for example, the increased share of container cargo in maritime traffic (see Sect. 4.2.3).

An increase in freight traffic can be observed as a result of the ongoing Europeanization which was essentially fuelled by the creation of a single European market from the 1990s onwards. Additionally, we are seeing changes in distribution structures which are characterized by setting up European central warehouses – primarily in the Benelux countries. Furthermore, a tendency towards relocating production facilities to Central and Eastern European countries, such as the Czech Republic, Poland, and Hungary, has become apparent and individual production stages are increasingly relocated to different countries. In the next few years, these developments will contribute to a significant intensification of cross-border traffic between the EU and accession countries and the bordering countries in Eastern and South Eastern Europe.

All of the aforementioned trends have varying ramifications for transport and logistics service providers, which will be analyzed in the following.

5.3 Modes of Transport and Transport Technologies

5.3.1 Transport Value and Transport Affinity

To roughly assess modes of transport and transport technologies, we can refer to the concepts of transport value and transport affinity. In the following, these concepts will repeatedly be employed in explaining transport systems.

Transport Value defines the individual quality features of the transport modes. Closely related to this is the concept of *transport affinity,* which describes the requirements of the transport object or of the consumer, respectively. Thus, the key requirements are[1]:

- *Mass transport capacity:* capability of a means of transport to transport large volumes at low costs
- *Speediness:* Transport duration, transport speed, capability of a means of transport to quickly transport goods
- *Network-forming capability:* capability of carrying out spatially inclusive and comprehensive transport
- *Predictability*: Measurement for transport time reliability (timeliness) of transports
- *Flexible schedules:* Frequency of transport services, capability to adapt to changed schedules and requirements
- *Spatial Flexibility:* Capability of spatially relocating/integrating means of transport and transport capacities
- *Safety:* Measurement for accident frequency of transports and the amount of damage
- *Environmental impact:* especially the use of energy, pollutant emission, and noise emission

The basis for the individual transport modes' quantitative performance measurement is the tonnage transported. Annual reports on this are published by the Federal Office of Statistics (see Fig. 5.1).

By linking the net tonnage to the distance covered, statements about the transport performance of the modes of transport can be made (see Fig. 5.2). Both of these statistics are also an indicator of the economic development and sector attractiveness. Furthermore, they serve as a basis for traffic planning.

The market shares of the individual modes of transport are visualized in Fig. 5.3.

---

[1] Cf. Ihde (2001), p. 197 et seq.

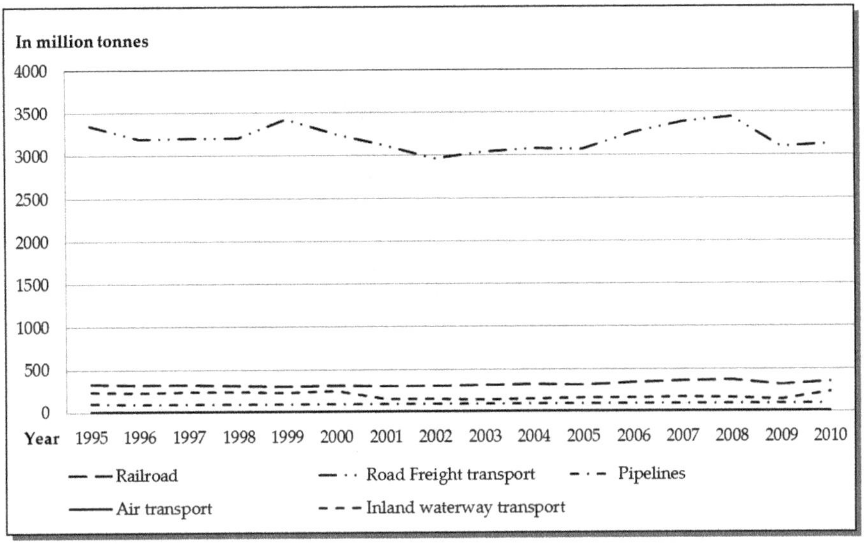

Fig. 5.1 Tonnage transported according to transport modes in Germany (Cf. BMVBS 2012, p. 240 et seq)

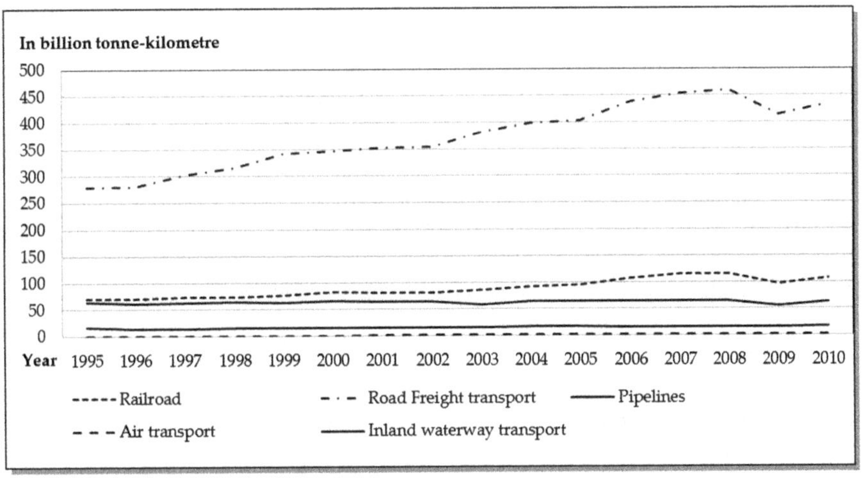

Fig. 5.2 Tonne-Kilometre performance according to modes of transport in Germany (Cf. BMVBS 2012, p. 244 et seq)

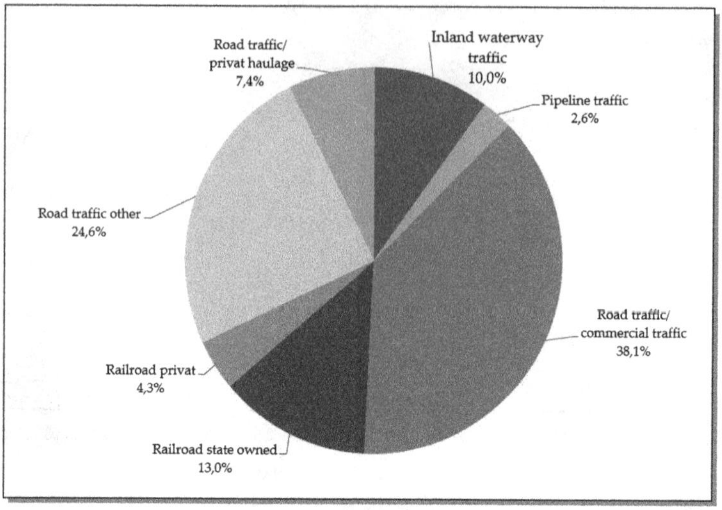

Fig. 5.3 Tonne-kilometer performance according to market share in 2011 in Germany (Cf. BMVBS 2012, p. 247; Bundesnetzagentur 2011, p. 19)

5.3.2 Road Freight Transport

Road freight transport is the most important mode of transport. With about 650,000 km of motor traffic roads, it constitutes the densest road network in the EU (see Sect. 4.2.1) and is therefore unrivalled for spatially inclusive and comprehensive logistic services. Road freight transport can be conceptualized by the characteristics listed in Table 5.1. Road traffic is highly controversial. For one thing, this is because of its environmental pollution through energy consumption, exhaust emission, noise generation, and traffic congestion. For another thing, state subsidies for traffic routes and the ongoing competitive distortion within Europe (despite liberalization) due to varying factor costs in the individual states (personnel, taxes, duties etc.) are cause for debate.

The most important distinguishing characteristics of road freight transport are shown in Fig. 5.4.

Commercial road transport is the transport of goods by means of motor vehicles in a business context or in return for payment. *Private haulage*, however, is transport carried out by industrial companies or commercial enterprises using their own vehicles. The distinction between short-distance and long-distance traffic harks back to the time when the governmental regulation of the transport market system will a significant impact on the supply and the market outcomes of road freight traffic. Capacity restrictions by granting concessions and tariff commitments to regulate pricing were supposed to increase or at least stabilize the train‘s market share compared to the truck's market share, which eventually yielded no success. Additionally, national companies were to be protected against foreign competition by means of restrictive cabotage regulations. *Cabotage* is a foreign carrier's permission to perform transports in a specific country, so long as they begin and end in the

Table 5.1 Important characteristics of road freight transport

Characteristics of Road Freight Transport
Good network-forming possibilites (i.e.. a good linkage between the primary elements of the traffic system is possible)
Speediness in conjunction with a relatively low transport risk (Direct delivery in door-to-door transport, transshipment and/or intermediate charging are only conditionally necessary or not necessary at all)
Rational utilization of the network (normally via groupage, delivery traffic and feeder traffic)
Utilization of vehicles specific to the volume of goods (silo truck and dump truck, refrigerated transport, tank transport, luggage transport, container transport, semi-trailer truck, high-capacity transport, heavy-load transport etc.)
Cost optimization and environmental-impact optimization by means of the piggy-back system with railway
In long-haul transport abroad, road freight transport is often used for the first and the last transport stages, e. g. pre-transport from sender to the sea-going vessel (sender) and onward transport from port of discharge to the buyer on the target market
Relatively short standstill periods and latency
High flexibility (with regard to acceptance dates, delivery deadlines, dates for transport and possibilities for new dispositions of goods and transport modes)
Complex competitive relationships (both within road freight transport and with other modes of transport)

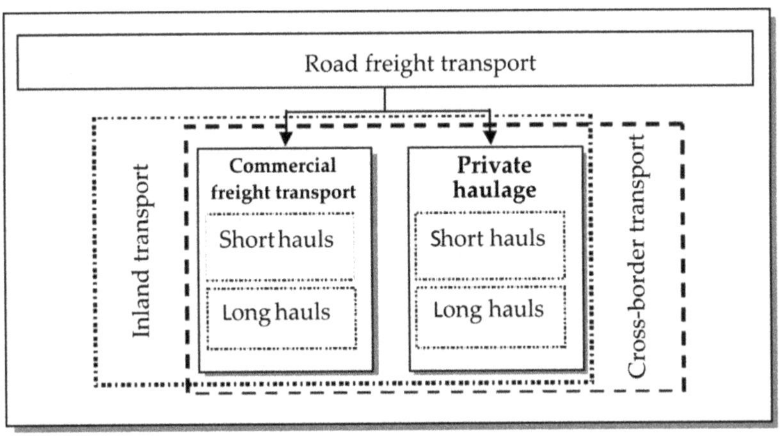

Fig. 5.4 Distinguishing characteristics of road freight transport

same country. Following the realization of free movement of services within the EU, the 1990s saw extensive deregulatory measures in the transport market, which resulted in free formation of prices and the abolishment of capacity restrictions.

High adaptability to the manifold and changing tasks of senders who service vast areas has made road freight transport the mode of transport which is most suitable to perform logistics services. In short-distance transport and long-distance transport up to 400 km, trucks are markedly faster than trains, both nationally and internationally. This is especially true for direct door-to-door transport relations.

Medium-sized service providers continue to be prevalent in road transport. Due to this, the transport services sector is having difficulties meeting the increasing requirements of senders. Individual carriers can only rarely service areas comprehensively, which is why they mainly work as subcontractors for major forwarders or co-operate with other providers.

5.3.3 Rail Freight Transport

The development of the railroad was essential for the industrialization and the establishment of raw-material-intensive base industries. In this respect, the advantages of rail transport in the form of scale advantages came into their own for bulk transports and long-range transports.

After the shift towards post-industrial economic structures, the railroad now has to meet different demands. The *volume-of-goods effect*, the *goods-structure effect*, and the *logistics effect* are increasingly changing the economic requirements compared to the traditional requirements the railroad had to fulfill. In addition, railroad transport has lost its erstwhile monopoly position on traditional mass-transport markets to inland vessel transport, pipeline transport, and road freight transport.

However, with rising environmental awareness and more bottlenecks in the road network, the railroad system is currently experiencing a comeback. This is supported by a trend towards containerization, which greatly enhances the possibilities of multi-modal traffic. An example of this would be the Hinterland traffic of sea ports. Furthermore, privatization has invited the development of innovative service offers, such as the Railport concept for cargo traffic suited to rail transport (see case study 4.1). Railports are simple and flexible transshipment points for general cargo, containers, bulk goods etc. between trucks and trains. A comprehensive network allows for main-run transports largely to be carried out by train, while pre-transports and onward transports are conducted by truck.

Table 5.2 summarizes logistically relevant and system-specific characteristics of the railroad in the context of freight traffic.

Figure 5.5 gives an overview of providers of rail freight transport services.

The efforts that have been made to privatize the so far dominant national railroads in Europe brought with them considerable changes for the rail freight transport market. These changes are part of a long-lasting process, since railroad transports have their legal basis in the German constitutional law and are therefore guaranteed. The privatization process of *Deutsche Bahn AG* and their efforts towards improving their capabilities of market cultivation have made great progress. A structural layout of *Deutsche Bahn AG* – Germany's main service provider of railroad transport – is depicted in Fig. 5.6.

Table 5.2 Important characteristics of rail freight transport

Characteristics of rail freight transport
Highly capable of performing mass transport resulting in low direct costs for production
Especially suited for long and direct mainland transport
Suitable for almost any kind of goods (for valuable, large-volume or bulky goods, for bulk-goods if there are no convenient connections via canals or rivers and for goods that cannot or may not be transported on the road)
Speediness in the case of block trains or direct trains without shunting
Timetables and rail-track allocation greatly ensure timely delivery. Relative independence from rush-hours in road traffic, holiday traffic and adverse weather conditions
Safe transport handling, especially when transporting hazardous materials, relatively environmentally friendly method of transport
Relatively low network density and thus limited possibilities for door-to-door transport, which entails costly and time-consuming transshipment activities. However, subsidies for railroad sidings to warehouses in industrial estates
Strict adherence to timetables and allocated train lengths
Low transport speed (more time required), especially in the case of single wagons due to the shunting procedures required for train formation and because of prioritized passenger service
Weak competition from foreign railroad competitors due to limited marketablity
Border delays due to varying technical requirements of country-specific railroad systems

By means of route allocation to third parties, other private service providers of railroad transport can enter the market. They mainly offer and run block train services. Figure 5.7 gives an overview of the different railroad transport services.

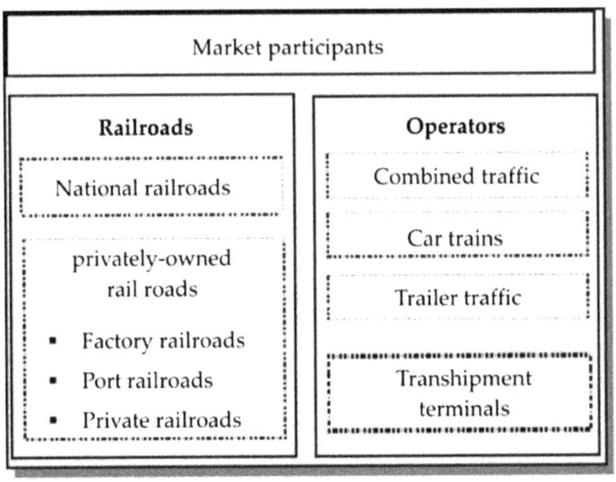

Fig. 5.5 Market players in rail freight transport

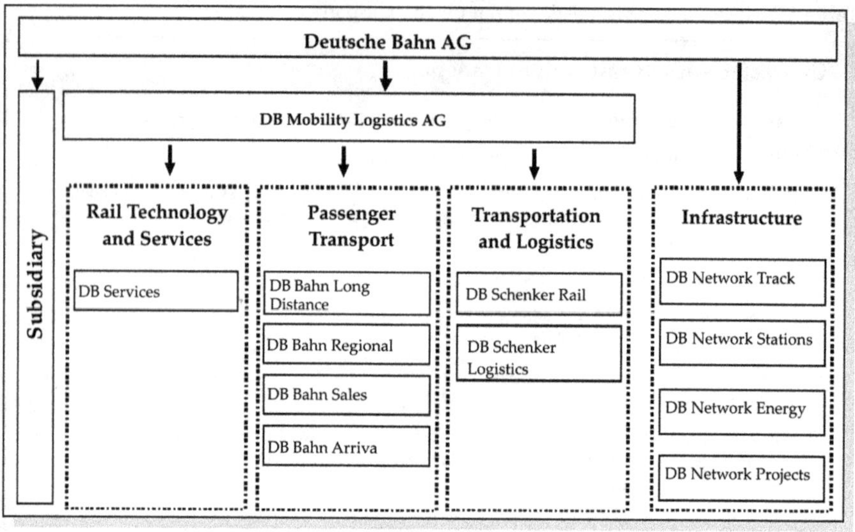

Fig. 5.6 Group structure of the Deutsche Bahn AG (Cf. Deutsche Bahn AG 2010)

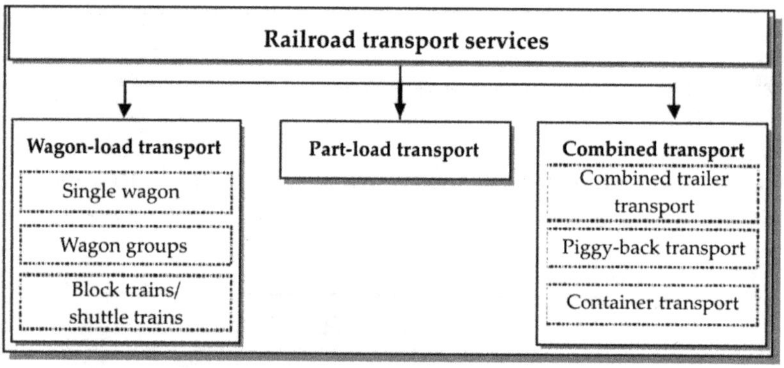

Fig. 5.7 Rail freight transport products

Wagon-load traffic is the transport of a major volume of goods as closed cargo by a means of transport. Characteristically, wagonload goods include goods such as agricultural and forestry produce, fertilizers, coal, ore, and quarrying produce.

Consignments below wagonload size are termed part loads and delivered by means of *part-load shipping*. This includes traditional general cargo above a package weight of 30 kg. Shipments like this are usually dispatched on pallets. Unlike road freight traffic, rail freight traffic is hardly used for part-load traffic any more. Such shipments are rather dispatched as consolidated cargo using combined traffic.

Single wagon transport is characterized by single wagons or small wagon groups which are loaded by customers who often have their own railroad siding and which the railroad service provider then transports to the destination station. Effective train formation can be achieved by grouping together several customers' wagons with the same destination. Today there are flexible service offers that allow for single wagon transports to be freely commissioned on an international level with regard to time, volume, and relation. As a part of these services, 1,400 freight traffic locations and 4,000 private rail track sidings are being operated. Freight traffic locations are goods stations, public loading stations, or border crossings. Amongst other things, they serve as a basis for tariff kilometer calculations within the German rail network.

Block train traffic enables the transport of a customer's high volume of goods from sender to receiver in complete trains without en-route handling. This kind of transport is usually employed for price-sensitive bulk goods and breakbulk cargo which do not require speedy transport or different wagon types, e.g., raw materials and products of the mining or mineral oil industry and building materials.

Block train traffic in the form of *shuttle trains* or *direct trains* are high-quality services for linking production sites as well as for procurement logistics and distribution logistics. This type of transport services are customized by means of individual timetables and thus guarantee on-schedule delivery. Shuttle trains provide point-to-point connections with fixed sets of wagons and without further addition of wagons during transport.

Besides its product range of bulk train services (*flextrain, variotrain and plantrain*), *DB Schenker* has developed industry-specific concepts catering to the needs of, for example, the paper industry, the chemical industry, the waste management industry, and the mineral oil industry.

Carriers provide *combined cargo* services, which come in the form of bulk trains or groups of wagons. Here, different carrier wagons are employed for containers, swap bodies and trailers. Combined cargo has the benefits of speediness by taking advantage of favorable traffic conditions at night. Furthermore, loading and discharge take place at the production site or at the warehouse and containers may serve as intermediate or buffer storage facilities.

Combined transport is utilized if trucks, trailers, and swap-bodies are to be transported using piggy-back transport. In Germany, the handling process for this is implemented by the company *Kombiverkehr Deutsche Gesellschaft für kombinierten Güterverkehr GmbH & Co. KG*. Currently, there are 230 limited partners holding stakes in *Kombiverkehr*, of which 50 % are held by forwarders and carriers and another 50 % by *Deutsche Bahn AG*.

Kombiverkehr operates over 50 transshipment terminals. Their services include the *Kombi-Netz 2000+* service, which is a system of 26 trains that service 60 relations on weekdays and the *Albatross Express* service of the freight forwarder *Transfracht*, which connects the ports in Hamburg and Bremen to 18 Hinterland terminals using nightly container trains.

The future market potential of rail freight services is largely dependent on the continued development of logistically relevant services and of the capacities to implement them. Constructing exclusive cargo routes and separating passenger transport and cargo transport will be an essential prerequisite for this. Not only should this happen on a national level, but also on an international, cross-border level, which calls for extensive deregulation in the EU-wide railroad sector. Furthermore, infrastructural, operational, and vehicle-related technical impediments need to be disposed of to enhance inter-operability.

Case Study 5.1: Railroad Usage in Industrial Logistics

BASF SE (*The Chemical Company*) is one of the world's leading chemical companies. Its subsidiary BASF Schwarzheide favors rail transport for the transport of their goods. The company's site in Lausitz produces a wide range of plastics, foam rubber, pesticides, and waterborne coatings. PU dispersions and Laromer® brands (coating raw materials) complement their portfolio. Feedstock, packaging, operating supplies, and technical goods are needed for the production process. The means of transport eligible for receiving and sending goods are tank wagons, covered freight wagons, road tank vehicles, trucks, various containers, and utility vans.

The development of economical, reliable, and environmentally-friendly transport concepts and their target-oriented implementation, as well as regular monitoring makes the site logistically attractive, even without direct connection to waterways. For the own transport volume and for other companies based on the same premises, BASF Schwarzheide GmbH uses an EU-wide block train railroad network provided by private railroad companies. This network connects BASF sites in Ludwigshafen, Antwerp, and Schwarzheide on a daily basis. The liberalization of the European rail freight traffic has allowed for competition to develop with *Deutsche Bahn AG* for these relations. This was the prerequisite for cost-effective rail freight traffic.

An international forwarder operates a public container terminal on the company premises in Schwarzheide (see Sect. 4.2.5). The terminal is being expanded to a hub for Eastern Europe with connections to Poland, Ukraine, and Russia. Plans to extend the connections to China are underway. The container terminal in Schwarzheide is connected to the German combi-terminal network, thus maintaining strong connections to Ludwigshafen and the whole of Europe. Various providers of combined transport services design and develop these networks in co-operation with partners and organize transports with different carriers.

A good railroad infrastructure and integration into the European block train and container network enable economical and high-quality rail transports. This provides immense potential for a shift from road transports to railroad transports. Customers, suppliers, the producer, and various service providers are included in the development of logistics concepts and transport concepts.

The transport system implemented at BASF Schwarzheide shows that cargo transport can purposefully be operated by rail (see Fig. 5.8). The transport shares in 2009 are:

Total traffic volume	100 %	1.8 million t
By rail	65 %	1.2 million t
By road	35 %	0.6 million t

In the future, the share of rail freight traffic is planned to be further increased at BASF Schwarzheide GmbH.

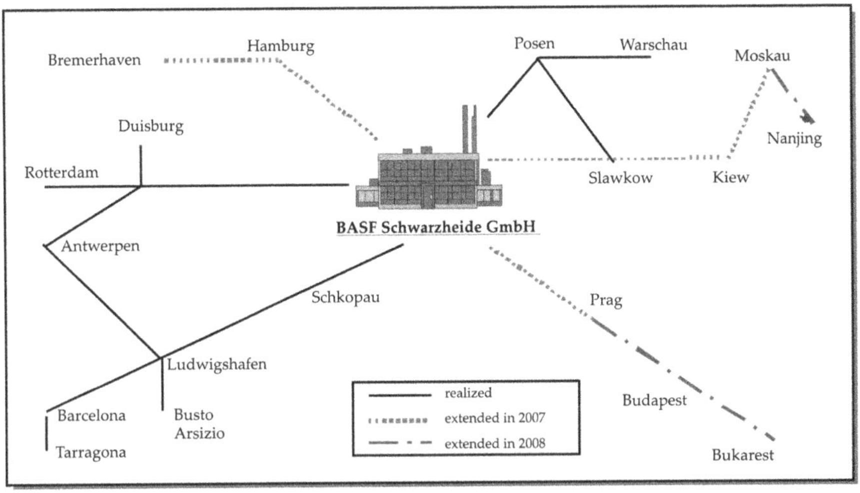

Fig. 5.8 Railroad transport network of BASF Schwarzheide GmbH (Cf. BASF Schwarzheide GmbH 2008)

5.3.4 Sea Freight Transport

International trade and international division of labor have only become possible through the development of maritime traffic. Transportation of goods by sea enables cost-effective transport of bulk goods over long distances. The most relevant characteristics for a logistical assessment of maritime navigation are shown in Table 5.3. The increasing containerization of consignments and shipments has been extremely conducive to the development of sea traffic. Containers bring with them the advantage of standardization (ISO standards), which may render the transshipment of goods unnecessary within a transport

Table 5.3 Important characteristics of sea freight transport

Characteristics of sea freight transport
Transport of large volumes of goods possible or usual
Especially suited for long-haul and intercontinental transport
Suited for almost any type of goods
Relatively cheap relation between transport cost and capacity (in € per tkm)
Rather high safety during transport
Relatively low transport speed (high time requirement)
Highly dependent on weather conditions

system and which makes it possible to change from one transport mode to another within a short period of time.

Thus, container shipping assumes a central role in maritime traffic and has greatly influenced the relevant transport and handling technologies as well as shipbuilding developments and ship sizes. High-performance container gantry cranes with handling capacities of over 55 containers per hour make it possible to load and discharge even the largest container ships within 1 or 2 days.[2] Quick discharge makes short laytime at the port possible. This reduces laytime charges in ports, enables speedy implementation of transport, and reduces the round-trip time of containers.

On the other hand, the development of goods volumes and ship sizes significantly influences the sea traffic and sea port infrastructures, leading to fewer ports in a specific area being navigable for large container ships (Main Ports) and to feeder services having to supply the main ports with consignments from smaller ports and Hinterland ports. Feeder services are small container ships capable of navigating canals and commuting between the large international container ports, smaller seaports, and inland ports accessible to coasters.

Some of the port infrastructure determinants are:

- Width and depth of fairway
- Wharfage, open space, and storage area
- Handling facilities, container gantry cranes
- Information and communication systems

Of major importance, alongside the infrastructure, is the port location. The following basic location factors of sea ports can be identified[3]:

- *Seafront location*: geographical location in relation to the main open sea traffic routes. Thus, for incoming ships via the English Channel the Antwerp-Rotterdam-Amsterdam range (ARA) is at a temporal advantage of about 1 day compared to the German North Sea ports.
- *Coast location*: proximity to open fairway. Thus, the ports of Bremen and Hamburg are at a distance of up to 100 km from the open sea.
- *Hinterland location*: economic area which operates its sea traffic via that specific port with varying expansion and depending on type of goods and transport modes. An example of this is the so called *Rheinschiene*, the route alongside the river Rhine.

Due to the large volume of exports, the ports with the world's greatest handling tonnage are located in Asia. The reception ports of Europe and North America are found among the lower ranks (see Table 5.4).

---

[2] Cf. Vis/Harika (2005), p. 58.

[3] Cf. Ihde (1991), p. 99 et seq.; Brinkmann (2005), p. 5 et seq.

Table 5.4 The world's largest container ports, 2010–2012 (Cf. Hafen Hamburg Marketing e. V. 2013)

Port	2010 (Volume handled in 1,000 TEU)	2011 (Volume handled in 1,000 TEU)	2012 (Volume handled in 1,000 TEU)	Change (compared to the previous year in %)
1. Shanghai	29,069	31,500	32,529	+3.3
2. Singapore	28,430	29,937	31,649	+5.7
3. Hong Kong	23,532	24,404	23,097	-5.4
4. Shenzen	22,510	22,570	22,941	+1.6
5. Busan	14,137	16,175	17,020	+5.2
6. Ningbo	13,144	14,686	16,830	+14.2
7. Guangzhou	12,550	14,426	14,744	+2.2
8. Quingdao	12,012	13,020	14,500	+11.4
9. Dubai	11,600	13,000	13,300	+2.3
10. Tianjin	10,080	11,582	12,300	+6.2

With regard to the organization of the sea traffic market and its market participants, the following performance patterns can be identified:

- Regular services with scheduled routes
- Charter traffic, one-time booking of an entire ship for a certain relation
- Occasional services (tramp shipping), booking for a specified volume of cargo according to specific transport needs (comparable to consolidated cargo transport in road freight transport)

Providers of scheduled services have formed alliances in so-called *liner conferences*. These are cartels with fixed agreements on prices and quotas. Their objective is to secure the clearing collateral in international sea traffic. Providers who do not participate in these conferences are called *Outsiders*. They try to position themselves on the market by means of cheaper freight rates.

For some time, shipping companies have been trying to influence cost structures through flagging-out. This means a deletion from international shipping registers and instead registering in so-called flag-of-convenience states (e.g. Liberia, Cyprus, Panama, Bahamas, Malta). National regulations as to manpower, safety requirements, and minimum wage do not apply in these states. Flagging-out rates have been on the increase for years (1994: 48 % world-wide; 2001: 62 %

world-wide).[4] Second registers and plans to introduce a European maritime shipping register are supposed to put a check on these developments.

The traditional procedure of *overseas container transport* can be described through the process chains of *Carrier's Haulage* and *Merchant's Haulage*.[5] With *Carrier's Haulage*, sea shipping companies offer door-to-door container freight transport. The sea shipping company takes the empty container to the sender's ramp, organizes pre-carriage of the container to the sea port, provides ship transport, organizes on-carriage to the receiver, and collects the discharged container. It assumes liability for the entire transport process and commissions other carriers with the forwarding of the goods. The carrier issues bills of lading, which are bankable documents of title confirming the rights to the goods transported and containing detailed liability regulations.

In the case of *Merchant's Haulage*, the sender instructs their forwarder to carry out transport. The forwarder orders a container from a sea shipping company and organizes the pre-carriage. The sea shipping company is responsible for the transport at sea. A corresponding forwarder receives the container from the sea shipping company at the port of destination and takes out customs clearance and transport in the country of destination. The container is returned to the sea shipping company after discharge.

Apart from sea shipping companies there are *Non-Vessel-Operating Common Carriers* (NVOCC). A NVOCC provides overseas transport and operates in a fashion similar to that of a sea shipping company. The only difference is that a NVOCC does not operate its own vessels and in most cases does not rely on its own containers. Freight capacity is chartered from sea shipping companies at their transport relation (also referred to as slot charter). Subsequently, the NVOCC offers these freight capacities to its customers on the market. The NVOCC is entitled to issue the necessary freight documents (bill of lading). In many cases it also operates its own offices both at the places of dispatch and destination or works with partners that organize dispatch (pre-carriage) and delivery (on-carriage) of the containers. This guarantees one-stop freight service for the sender.

Case Study 5.2: Container Transport

FCL MARINE AGENCIES GmbH in Bremen is a Non Vessel Operation Carrier (NVOCC) and conducts world-wide container transports. One of their transport operations involved the shipment of 24 off-road racing vehicles to the USA for testing purposes. The value of the vehicles ranged from 175,000 to 250,000 Euros per vehicle. They were loaded into containers in Stuttgart and subsequently shipped to Houston, Texas. Customs clearance for temporary import was taken out by a partner in Houston. The vehicles underwent test drives in the Arizona desert and through the Southern US states (New Mexico, Texas, Louisiana, Mississippi, and Alabama). After 4 weeks the client ordered the containers to be brought to Tuscaloosa, Alabama for return transport. Loading

[4] Cf. Aberle (2009), p. 268.

[5] Cf. Ihde (2001), p. 143 et seq.

the containers out of town so that the vehicles would arrive in Germany in good condition was especially challenging. Following this, the containers were transported to the sea port by truck and subsequently shipped to Germany. The import formalities had then to be taken care of in Bremerhaven and on-carriage to the final destination needed to be arranged. Each step in this high-value transport process was coordinated by FCL who were also the point of contact for the client at the same time.

5.3.5 Inland Waterway Transport

The most important forms of inland waterway transport are bulk goods waterway transport, general cargo waterway transport, and river and maritime navigation.

Bulk goods waterway transport mainly involves the transportation of raw materials and primary products in solid and liquid form. The development of tugboats brought about increased flexibility in the loading and discharging process and in the organization of routes since the transport units (lighters) are decoupled from the motorized push units.

General cargo waterway transport is primarily used for finished goods from the automobile, engineering, and construction sector. Roll-on-roll-off transports are ferry services allowing new vehicles or trucks to drive on and off the vessel, which greatly facilitates handling.

In *river and maritime navigation* there is an overlap between inland shipping – i.e. transport carried out exclusively on rivers and canals – and sea traffic. Owing to their size, vessels used in river and maritime navigation are capable of navigating both at sea (along coastlines) and on large, developed inland waterways (e.g. on the Rhine river up to Duisburg). Thus, by using appropriate navigable units and coastal motor vessels, additional shipping connections may be established, such as *short-sea shipping* (see Sect. 4.2.1).

Further distinctions can be drawn according to product groups transported in inland waterway transport (see Fig. 5.9).

As is the case in sea freight transport, *container traffic* is becoming more and more important for inland waterway transport. The increase in turnover by 2015 is estimated to be 10 % p.a. Accordingly, the significance of inland ports and handling facilities in the Hinterland for groupage in pre-carriage and on-carriage is on the rise. To create high-performance transport chains and to support multimodal traffic, there is increasing demand for multimodal (road/rail) and trimodal (road/rail/inland waterway) port facilities. Moreover, inland ports are increasingly being converted into *logistics service centers* (see Sect. 4.2.6), where procurement-logistical and distribution-logistical services are offered by logistics service providers alongside purely port-logistical services. These developments are largely determined by the characteristics of inland waterway transport, which can be summarized as follows (see Table 5.5).

The supply structure of inland waterway transport market in Germany is essentially dominated by few shipping companies, such as Imperial, Rhenus Partnership, Haeger&Schmidt, Lehnkering and the Deutsche Binnenreederei, which in most cases have emerged from formerly company-owned shippers. Some larger shipping

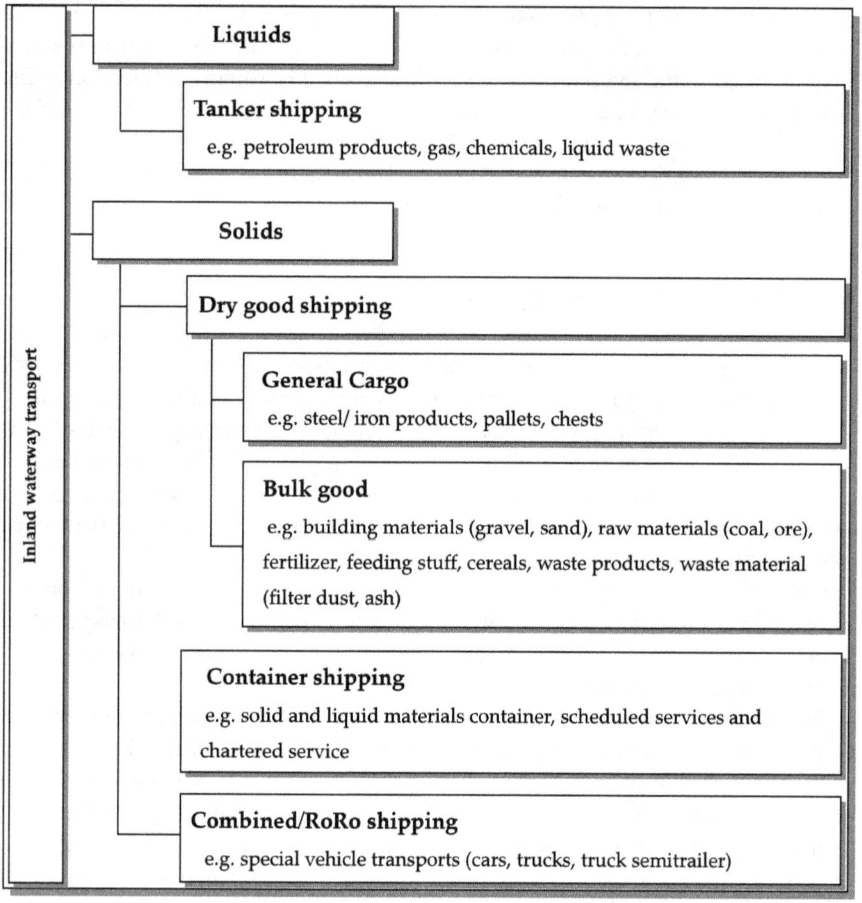

Fig. 5.9 Most important forms of inland waterway transport

companies, as well as many small independent owner operators (private shippers with up to three boats) complement this supply structure. Charterers do not own their own shipping space but merely organize ship transports on behalf of senders. They could be called forwarders for inland waterway transport. The role of shipping companies has changed significantly. While traditionally their responsibilities included the organization of shipping space, the provision of freight, and the disposition and implementation of transport, they primarily only assume the role of shippers nowadays.

The future development of inland waterway transport mainly depends on whether freight traffic forecasts will hold true and on the influence they will have on the modal split. By now, the modal split – i.e. the share that the different modes of transport have in the total volume of transport – is largely determined by the transport mode's capacities and its ecological relevance. Inland waterway transport enjoys a competitive advantage in this regard.

Table 5.5 Most important characteristics of inland waterway transport

Characteristics of Inland Waterway Transport
High mass-transport capacity (low costs at large goods volumes and large distances)
24hr service due to radar navigation
Free capacities on waterways
High reliability and transport safety (also for hazardous goods)
Environmentally friendly (low energy consumption per transport unit)
Low network density
Low transport speed
Dependent on weather conditions (high/low water levels, icing)

Improvements need to be made with regard to consistent navigability by means of expanding inland waterways and upgrading locks (see Fig. 5.10 and Sect. 4.2.1). The purchase of more efficient inland vessels would bring about a general efficiency enhancement and thus increased competitiveness.

Analogously to the situation in rail freight transport, it will be crucial for inland waterway transport to adapt its range of services to market requirements and thus render it attractive for shippers. The rail's competitive behavior will continue to have an influence on inland waterway transport as well as on the persisting competitive gaps between Western European countries (esp. between Germany, Belgium, and the Netherlands).

Case Study 5.3: Service Offers in Inland Waterway Transport: Vehicle Transport

Apart from the transportation of bulk goods, inland waterway transport may also serve to transport high-value goods such as brand-new passenger cars. Ever since the development of an innovative logistical solution by the charterer *E.H. Harms-Terminal Kelheim GmbH & Co. KG*, cars of *Mitsubishi*, *Ford*, and *Renault* have been transported on the Danube from Kelheim or Vienna to Budapest.

Suzuki cars are transported from Budapest to Kelheim on the return trip. Thus, empty trips can be avoided and the cost efficiency of the transporting vessel can be increased. The vessels run on a non-regular schedule according to peak seasons over the year. Trips can also be scheduled or cancelled on short notice in order to react to sector-specific fluctuations in demand in a prompt and flexible manner. This liner service is carried out by the small-scale German shipping company *Trödel* and by *Lehnkering GmbH*. The waterway motor vessel *Heilbronn* is used for the service. The vessel is furnished with three decks with grid floors designed for vehicles up to 2,000 kg each. The *Heilbronn* has a capacity of 205–270 vehicles, depending on vehicle type. The handling of the

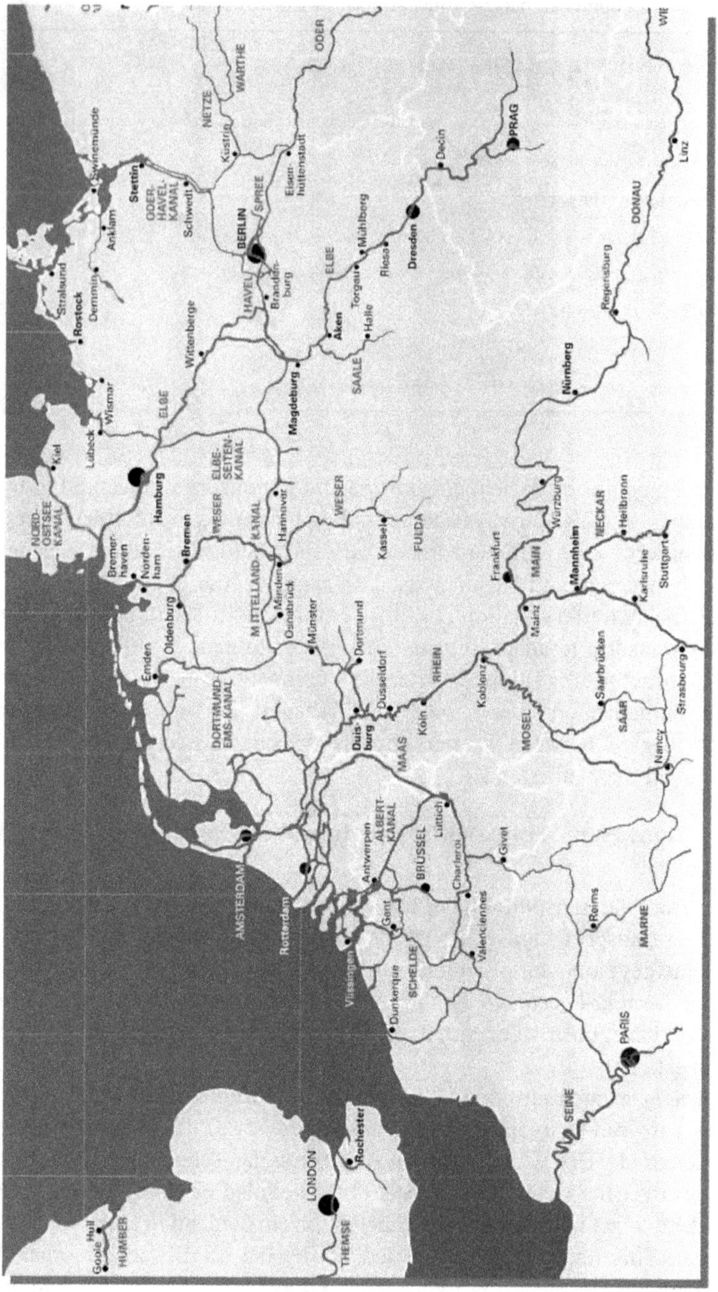

Fig. 5.10 Network of North European inland waterways (Deutsche Binnenreederei 2007)

cars is facilitated by means of a bow ramp which connects to the concrete RoRo ramp of the port.[6]

Case Study 5.4: Service Offers in Inland Waterway Transport: Transport of Large-Dimension Goods

The *Deutsche Binnenreederei AG* in Berlin transports rotor blades and equipment for the wind power station manufacturer ENERCON GmbH in Aurich. The rotor blades for the wind power stations are produced in Magdeburg. Using 480 km of inland waterway routes, the rotor blades are then shipped to Emden by means of a liner service, where they are loaded onto overseas vessels. Using a special, double-layered transport rack, six rotor blades can be transported per push-barge. From Magdeburg to Bergeshövede (Dortmund-Ems-canal), two push barges can be coupled together for transport. The barges, however, need to be de-coupled in Bergeshövede to carry on transport individually due to the underdeveloped state of the waterway route. Furthermore, additional ballasting of the barges with water is necessary at the lock in Sülfeld (Mittelland Canal), due to a clearance height of 4.20 m between the water surface and the bottom edge of the bridge.

The barges therefore need to be double-hull vessels. To lower the vessel, the space between the hulls is filled with water, which is subsequently pumped out again.

5.3.6 Air Freight Transport

Air freight transport is becoming more and more important. For one thing, this is due to the creation of global and intercontinental logistics chains with decentralized structures of production and distribution. For another thing, more and more high-value products from the engineering sector (machine parts), the electrotechnical sector (PCs) and the automobile sector (modules, spare parts) are being transported by air freight.

The characteristics of air freight transport are summarized in Table 5.6.

Air freight services fall into the organizationally and institutionally separate partial performances of:

- Air transport services, rendered by airlines
- Air traffic control systems, provided by the air traffic control authorities
- Airport services, provided by the airport.

The supply of air transport services is essentially determined by the market and production conditions in air traffic. These conditions are characterized by oligopolistic competition with market access restrictions, national regulations and traffic laws, high capital intensity, and adherence to flight safety requirements.

Air traffic distinguishes between two institutions, namely the *ICAO* (International Civil Aviation Organization), which includes all countries operating air traffic, and the *IATA* (International Air Transport Association), which organizes

[6] Österreichische Wasserstraßen-Gesellschaft mbH (2007), p. 27.

Table 5.6 Important characteristics of air freight transport

Characteristics of air freight transport
Short transport times in the air
Especially suited for long distances (intercontinental transport)
High transport costs
Small transport capacities compared to other tranport modes
Low network density due to dependence on airport locations
Dependent on weather conditions, especially during take-off and landing

all air-traffic companies. The objectives of the ICAO include the development and promotion of civil aircrafts, international airports, and shared air traffic control systems. One of its most important agreements is the treaty about the *freedoms of the air*, which stipulates universally binding regulations on air sovereignty and technical requirements. The IATA is an antitrust organization whose main goal is to reach agreements on pricing, on-board service, the standardization of free baggage allowance, and on the requirements for registration as an IATA agency.

The logistically relevant air freight services are divided into three segments:

- Scheduled flights and charter flights in passenger transport (belly cargo)
- Air cargo traffic
- Air cargo express traffic, messenger service air mail traffic (CEP services, integrators)

The individual product offerings are subject to restrictions in time, volumes, and price. The decisive factor in opting for air freight transport are the total costs since the high freight rates make it possible to cut costs in other places of the transport chain (storage costs, packaging).

Air freight transport chains may be formed in different ways.[7] Terminal to terminal transport of goods may be carried out by co-loading them onto a passenger plane (Lower Deck) or using a cargo plane. Besides all-cargo planes such as the Boeing 747 or the Mc Donald Douglas MD 11F, there are the so-called mixed versions or *quick changes*, which can be used flexibly as either passenger or cargo planes. In most cases, pre-carriage from the sender to the airport and on-carriage from the airport to the receiver are carried out by truck. In many cases, main carriage is also performed as ground transport on behalf of the airline. This is referred to as Road Feeder Service (RFS) or trucking. In these cases, the freight retains its status as air freight and is billed accordingly.

Airports are bases or start and end points of air traffic within air freight transport chains. They serve as an interface between air and ground transport. While formerly pure infrastructure enterprises, they are now developing into modern service

[7] Cf. Mensen (2007), p. 18 et seq.; Cf. Vahrenkamp (2005), p. 293 et seq.

companies for air cargo handling (Airport Industrial Parcs) (see Sect. 4.2.2). An airport operator's range of services with regard to the air freight sector should include the provision of suitable transport facilities with adequate space and capacity, and quick and direct handling facilities. In addition, minimum ground times as well as 24-h and all-weather operation should be guaranteed.

In order to provide air freight services and to operate air freight transport chains, it is vital that *air freight information and communication systems* be in place. Examples include customs and document processing (electronic air waybill), disposition of holds and flights, tariff and rate calculation, as well as billing and tracking.

Further information and cargo handling systems offer features such as data collection and synchronization for drop-off and receipt processing, inventory management, charge preparation, and generation of loading lists at initial clearance. Air freight forwarders rely on the systems of airlines and other carriers in pre-carriage and on-carriage to view flight schedules, match transport terms, and make bookings. Internet-based applications enable cross-over transactions between agents, countries, and companies.

5.3.7 Pipeline Transport

Another transport technology is pipeline transport. It is an independent mode of transport. However, unlike other transport modes, transport route, transport container, and means of transport form a unit in this instance. Pipelines can be categorized into four groups: oil, gas, and product pipelines, as well as other energy pipelines.

Oil pipelines distinguish between feeder pipelines and distance pipelines. Feeder pipelines only have a rather small diameter of about 10 cm and are operated under low pressure produced by a pump. They transport the oil from the well to central points where it is first processed in order to free it from gases and water. Subsequently, the oil is stored in tanks. Distance pipelines start from these tank farms. These pipelines can have a diameter of up to 1.20 m and are operated under higher pressure. Pumps along the route generate pressure in the pipes and are located according to the relief (altitude difference between ground and sea level). The oil travels several hundred kilometers through the pipeline until it arrives at the port or at a refinery.

The transport of gas in *gas pipelines* is similar to the transport of oil but differs as to the operating conditions and the equipment used. Smaller pipelines lead from a gas field to a gasworks where the gas is cleared of impurities such as water, sulfur, toxic gas, sulfides, and dioxides. After this cleansing process at the gasworks, the gas is fed into the distance pipeline. In contrast to oil pipelines, compressors along the route generate sufficient pressure in the pipeline.

Product pipelines transport refined mineral oils such as gasoline, diesel fuel, and heating oil but also secondary products such as vaseline, paraffin, bitumen, petroleum coke, lamp oils, and lubricants from refineries to storage and distribution centers. Additionally, gases filtered out in the oil and natural gas production process (butane, ethane, propane) are transported in liquid aggregate state to refineries where they are separated again. As opposed to oil and gas pipelines, product pipelines do not solely transport one specific type of goods. Moreover, they serve to carry several different products consecutively or even simultaneously.

To avoid an intermingling of individual components when transporting different products, three basic procedures may be utilized[8]:

- The valves necessary to separate the batches at the infusion stations briefly remain closed
- *Balloons* which are marginally smaller in diameter than the pipeline are inserted between the individual batches
- A buffer batch is interposed in between two products which may intermingle with both products

Mineral oils and liquid gases require higher pressure for transport since they are lighter than crude oil or natural gas. Furthermore, higher pressure is required to prevent mixtures of liquids and gas as such a mixture could permanently damage the pumps in the pipeline.

Energy pipelines comprise so-called two-phase pipelines which are mainly used as feed pipes between oil or gas fields. They simultaneously transport liquids and gases resulting from the extraction process. These two products are separated in refineries and gasworks close to the production area. Another type of energy pipelines are liquefied-natural-gas pipelines which – after liquefaction in special plants – carry the natural gas to a port for transport by tankship or collect it from the tankship at the port of destination. Energy pipelines also include the rarely used type of coal-slurry pipelines. These pipes carry finely ground coal blended with water.

Transcontinental pipelines stretching from production areas to consumption areas or to sea ports attract the most attention. In these cases the pipelines span long distances and are installed either above ground, underground, or on the seafloor. Since they are to a greater or lesser extent taken for granted, less attention is attracted by shorter pipeline networks used as an integral part of production, as can be found in the pharmaceutical or chemical industry or in the communal supply and disposal networks for gas, water, and sewage. The characteristics of pipeline transport are summarized in Table 5.7.

However, owing to its investment volume and high fixed costs, pipeline transport is not very wide-spread. Thus, pipeline traffic only accounts for a small share of

Table 5.7 Important characteristics of pipeline transport

Characteristics of pipeline transport
High mass-transport capacity, depending on pipe diameter and conveyor speed
High network-forming capability (mostly direct connection from sender to receiver)
High reliability and thus planability
Low risk of air or water pollution (in case of flawless construction and smooth operation)
Environmentally friendly by avoiding emissions

[8] Cf. Brecht et al. (1982), p. 31.

the total traffic volume (see Figs. 5.1 and 5.2). In most cases pipelines are constructed and provisioned by the users themselves – normally raw material producers.[9] In the wake of the Cold War, large capacities of pipelines previously

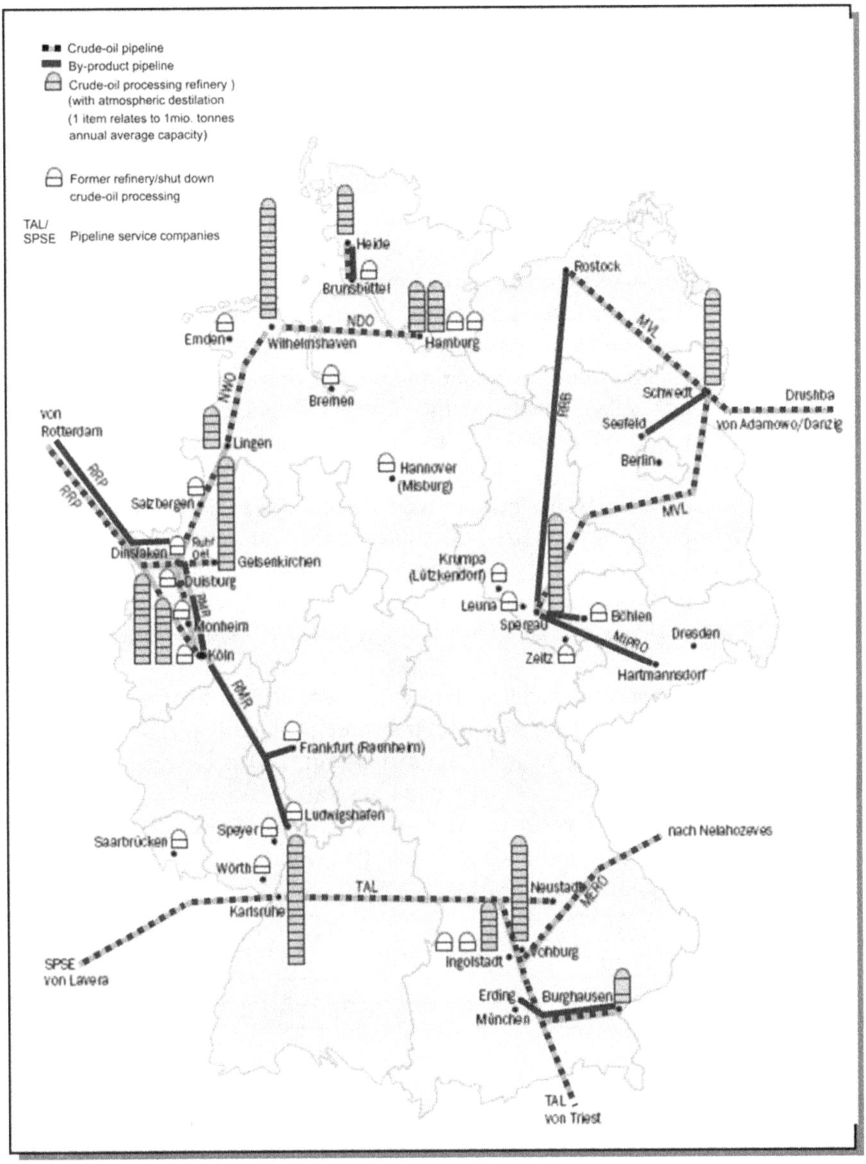

Fig. 5.11 Pipeline network and refineries in Germany (Mineralölwirtschaftsverband e. V., Hamburg 2011, p. 10)

[9] Cf. Ihde (2001), p. 190 et seq.

used for military purposes could additionally be harnessed for civil utilization. The main traffic routes of the German pipeline network with European connections are depicted in Fig. 5.11.

5.4 Transport Systems and Means of Transport

5.4.1 Transport Chain

A transport system can be designed as a *transport chain*. A transport chain is characterized by the technical and organizational linkage of the stations which a goods transport passes, starting from the place of dispatch (source) and ending at the receiving point (drain). The transport from sender to receiver can generally be broken down into pre-carriage, main carriage, and on-carriage. This distinction is mainly made for general cargo transports. Here, pre-carriage describes the process of collecting the goods from the sender and transporting them to a transshipment point. The shipment is then bundled and transported to another transshipment point during main carriage. What follows is the dispersion to the recipients during on-carriage (see Fig. 5.12).

Transport chains may be designed as a single-link chain. In this case, transport objects are transported directly from the source to the drain, whereas in multi-link transport chains, transport may be effected using different means of transport (disrupted transport) or using different transport modes (combined transport). Figure 5.13 shows single-link and multi-link transport chains as well as application examples.

With *direct transport*, no handling between sender and receiver takes place but an entire loading unit is exchanged. Direct transport is primarily used in road freight transport, to a lesser extent in rail freight transport, and occasionally in inland waterway transport in the form of combined transport. In cases of full chartering of cargo aircrafts and ocean vessels, direct transport may also constitute a viable solution in air freight transport and sea freight transport.

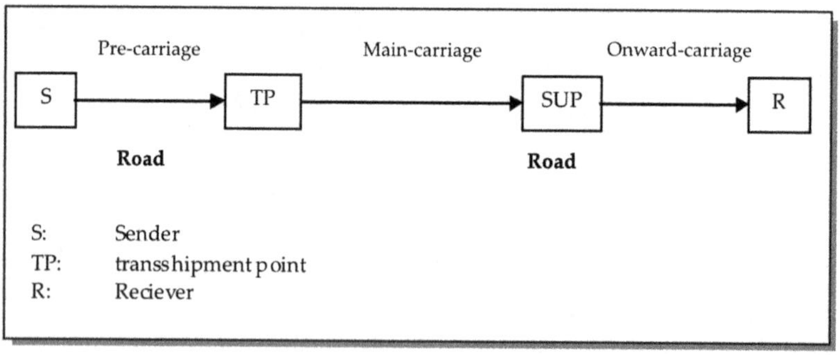

Fig. 5.12 Pre-carriage, main carriage and on-carriage

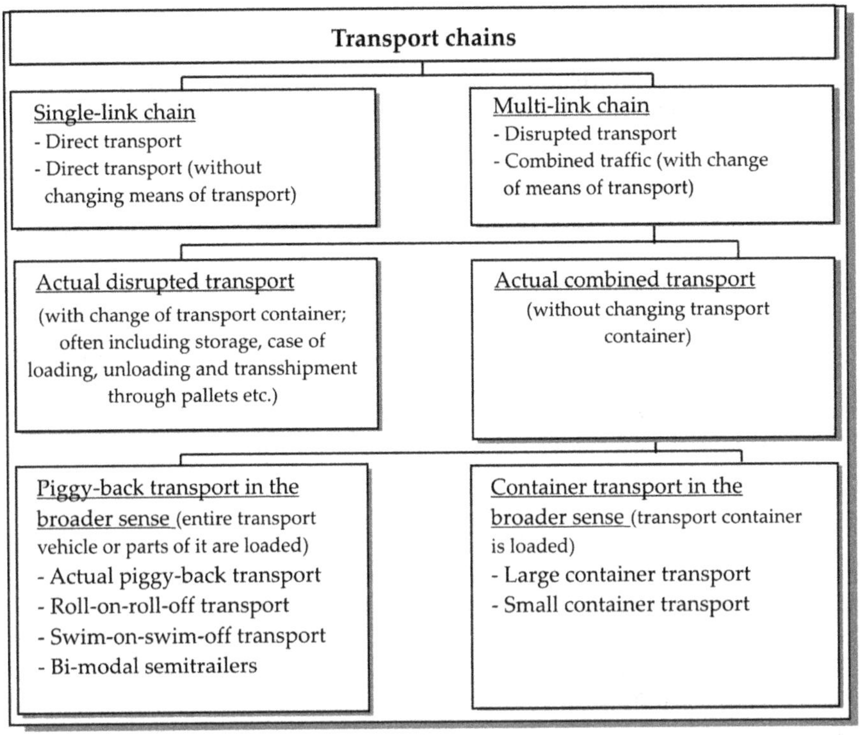

Fig. 5.13 Setup of transport chains (Cf. Jünemann and Schmidt 1999, p. 328)

Table 5.8 Important characteristics of combined transport

Characteristics of combined transport
Benefits from the individual transport modes advantages (costs, safety, adherance to delivery dates, environmentally friendly etc.)
Favorable transport times for long distances
Relatively environmentally friendly
Disrupts the transport chain
Rather time consuming due to transshipment of goods
Increased need for spatial and technological capacities due to transshipment

In order to benefit from the advantages of the individual transport modes, especially on long distances and under ever-changing transport conditions, systems have evolved which make use of two or several different transport modes. The main differentiators of *combined transport* are listed in Table 5.8.

Combined transport is a way of conducting transport in which goods are carried from the drain to the source in one loading unit without changing the load carrier, by using several transport modes consecutively within a transport chain. To this end, entire vehicles are loaded onto carrier vehicles, or loading units – if suitable – are loaded from the carrier vehicle of one transport mode onto the carrier vehicle of another transport mode.

Types of combined transport are:

- *Piggy-back transport:* Trailer trucks and semi-trailers are loaded onto special railway carriages which perform the main carriage to the destination area; on-carriage to the point of destination is carried out by truck
- *Loading unit or container transport:* Containers are used as rationalizing transport aids, which can be transported in any transport mode
- *Roll-on-Roll-off transport:* trucks or semi-trailers are driven onto RoRo ships using ramps
- *Trajectory transport:* rail freight wagons are rolled onto rail-equipped ferries

5.4.2 Means of Transport

In order to provide transport services, *means of transport* need to be utilized. We distinguish between stationary and mobile means of transport. Stationary means of transport are means of conveyance (see Chap. 6) which are only used in hubs within transport chains, such as warehouses or transshipment sites. Mobile means of transport are vehicles which are used for transport between hubs.[10] A variety of vehicles is employed in *road freight transport.* They can be categorized according to net load class and volume capacity and include utility vans, trucks (motor vehicle and trailer), and semitrucks (tractor unit and trailer). The most common transport means in road freight transport are shown in Fig. 5.14.

The broad spectrum of usage possibilities is complemented by *loading aids*, which can be fixed or interchangeable. The former type comprises superstructures and roadbeds providing more transport safety than, for example, covers or roof arches. They are also a technical prerequisite for refrigerated cargo transport or hazardous goods transport. Interchangeable loading aids are swap bodies in the form of standardized containers which are equipped with foldable supports. Trucks are able to drive below these swap bodies to pick up cargo, which renders the loading and unloading process independent of the transport vehicle.

Rail wagons are the means of transport in *rail freight transport.* Apart from the engines, there is a variety of goods wagons in standard and special design (depending on the transported goods) and are categorized into various types (see Table 5.9 and Fig. 5.15). The goods wagons are owned by the railway

[10] Cf. Isermann (1994), p. 1095.

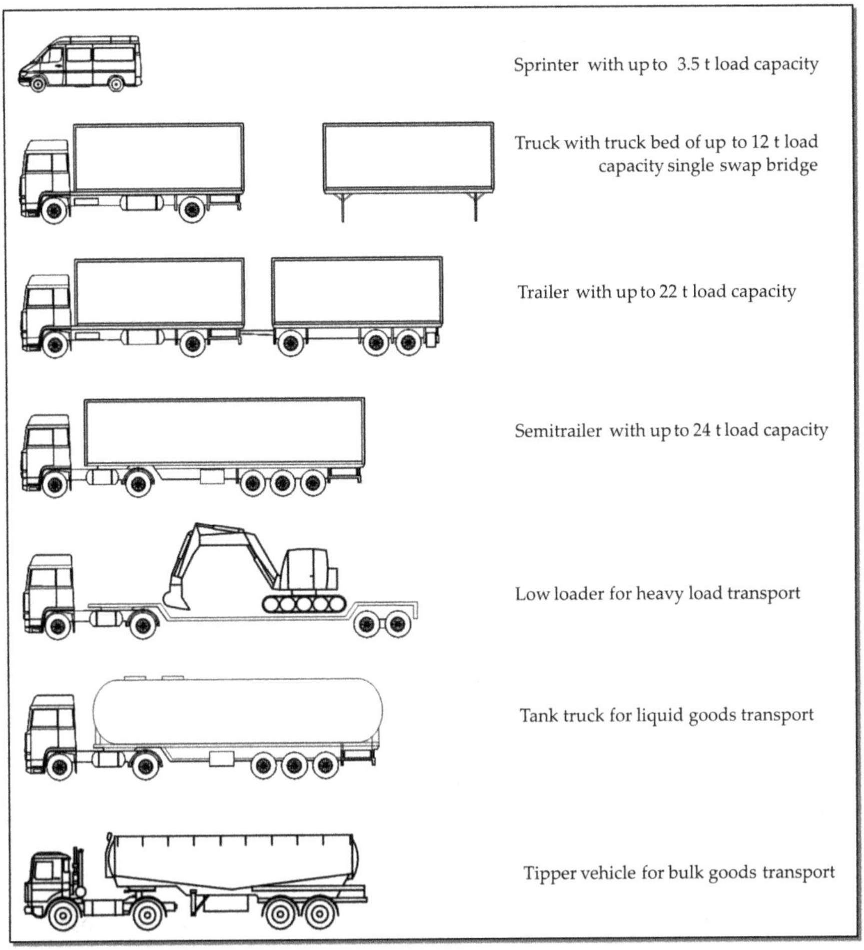

Fig. 5.14 Transport means in road freight transport

companies as well as by the dispatching companies. Furthermore, there are rental companies offering wagons, especially tank wagons, for hire.

The loading gauge determines up to what height an open wagon may be loaded. This, in turn, is dependent on the so-called minimum clearance outline along railway routes. The minimum clearance outline is the same in most European countries. Exceptions are France, Italy, and Switzerland, where a smaller loading gauge applies accordingly.

Container ships are the prevalent means of transport in *ocean shipping*. Depending on the goods to be transported, other types of ocean vessels include crude oil tankers, gas tankers, and bulk ships. Bulk ships are equivalent to traditional freight ships which can be loaded with individual bulk cargo. However, bulk

Table 5.9 Wagon types of DB Schenker Rail Deutschland AG (Cf. DB Schenker Rail Deutschland AG 2011)

- Type E: Open wagons
- Type F: Open hopper wagons
- Type G: Covered wagons
- Type H: Covered, spacious sliding wall wagons
- Type K: Flat wagons with double wheel sets
- Type R: Bogie flat wagons with four wheel sets
- Type S: Bogie flat wagons for coil transports
- Type S: Bogie flat wagons for metal plates
- Type S: Bogie flat wagons for stakes
- Type S: Bogie flat wagons with six wheel sets
- Type T: Covered hopper wagons
- Type T: Wagons with openable roof
- Type U: Low-floored wagons with/without special equipment
- Autoracks
- Wagons for combined transport
- Tank wagons for fuel, chemicals, or other types of liquid

ships are becoming less important since container transport is gaining more and more popularity due to its efficiency.

There are different types of containers, such as:

- Open-Top-Container (with detachable roof),
- Open-Side-Container (without sidewalls),
- Platform Container (without walls),
- Special containers (tank, bulk, refrigerated).

The most important types of containers are 20- and 40-ft containers, whose sizes allow transport of 14 or 29 pallets, respectively. The measurements of a 20-ft container are 6.06 × 2.44 × 2.59 (L × W × H in m; external); 40-ft containers measure 12.19 × 2.44 × 2.59 (L × W × H in m; external).

Container ships have become significantly more efficient in the past few years with regard to size and loading capacity. We can differentiate between the following types or generations of all-container ships[11] (see Fig. 5.16):

- Fourth, outdated generation with about 4,500 TEU loading capacity (so-called post-Panamax class – since 1988),
- Fifth, current generation with up to 10,000 TEU loading capacity (so-called super-post-Panamax class – since 1997),
- Sixth, future generation with up to 13,000 TEU loading capacity (so-called Suezmax class – since 2006).

[11] Cf. Brinkmann (2005), p. 66 et seq.

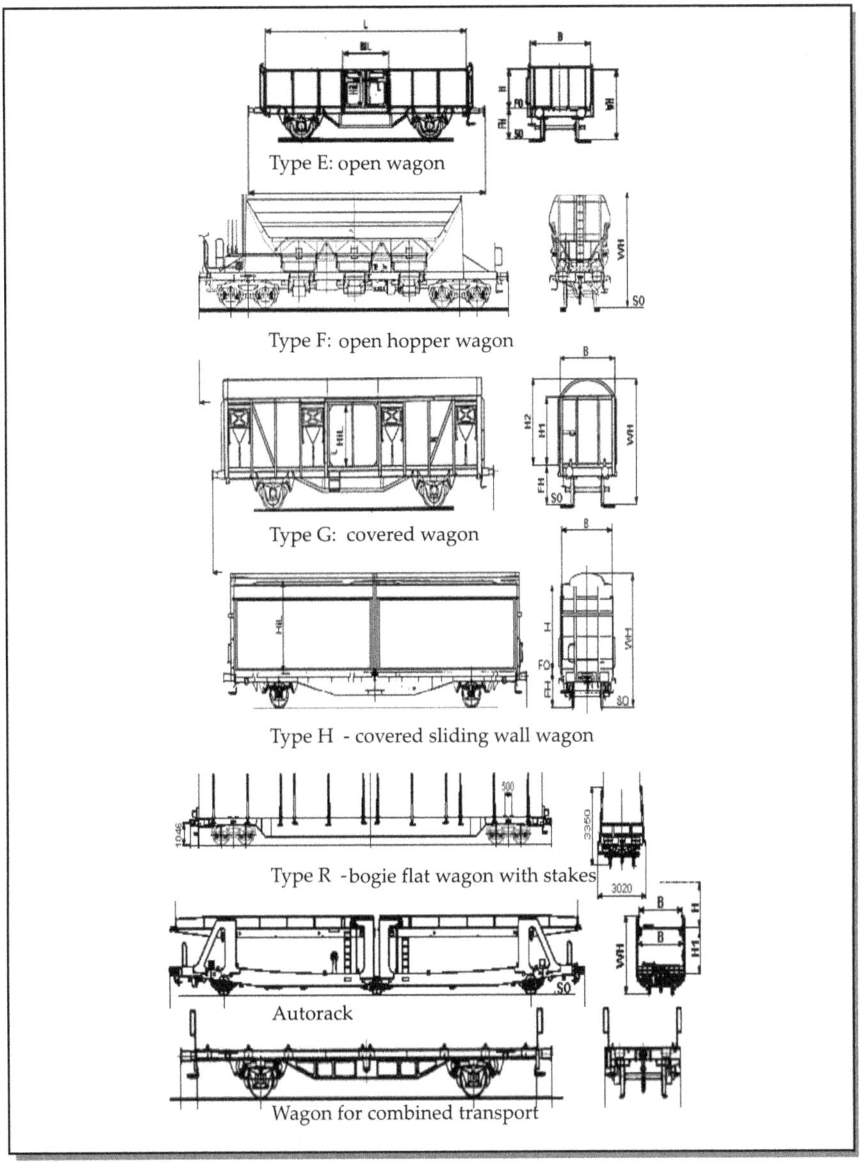

Fig. 5.15 Railway wagons (Cf. DB Schenker Rail Deutschland AG 2011)

TEU stands for Twenty-Foot Equivalent Unit and is the measurement most commonly used in container transport based on the measurements of a 20-ft container. This means, for example, that a 10,000 TEU container ship can transport the equivalent of five thousand 40-ft containers. The terms Panamax and Suezmax

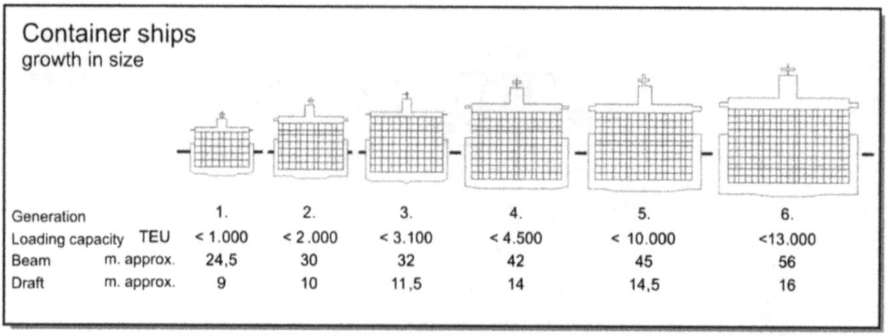

Fig. 5.16 Ship generations in container transport (Cf. Nuhn 2005, p. 113)

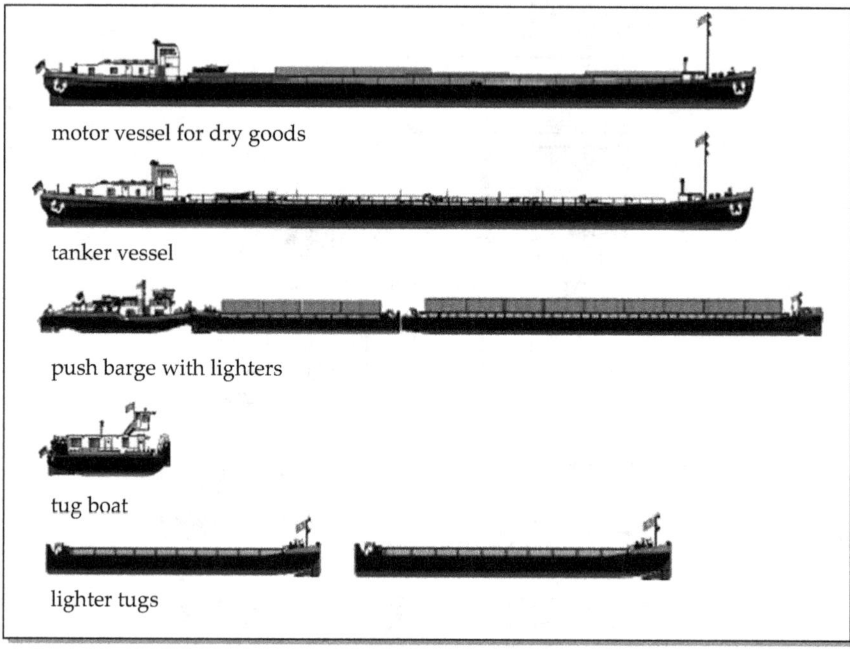

Fig. 5.17 Transport means of inland waterway transport (Deutsche Binnenreederei 2007)

relate to the maximum ship size for navigating the respective canal. The largest container ship that can barely navigate the most important shipping routes is the *Emma Maersk* of the Maersk shipping company in Copenhagen.

Ship types used in *inland waterway transport* include motor vessels, push barges, lighters, tug boats, lighter tugs, and towed barges. Other types of vessels include sea-going inland waterway vessels, ferries, and RoRo schips. Figure 5.17 shows the individual transport means typically used in inland waterway transport.

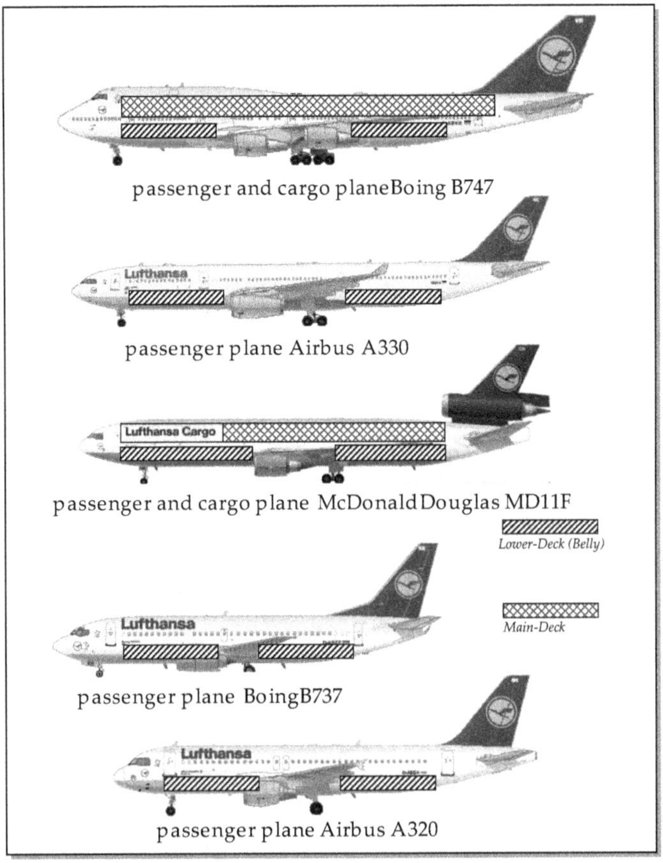

Fig. 5.18 Airplane types (Deutsche Lufthansa AG 2007)

With *air freight* transport services, the airplane as a means of transport is differentiated into passenger plane, quick-change plane and all-cargo plane. To a limited extent the lower deck loading capacity (belly capacity) of passenger planes can be utilized for goods transport.[12] More lower deck capacity is available on combined passenger/cargo planes. All-cargo planes are solely used for freight transport. Figure 5.18 shows the different airplane types and their basic freight capacities.

The distinct shape of airplanes requires the use of special loading aids for air transport. A selection of commonly-used air freight containers which are fitted to airplane shapes is depicted in Fig. 5.19.

[12] Cf. Mensen (2007), p. 52 et seq.

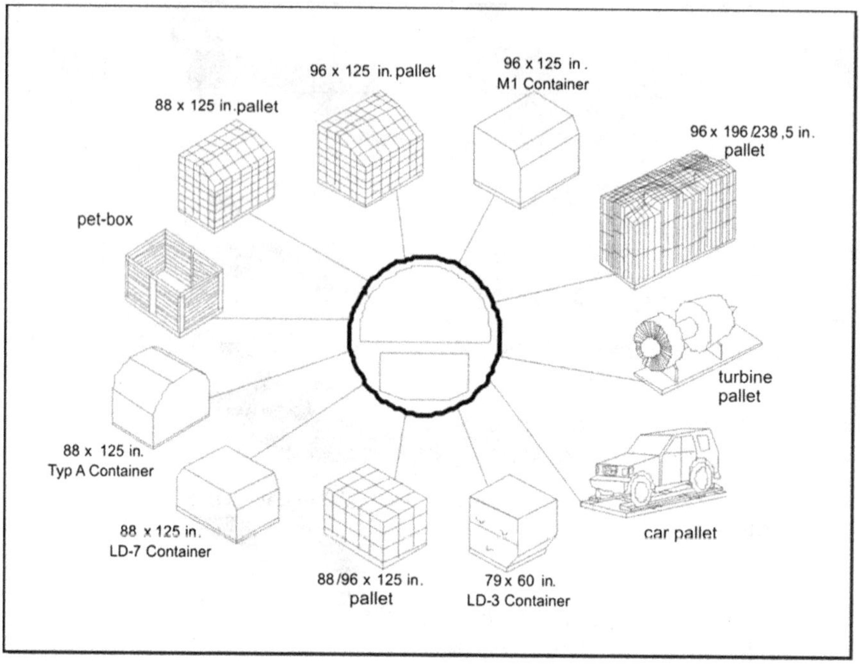

Fig. 5.19 Air freight containers (Deutsche Lufthansa 2005, o. S.)

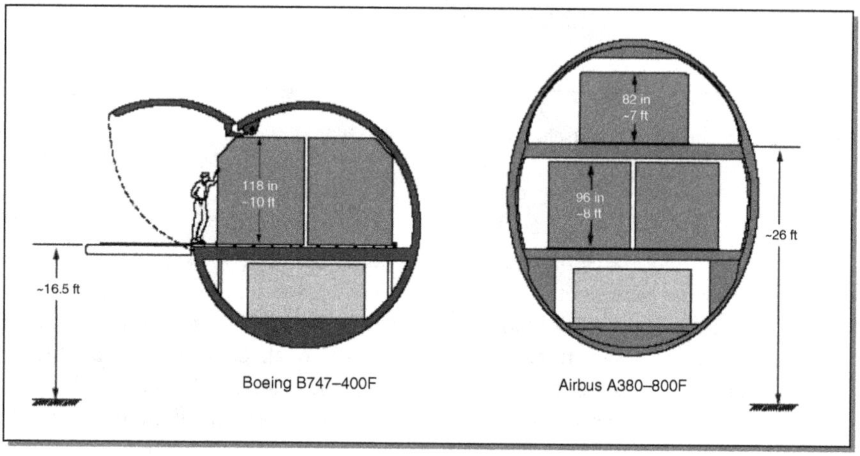

Fig. 5.20 Loading profiles of all-cargo planes (Boeing 2006)

Figure 5.20 shows the different loading profiles of cargo planes on two and three levels. The planes are loaded either from the side, in front, or from the rear.

Case Study 5.5: Loading of Air Cargo

In accordance with the loading and booking list for a flight, the consignment is first checked against prohibitions on co-loading certain goods. This is especially relevant for shipments of living animals and hazardous goods. Furthermore, package volumes are checked to ensure efficient space utilization on air pallets or in air containers.

Depending on airplane type, there are different restrictions in place as to the loading height on air pallets or the use of certain kinds of containers. A so-called counter frame is used to indicate the exact height and inclination of the external walls on the cargo plane's main deck.

Type MD 11 all-cargo planes provide a loading height of 1.6 m on the lower deck. Passenger planes transporting cargo on the lower deck offer loading heights of 1.16 m (type A320/321), 1.09 m (type A319), and 0.86 m (type B737).

If the goods contain prefabricated elements which are firmly attached to the pallets, additional straps are used to secure heavy and wide load parts. Finally, a cover sheet for weather protection (rain) and a net to secure the whole pallet are wrapped around the goods.

Subsequently, the pallet is weighed. The data is then transmitted to the operations department at *Weight & Balance*, where a trim calculation is carried out. This way, pallets with different weights can be loaded and spread evenly across the airplane's loading space.

5.5 Logistics Service Providers

5.5.1 Carriers and Forwarders

Apart from transport means and transport chains which represent the technological and organizational aspect of transport systems, we can draw institutional distinctions according to the stakeholder involved in the transport system. More specifically, we can distinguish between carriers, haulage contractors, forwarders, couriers, express service providers, parcel service providers, and logistics service providers.

Transport system operators, in turn, can be differentiated according to the technologies they use. For every transport mode there is a corresponding *carrier*. These are companies that carry out transport autonomously, such as airlines, shipping companies, independent ship owners, and railroad companies. In ocean shipping, carriers are also referred to as consignors. In § 7 par. 1 of the German Code of Commercial Law, a haulage contractor is defined as somebody who is bound by the haulage contract to transport a good to the point of destination.

According to the German Code of Commercial Law HGB § 453 par. 1, *forwarders* are companies that are contractually obliged to organize transportation of goods. This, however, usually implies disposition services, as opposed to actual transport services. Forwarders act as mediators between the party demanding the transport service (sender) and the party providing the transport service (commercial goods traffic). Thus, a forwarder organizes mode-neutral transportation of goods. For this reason, forwarders are sometimes called the *architects or organizers of transport chains*. Through own-name transactions (Code of Commercial Law HGB § 458), transport services and other services may also be rendered by the forwarders themselves. Forwarders focus on certain transport modes or service areas, such as truck, railroad, air freight, or sea freight. Table 5.10 provides an overview of the different service areas and the corresponding focus areas.

Consolidated transport or groupage transport on the road assumes a prominent function. The table shows that 41.3 % of all forwarders surveyed carry out this kind of transport. Such operations were the main focus area for 21.6 % of the companies

Table 5.10 Service areas and service focus of forwarders (Benchmark: 2,900 surveyed companies, multiple answers possible) (Cf. DSLV (2010), p. 2)

Service area of forwarders	service area	service focus
Consolidated road freight transport	41.3 %	21.6 %
Consolidated rail freight transport	4.4 %	1.4 %
Parcel and express services	11.1 %	4.6 %
Freighting of external trucks	63.3 %	21.9 %
Long-distance truck transports (own-name)	47.4 %	29.4 %
Short-distance transport / cartage	51.8 %	21.4 %
DB general cargo private haulage	7.8 %	3.9 %
International forwarder	55.1 %	24.1 %
Air freight transport company	23.9 %	11.9 %
Sea port forwarder	17.1 %	7.8 %
Customs clearance	45.2 %	12.2 %
Inland waterway forwarder	6.6 %	2.1 %
Inland waterway transshipment	5.6 %	2.6 %
Moving company	9.5 %	4.6 %
Distribution warehousing	39.5 %	4.6 %
Bulk commodity warehousing	8.3 %	3.3 %
Cereal and feedingstuff warehousing	3.7 %	2,0 %
Hazardous goods handling	31.2 %	5.9 %
Distribution logistics	24.0 %	11.0 %
Procurement logistics	19.3 %	7.8 %

surveyed. With these operations, the forwarder organizes dispatch of goods shipments from different senders as consolidated cargo (consolidated shipment or groupage consignment). Apart from consolidated cargo traffic there is combined transport. Combined transport is used for goods that either do not take up the full capacity of a means of transport (part-load shipping) or that take up a transport means' full capacity (full-load shipping). The term consignment comprises all goods which have been handed over to a forwarder by a sender together with a shipping order and which are meant to be shipped to a receiver. A consignment may consist of several individual shipping orders. Conversely, one shipping order may be carried out through several consignments. In groupage traffic, a transport chain involves:

- The collection of single consignments within a local/regional area (pre-carriage)
- The spatial and temporal consolidation of grouped consignments of as many senders as possible and for receivers in the same destination areas
- The joint transport of those consignments in one means of transport over as long a distance as possible
- Discharge at the points of destination
- Delivery of the consignments to the receivers (onward carriage)

One of the benefits of groupage traffic is economies of scale, which are brought about through increased capacity utilization of the means of transport, thus reducing transport costs per shipment.

Groupage transport is very common for the transportation of *general cargo* shipments. *General cargo* consists of individual goods in solid form, which can be separately handled and informationally registered. These include unpacked goods, packages, and loading units. They can weigh from 25 up to 3,000 kg. *General cargo* shipments are usually very heterogeneous and consist of several shipping units. Groupage transport is an important submarket of *small goods transport*, which also includes courier services, express services, parcel services, and mail services.

5.5.2 Courier, Express, Parcel and Mail Service Providers

In contrast to forwarders of groupage cargo, courier messenger, express and parcel (CEP) service providers focus on specific transport objects, such as documents or parcels with weight restrictions, whose shipment usually only comprises one shipment unit. These services require speedy transactions and high reliability. Express services provide transport of consignments with or without weight and size restrictions. Speediness and reliability are again paramount for these kind of services. Parcel services offer transport of one-piece parcels with weight and size restrictions (e.g. maximum weight 31.5 kg; maximum girth measurement 3 m).

In many cases, *CEP service providers* are integrators (Integrated Service Carrier) which are providing the entire transport chain. They produce and sell

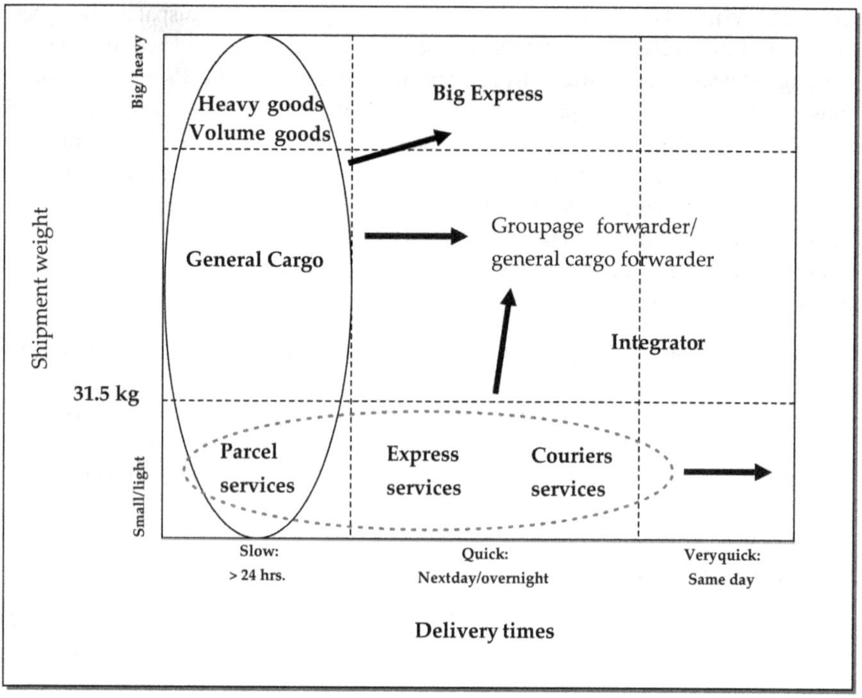

Fig. 5.21 Competition matrix: forwarder/integrator (Vgl. Bjelicic 2005, o. S.)

their services autonomously by means of using their own airplanes, transshipment facilities, and vehicles. Integrators continuously expand their world-wide networks. Thus, they are also expanding their business field by offering additional logistics services. Extended or suspended weight restrictions are leading to an increased overlap between the business fields of integrators, groupage cargo forwarders, and CEP service providers and to growing (substitution) competition among these stakeholders, as can be seen in Fig. 5.21.

Mail services include the transport of letters, small goods, and mail items. In Europe, mail service providers often used to be state-owned. Meanwhile, however, many of the formerly state-owned logistics companies have been privatized or are undergoing privatization processes, as is the case in Germany, for example. Mail service providers also frequently offer monetary transactions, as well as fax, email, and other kinds of services. In principle, mail services are a highly specialized form of goods transport dispositioned by the state. The fact that mailboxes are emptied on a regular basis – regardless of how many mailings are in it – serves as an example to differentiate mail services from other logistics and forwarding services. Apart from the formerly state-owned mail companies, there are privately-run mail service providers. These are going to broaden the range of services on the mail markets, especially due to the abolishment of the letter monopoly.

Case Study 5.6: Express Service Provider as Integrator

The term *integrator* comes from the English verb *(to) integrate*. More specifically, this refers to the logistical service structure of a transport service provider. This means that an integrator unifies or integrates procurement, production and sales of a service in their portfolio on an international level. To this end they make use of their own networks, hubs, vehicles, and airplanes. For example, with 672 airplanes the American express service provider FedEx runs the world's largest air fleet. In parallel to air freight companies, integrators also participate in the international freight market by offering loading space to other airlines or by requesting additional loading space in the case of a bottleneck situation. The product portfolio of these companies not only comprises CEP services but by now also includes aspects of contract logistics (especially storage, order-picking, billing), consulting services, mail services, and financial services. This underlines a strong tendency towards the formation of integrated logistics groups. Additionally, the logistics service provider's extended range of services becomes apparent by offering a number of value added services.

World-wide there are four integrators which generated an aggregate turnover of about € 120 bn in 2010 (see Table 5.11). With a turnover of roughly € 11 bn, TNT is the smallest company out of the four. TNT operates their own air freight and road network in Europe to provide their services (see Fig. 5.22).

These networks are based on the hub-and-spoke principle, which connects superordinate locations with subordinate ones in a radial way. TNT's European air network is a single-tier system with one central airport in Lüttich (Belgium) and 58 associated destinations in 23 countries. The European road network comprises 414 depots in 33 countries and has a multi-level structure since several smaller hub-and-spoke systems are integrated into an overarching hub-and-spoke system. Thus, TNT's Express division operates the most extensive air and road network in the whole of Europe.

5.5.3 Systems Service Providers and Contract Logistics Providers

Over the past few years, there has been a significant trend towards *outsourcing* logistics services to external service providers (see Sect. 8.5). Several forms of outsourcing have developed on the market. These developments include carriers and forwarders outsourcing freight services as well as allocating entire logistics

Table 5.11 The most important integrators and their turnover in 2010 (Cf. Annual reports of TNT, UPS, FedEX, DPWN 2010 ($ exchange rate as of 31.12.2010))

	TNT	UPS	FedEX	DPWN
Turnover (across the group in 2010)	€11.3 bn	€37.3 bn	€26.1 bn	€51.5 bn
Market share	8.9 %	29.6 %	20.7 %	40.8 %

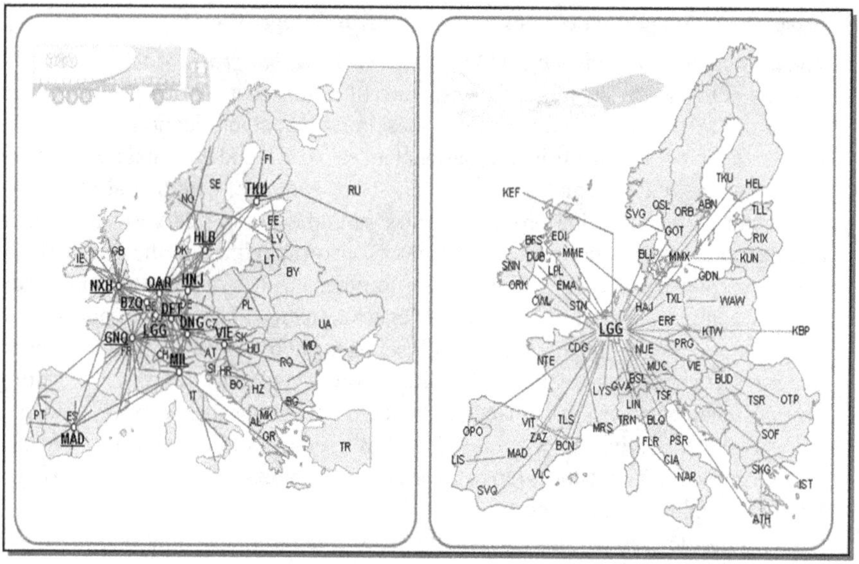

Fig. 5.22 Air and road network of TNT Express (TNT Express GmbH 2007)

locations to logistics service providers. Comprehensive services like these are also termed *contract logistics.*[13] In such cases, logistics service providers become *contractual partners* for systems services and are responsible for the organization of complete supply chains (see Sect. 12.4).

In particular, contract-logistical services may include:

- Central warehousing management for procurement and distribution logistics
- Internal and external production-logistical functions or job order production
- Comprehensive pickup and delivery systems

One of the main characteristics of contractual performance – together with the scope of services provided – is the sustainability of co-operation between the outsourcing company and the service provider. For this reason, contract periods of 3 years or longer are not unusual, especially if the outsourcing process requires investments on the part of the service provider.

Contract logistics opens up a market with high profit margins and rapid growth for the logistics service sector. Thus, double-digit growth rates are forecast for the next few years. In this context, the German logistics market can be counted among the largest ones in Europe with a turnover of 205 billion Euros in 2007. The logistical expenses comprise all transport, warehousing, transshipment, and value-added services as well as all activities associated with logistical co-ordination. Transport services account for a share of over 40 % of the overall

[13] Cf. Weber et al. (2007), p. 37 et seq.

logistics volume, while storage and transshipment services make up about one quarter. The remainder includes services related to order processing, inventory management, and supply chain management.[14]

There are, however, numerous requirements to be met by providers of contract-logistical services. These include a minimum enterprise size, corresponding availability of capital and, most importantly, know-how in logistics and market cultivation. In addition, certain company functions, such as marketing, production, quality management, controlling, and IT, are confronted with increased challenges.

[14] Cf. Klaus and Kille (2008), p. 45; BVL e. V. (2011), p. 1.

Review Questions

1. Differentiate between transport services and logistics services.
2. What is meant by volume-of-goods effect and goods-structure effect?
3. What is containerization?
4. What are the advantages of inland waterway transport as a mode of transport?
5. What are the differences between wagonload transport, part-load transport, and combined transport?
6. What do the acronyms ICAO and IATA stand for?
7. Explain the structure of an air transport chain.
8. What is the significance of road traffic?
9. What is the difference between a transport chain and a means of transport?
10. What is contract logistics?

Additional Literature

Aberle, G. (2009). *Transportwirtschaft: Einzelwirtschaftliche und gesamtwirtschaftliche Grundlagen* (5th ed.). München/Wien: Oldenbourg.

Biebig, P., Althof, W., & Wagener, N. (2004). *Seeverkehrswirtschaft: Kompendium* (3rd ed.). München: Oldenbourg.

Klaus, P., & Kille, C. (2008). *Die Top 100 der Logistik*. Hamburg: Dt. Verkehrs-Verl.

Korf, W. (Ed.) (2005). *Lorenz-Leitfaden für Spediteure und Logistiker in Ausbildung und Beruf, Vol. 1*, (20th ed.). Hamburg: DVZ.

Kummer, S. (2010). *Einführung in die Verkehrswirtschaft*. Wien: Facultas.

Liberatore, M. J., & Miller, T. (1995). A decision support approach for transport carrier and mode selection. *Journal of Business Logistics, 16*(2), 85–116.

Merkel, H., & Bjelicic, B. (Ed.) (2003). Logistik und Verkehrswirtschaft im Wandel: Unternehmensübergreifende Versorgungsnetzwerke verändern die Wirtschaft, Festschrift für Gösta B. Ihde. München: Vahlen.

Schubert, W. (Ed.) (2000). Verkehrslogistik: Technik und Wirtschaft. München: Vahlen.

Sterzenbach, R. (2009). Luftverkehr: Betriebswirtschaftliches Lehr-und Handbuch, 4. Edition, München: Oldenbourg Verlag.

6 Warehousing, Handling, and Picking Systems

Learning Objectives

Storage and warehousing are central logistics tasks. They require logistics systems which are to a great extent technology-driven and which any logistics expert needs to work with in everyday practice. This chapter aims to familiarize the reader with the basics of warehousing, handling and sorting systems. In so doing, it will address the basic methods to organize storage areas and storage processes. Furthermore, tried and tested solutions for warehouse equipment will be explained. The main focus will be to elucidate the business-economical contexts and effects associated with different solutions in order to point out the diverse options to the readers and to assist them in the assessment thereof.

Keywords

- Functions of warehousing
- Storage systems and conveyors
- Technological and organizational warehousing designs
- Task areas within a warehouse
- Transport and handling of goods
- Integrating a warehouse into transport chains

H. Gleissner and J.C. Femerling, *Logistics*, Springer Texts in Business and Economics,
DOI 10.1007/978-3-319-01769-3_6,

6.1 Warehousing Basics

Storage is a key element of the logistical system of services and occurs in all stages of a supply chain and on all levels of the value-adding process and production process. Storage is of particular importance for distribution logistics in the context of end-customer delivery of goods. Apart from the actual storing of goods, additional functions, such as picking, packaging, or return processing need to be fulfilled in this area.

There is a direct link between warehousing and inventory management (see Chap. 7). At a warehouse, goods are *prepared* or *transshipped* for on-carriage to the ultimate consumer and may undergo *quality-checks* and *maintenance* in cases of long storage periods.

Warehouses also fulfill *allocative functions* within the distribution process to direct shipments to their destinations in the right composition. Accordingly, warehouses serve to bring together flows of goods from different sources and places, e. g. production sites or suppliers. In one warehouse, an entire range of products may thus be available, which can be transported to the respective distribution area or from which actual customer orders can be compiled. The on-carriage may be organized in a way which utilizes the capacities of the outgoing loading or transport unit most efficiently (see Chap. 5).

The *Deutsches Institut für Normung, DIN e. V.* (German Institute for Standardization) and the *Verein Deutscher Ingenieure* (Association of German Engineers) have developed several definitions with regard to the most important warehousing terminology. *Storage* is any *planned* waiting time for goods or work objects in the flow of materials and goods.[1] Throughout the literature, the term *flow of materials* is more commonly used for production processes in the industry sector. In the distribution and trading sector, however, the term *flow of goods* is more common.

A warehouse is a room or an area in which goods are stored in solid (general cargo or bulk cargo), liquid, or gaseous state and where their quantity and value is usually recorded. In this context, the term general cargo characterizes goods which can be handled or carried by conventional conveyors either directly or in their respective packaging. Bulk cargo consists of very small, loose, and pourable conveying goods. Storage, in the economic sense, is a deliberate process. However, only if the goods stored are documented the term storage actually applies.

Conveyance occurs if goods are moved within a storage area or plant, e.g. from or towards the productions site. Thus, conveyance mainly refers to in-plant flows of materials and goods. Within the confines of such a system, conveyance may also comprise the movement of people, besides the movement of goods. The conveyance of goods from one system to another (warehouses, production sites etc.) over long distances is referred to as transport.

---

[1] Cf. VDI code No. 2411.

A *package* is the most frequently occurring element in both warehousing and transport. A package consists of the packaged good (the commodity) and its packaging, which should also be suitable for conveyance and transport.[2] Several packages or goods with the same destination may be compiled to *loading units* for the purpose of efficient conveyance, transshipment, and transport. *Loading aids* are usually used to form loading units. Pallets are the most commonly used loading aid.[3]

6.2 Storage Facilities

Apart from the basic storage function that warehouses assume, they also fall into different operational categories for storage, buffering, and distribution. Depending on warehouse type, they primarily serve to bridge the time or to change the composition of a package, loading unit, or consignment between receipt and dispatch at the warehouse. If there is a focus on offsetting fluctuations in demand over more or less long periods of time, the warehouse is referred to as a *storage warehouse* (procurement logistics). Innovative logistics solutions aim to minimize the inventory in such storage warehouses (see Chap. 7). A *buffer warehouse* serves to offset imbalances between receipt and dispatch for relatively short periods of time. This type of warehouse is especially important in production logistics, where the output and input for individual, consecutive production steps need to be balanced. It is the task of production planning to minimize these buffer stocks in order to reduce the amount of fixed capital in the production process. *Distribution warehouses* (distribution logistics) serve to direct the flow of goods to the end customer, to make goods quickly available in the respective sales regions, and to compose a consignment according to the customer's specifications.

Different storage systems are available to render the storing process efficient. Which kind of storage system is utilized is dependent on the goods and on the market's requirement for availability (service level). Furthermore, costs are an important factor when choosing a storage system. Maximum utilization of storage area or space volume, small investments for warehouse equipment, as well as simple and cost-efficient processes for storage, dispatch, and picking with high output are the main factors to consider. Thus, storage systems fall into three categories. In a *static storage system* the packages or loading units remain in one place without being moved between admission to the warehouse and taking out of the warehouse. This does not include movements as a result of a temporary withdrawal to remove individual parts of a loading unit or to re-stock the loading unit in another place. In *dynamic storage systems* the goods are moved within the storage area during their storage time. These movements are triggered by specific storage techniques, which, for example, enable goods to move forward as soon as

[2] Cf. DIN code no. 55405.

[3] Cf. DIN code no. 30781.

one unit is removed at the front. Another distinction can be made according to the accessibility of storage goods. In cases where *line storage* is employed, all loading units can be accessed at any time, which is not possible if the loading units are stacked in compact *block storage*. Finally, we distinguish between storage systems *with storage racks* and *without storage racks*.

6.2.1 Static Storage Systems

Block storage is the simplest form of storage. The packages or loading units are in this case stored in large blocks on the floor. If they are stable enough they may also be stacked on top of one another. Only those loading units which are on top along the aisles are directly accessible, which corresponds to the system of stacking. Most block storage systems work this way and thus achieve high storage space utilization. The greatest disadvantage here is the limited accessibility of all storage goods, which precludes the use of storage strategies such as First In-First Out (FiFo) (see Sect. 6.6). Block storage may also be designed as a line storage system, to provide better accessibility. In this way, however, much storage space is taken up by the aisles between the lines. Block storage is mainly used for storing large-volume, heavy and sturdy goods and is often found in outdoor storage areas (see Fig. 6.1).

Since block storage does not necessitate any rack and shelving systems, no further investments need to be made and the storage space can be used flexibly for various storage goods and transshipment or retrieval options. The possible stacking heights vary according to the goods being stored. In most cases, however, they do not reach the same level of space utilization as rack and shelving systems do. The level of space utilization results from the share of the storage volume (including operating aisles between racks) in the overall warehouse volume.

Pallet racks are the most commonly used storage system. Using the loading aid of the pallet, goods and loading units are stored on especially designated shelves. We distinguish between one-unit and multi-unit space systems. Multi-unit systems

Fig. 6.1 Block storage figure

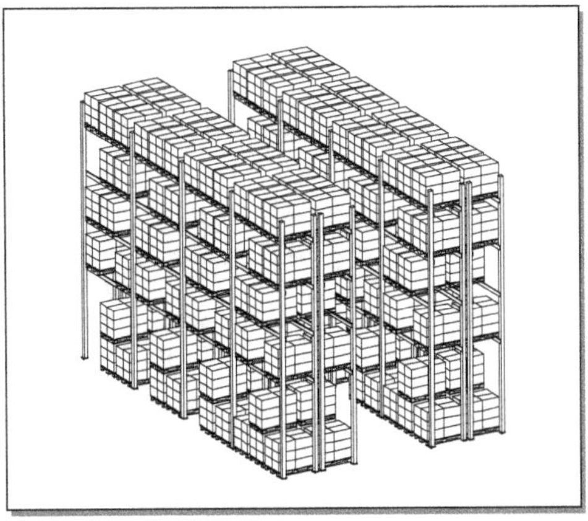

Fig. 6.2 Pallet rack

are designed to store more than one pallet next to each other on the shelf. Pallets can have varying measurements, but in most cases they are rectangular. Therefore, they can either be stored deep (mostly three next to each other), which results in greater capacity utilization, or they may be stored horizontally if goods are picked directly from the pallet. Pallets can also be stored one behind the other. If a pallet rack exceeds a height of approximately 15 m, it is called high rack.[4] Depending on type of goods and picking frequency the pallet racks may be separated by broad operating aisles or they can be designed as a narrow-aisle construction (see Fig. 6.2).

In contrast to block storage systems, pallet racks provide direct accessibility to each loading unit, as long as they are stored next to each other on the shelf. If standard pallet types are used, pallet racks also guarantee relatively high flexibility and transshipment performance with regard to goods types and picking requirements. The high storage volume utilization which can be achieved using this type of storage system is especially advantageous in metropolitan areas where storage space is costly. The ground on which a pallet rack is set up needs to meet certain constructional requirements irrespective of whether the construction is located outdoors or in a warehouse. In comparison to a block storage system, additional costs may arise for investments in shelving technologies and possibly also for technologically more advanced conveyor techniques used in the storage and retrieval processes.

Instead of using pallets, rack shelves may also serve to store small parts in containers of different sizes. To this end, the space between the shelving levels can be reduced. This storage type is termed *container shelving*. The storage of goods on flat, trough-like plates (trays) is referred to as *tray storage*.

Drive-in racks or *drive-in shelves* combine pallet shelving and block storage and are used for storing palletized goods. In principle, they are similar to one-unit pallet shelves. With this storage system, however, several pallets can be stored deep one

[4] Cf. Schulte (2009), p. 234.

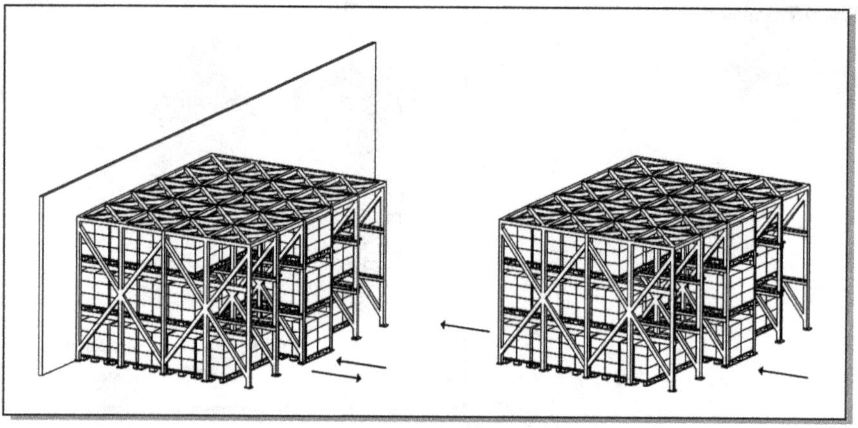

Fig. 6.3 Drive-in rack and drive-through rack (Cf. Jünemann and Schmidt (2000), p. 52.)

behind the other. This type of shelf can only be operated from one front side. Storage and retrieval is carried out by forklifts, which drive between the rack supports.

The storage principle of drive-in racks greatly limits the accessibility of the pallets (see Fig. 6.3). Beginning at the back, the loading units need to be stored from the top to the bottom, while they have to be retrieved the other way round. This means that merely the LiFo (Last in-First out) principle can be implemented. At first glance, the storage space utilization seems to be higher in comparison to pallet rack systems. In practice, though, this can only be achieved with a high number of articles coupled with a small range of articles.[5] Accordingly, drive-in racks are used for articles which are heavy or pressure-sensitive and therefore cannot be stored in a block storage system.

If the articles can be taken out on the other side, the drive-in rack is referred to as a *drive-through rack*. These racks allow for the realization of the FiFo principle, which is in many cases more cost-effective.

Bay shelves are made up of vertical posts and shelves which can be mounted or screwed into the construction. They are primarily operated manually without the use of ground conveyors (see Fig. 6.4).

Since the shelves can be easily mounted on the posts by means of hole pins, this storage system offers great flexibility. Additionally, it provides high storage space utilization as the stored articles determine which kind of shelf will be used. This type of shelf is mainly used to store loose small parts or small parts in boxes or outer packing. Outer packaging is an additional piece of packaging – such as plastics or straps –, which serves to keep the goods together in their original wrapping. Since each of the shelves is accessible, a broad range of articles can be stored. Picking position and distance are decisive for quick accessibility in the picking process.

[5] Cf. Jünemann and Schmidt (2000), p. 51.

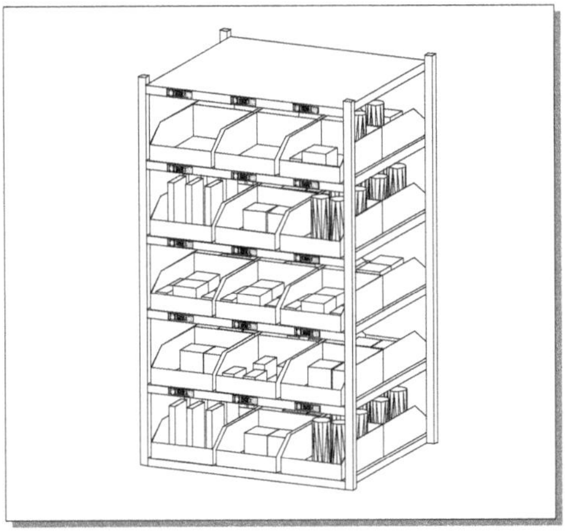

Fig. 6.4 Bay shelf

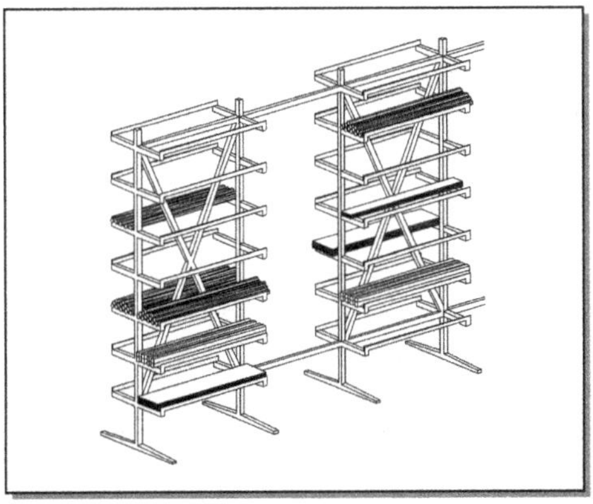

Fig. 6.5 Cantilever shelf

Articles that are stored on top or at the bottom are more difficult to pick than those stored at breast height.

Cantilever shelves consist of vertical posts on arms jutting out laterally. Long loads, such as tubes, profiles, bars, beams, and boards are stored on these arms. The number of arms on top of each other and the arm length are determined by the measurements and weight of the storage goods (see Fig. 6.5).

If the articles are supposed to be individually accessible, the number of articles per cantilever arm is limited.

6.2.2 Dynamic Storage Systems

Live storage shelving is the most commonly used dynamic storage system. The loading units are stored on slightly inclined roller lanes and are fed in at the higher

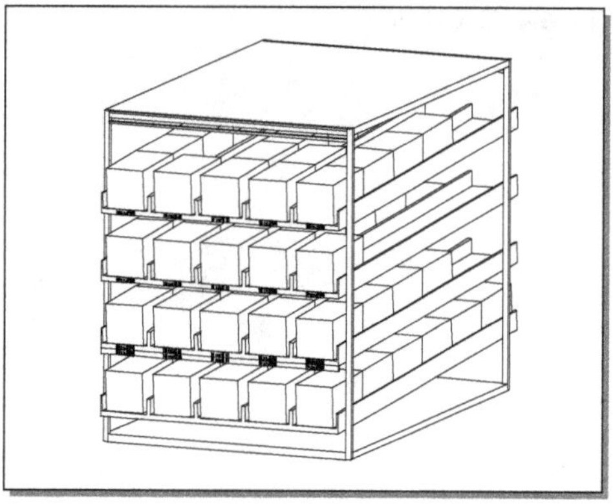

Fig. 6.6 Live storage shelf

end of the lane and taken out at the lower end. As soon as one loading unit is removed, the remaining units move forward automatically. Live storage shelves require the articles to be stored on suitable load carriers or to be packed in stable cardboard boxes (see Fig. 6.6).

The moving load carriers enable the implementation of a FiFo principle. High picking and withdrawal performance can be achieved if each shelf stores one article. The main advantages of this storage system are the separation between the feeding and exit aisles as well as its high storage space utilization if the lanes are full.

Push-back shelving uses the same construction as live storage shelving. The articles, however, are both fed in and taken out at the lower end of the lane. This kind of shelf is only used for palletized articles. The first pallet is put on the roller lane and pushed back as the second pallet is fed in (see Fig. 6.7).

In line with the LiFo principle, the goods are fed in and taken out at the same end. The pallets are supposed to push back those pallets which have already been

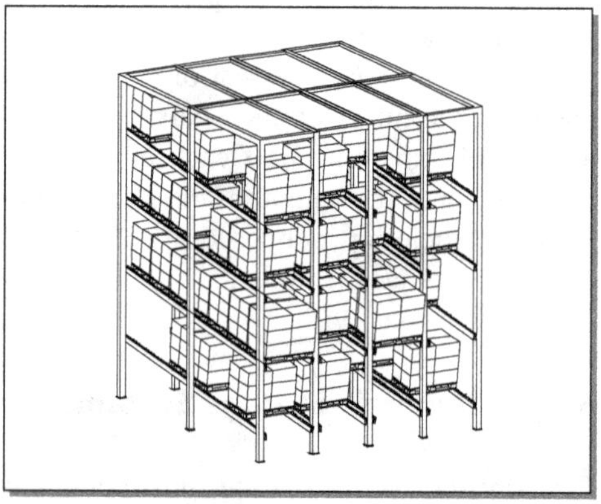

Fig. 6.7 Push-back shelf

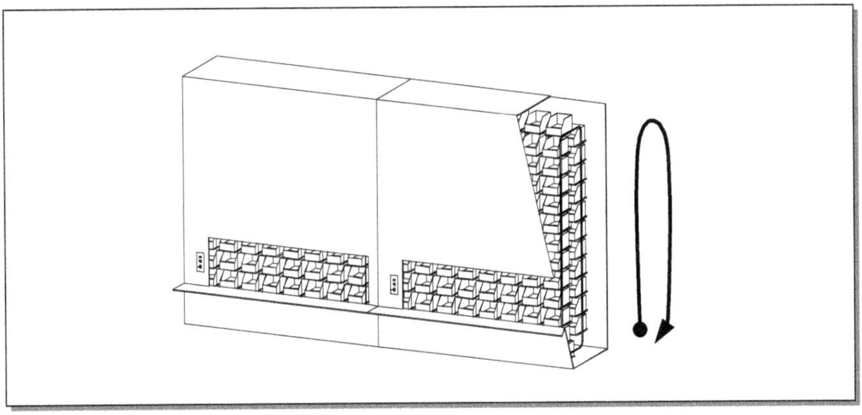

Fig. 6.8 Vertical carousel

stored on the shelf. This needs to be taken into account when opting for a shelving system. Furthermore, the pallets are prone to canting.

Satellite storage systems are a variation to push-back shelving. With this storage system, the pallets are put on self-propelled carts which independently drive into the lanes. Subsequently, the carts drive back to the point of loading. This system's advantage over push-back systems is the fact that forklifts or shelving operating devices can proceed with other tasks while the pallet is still being stored.

A *vertical carousel (paternoster)* is a type of shelving storage system in which the individual shelves are attached to two vertically circulating chains. Using a control panel, the required shelf is moved to the discharge opening (see Fig. 6.8).

In order to avoid full circulations when retrieving goods, the same articles should ideally be grouped together and stored on different levels. To this end, the shelves are partitioned and allow for separation according to article type. The handling performance is dependent on the height of the carousel and on the range of articles. Paternosters offer a higher percentage of storage space utilization than bay shelving systems while providing the same accessibility. The system is lockable and is well suited for the storage of small to medium-sized goods and high-value goods. Manual operation, e.g. in the case of a system failure, is not possible.

A *horizontal carousel* is another type of shelving storage system. A horizontal overhead conveyor moves the individual shelves, which are attached to the conveyor. Using a control panel, the requested shelves are moved to the side of discharge (see Fig. 6.9). Thus, the *goods-to-man* principle (see Sect. 6.5) applies both to vertical and horizontal carousels.

This shelving type is primarily used in picking warehouses. Employment of this solution begs the question whether the picking person would have access to the needed article more quickly on foot or using this type of system. Ideally, several carousels should be inter-connected in one picking zone so that the requested article can be retrieved during the picking person's transit time. Transit time describes the time that is needed for a person to move from one picking point to another.

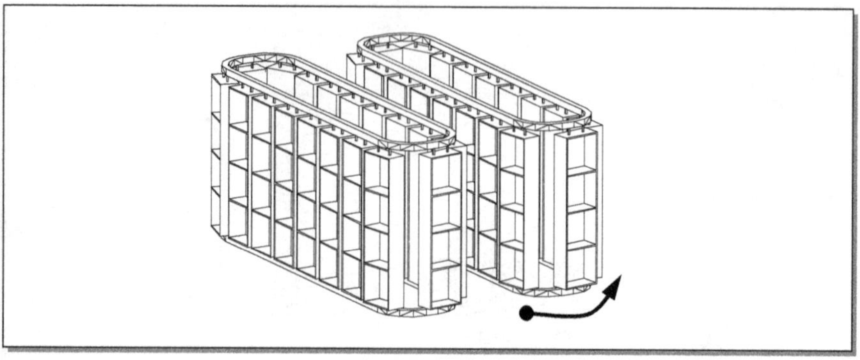

Fig. 6.9 Horizontal carousel

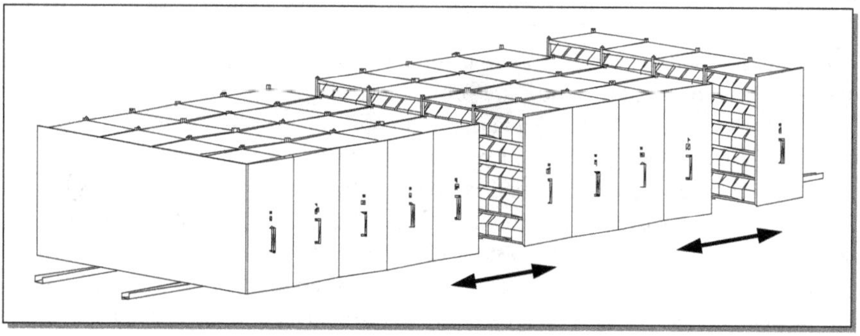

Fig. 6.10 Slide shelves

With *slide shelving*, the pallet shelves or bay shelves are mounted on slide racks, which are operated either manually or electrically. Only one or two aisles are open along the entire shelving block. The remaining shelves are densely put together. If goods are needed from one specific aisle, the shelves are moved sideward to open up a new aisle (see Fig. 6.10).

Highly cost-effective storage and warehouse space utilization is contrasted by limited simultaneous accessibility to goods stored in different aisles. For this reason, this type of storage system is predominantly used for archiving or for storing spare parts, e.g. in workshops. Among the system's advantages are optional lockability and accessibility to individual goods.

Case Study 6.1: Alternative Storage Systems

A mail-order company is planning to set up an additional auxiliary warehouse for cardboard boxes palletized on euro pallets. This is to be implemented on the existing company grounds, which cannot be expanded any further. The company offers a large range of articles and each article must be accessible at all times. Furthermore, it is imperative that opened pallets can be re-stored in their open state. The average height of one pallet is 1 m and the pallets are not stackable. To cover the long-term storage requirements, the new warehouse is supposed to have more than 6,000 pallet spaces available. Constructional restrictions by the responsible building authority allow an indoor clearance of 13 m.

The planning department is discussing three alternative storage systems:

(a) Pallet shelving (storage space utilization 45 %),
(b) Live storage shelving (storage space utilization 65 %),
(c) Block storage (storage space utilization 80 %).

The level of storage space utilization is the share of the storage area (including operational aisles between shelves but excluding main traffic areas and zones for receiving, issuing, and provisioning goods) in the total area of the warehouse. For safety purposes and to ensure sufficient free scope for storing and retrieving, a clearance height of 0.5 m per pallet needs to be factored in for pallet shelving. Other areas, such as goods receiving, goods issuing, and main traffic areas may be disregarded in the calculation.

Calculate the space requirements for each of the three alternatives.

Considering the circumstances, which storage system would you recommend?

6.3 Conveyors

To enable conveyance, i.e. the movement of goods within systems over short distances, technological aids – so-called conveyors or conveyor systems – need to be employed.[6] Conveyors transport goods within locally determined operational units, e.g. within production sites, warehouses, transshipment terminals or, airports. Besides their main task of conveying, conveyor technologies serve additional purposes such as *distributing or sorting* (sorting out of shipments), *collecting* (merging of individually incoming packages), *buffering* (transit time from dispatching location to destination or accumulation zones at the end of conveyor), or *picking* (sorting of goods for outgoing shipments or customer shipments). If some of the conveying goods can be discharged at certain points along the conveyor, we also talk about sorters or sorting technology.

Conveyor systems can be distinguished according to the following criteria:

- *Discontinuous conveyors* produce an intermittent stream of conveying goods and operate in individual working cycles (so-called intermittent conveyors).
- *Continuous conveyors* convey bulk or general cargo continuously.
- *Floor-bound conveyors* drive on the floor or on installations that are flush-mounted in the floor.
- *Supported conveyors* move above the floor on supports or on supported rails.
- *Suspended conveyors* move along rails on the ceiling.

[6] Cf. DIN 30781.

6.3.1 Discontinuous Conveyors

Manual lift trucks are manually-operated conveyors for the movement of single pallets on short distances. They are used for loading and un-loading vehicles (trucks or trains wagons).

Manual lift trucks can only lift a pallet to a height which enables them to drive carrying the pallet. For this reason they cannot be used to stack goods. Due to their technical simplicity, sturdiness, and good maneouvrability, manual lift trucks are suited to cover short distances in cramped spaces (see Fig. 6.11).

Lift pallet trucks, which belong to the category of *stackers*, are a variation of the manual lift truck. The base lift functionality as a so called low lift pallet truck enables them to stack pallets at a low height, around 1 m, which is why they are frequently used in the goods receiving and goods outgoing departments where goods are stacked up. Lift pallets trucks may also operate in storage areas. There are also high lift pallet trucks available to reach stacking highs up to 5 m (see Fig. 6.12).

Both manual lift trucks and lift pallet trucks can be fitted with an engine. This may serve to support staff members in pulling or pushing the truck, or the truck may be designed as a driver-seated lift truck. Both trucks are mainly used indoors since they cannot be driven on uneven surfaces due to their small wheels.

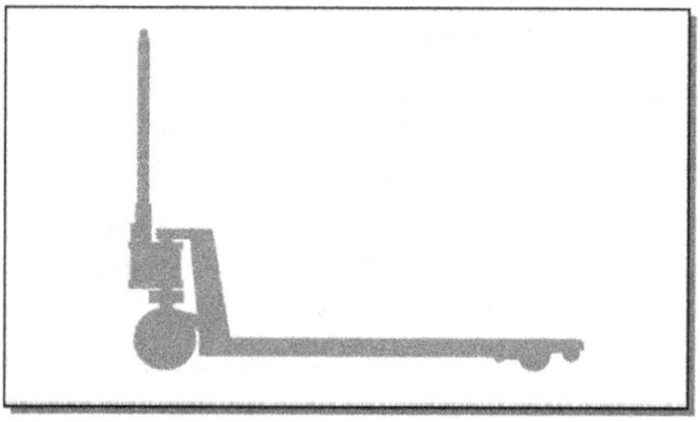

Fig. 6.11 Manual lift truck

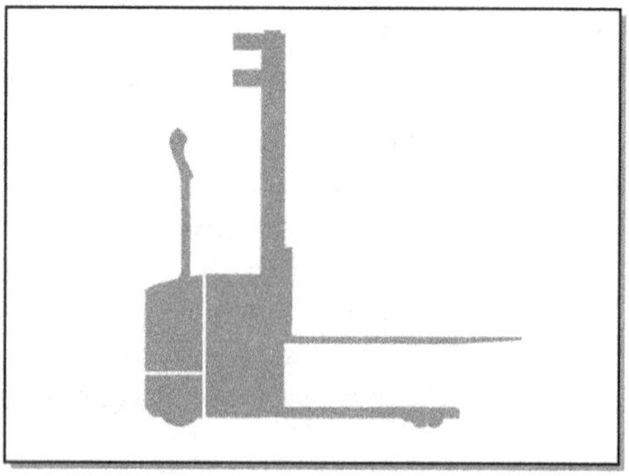

Fig. 6.12 Lift pallet truck

Fig. 6.13 Counterbalance forklift or frontload truck

There are different types of stackers to be utilized according to goods requirements and varying weights, heights, or ground conditions. Their main differentiators are their engine type (diesel, LPG, or electric), add-on devices (e.g. fork, clamp, mandrel, gripper, and lifting platforms), steering (four-wheel drive and three-wheel drive) and tires (large and small pneumatic tires or solid rubber tires).[7] Examples of the most commonly used types of stackers are explained below.

A *counterbalance forklift* or *frontload truck* carries the load unsupportedly in front of the front wheels. These kinds of stackers therefore require a certain overall length to counterbalance the weight. The reachable lifting height depends on the construction of the stacker but is in most cases higher than with a lift pallet truck (see Fig. 6.13).

Due to their robust construction and relatively large pneumatic-tired wheels, counterbalance forklifts can be operated outdoors to transport medium-sized loads at high speeds. Since these characteristics render it a truck for various purposes, its lifting height and maneuverability are limited, which makes it unsuitable for narrow operating aisles. The most common types are the usually smaller and more versatile three-wheel stacker as well as the four-wheel stacker, which is more robust and hence more suited to transport heavier loads.

Stackers with outriggers counterbalance the transported load by means of outriggers. Among these, the most important type is the *moving-mast reach truck*. It is the most frequently used stacker for in-house storage of pallets since the required aisle width is smaller than it is for counterbalance forklifts. This type of stacker is a mixture of a counterbalance forklift and a forklift with outriggers. For loading and unloading, the mast moves forward to the front wheels. During transport the mast is retracted which reduces the length of the stacker. Moving-mast reach trucks are used for medium-sized loads. Their load-carrying capacity decreases with increasing lifting height (see Fig. 6.14).

[7] Cf. Schulte (2009), p. 161.

Fig. 6.14 Reach truck

Their protruding outriggers render these trucks unsuitable for lifting up pallets from the floor since the pallets must be hoisted over the outriggers and the mast needs to be retracted towards the driver's cabin before driving. Neither are moving-mast reach trucks very well suited for outdoor conditions due to their small wheels. They are mainly employed for stacking rather than for horizontal load handling.

Other types of stackers with outriggers include the *pantograph reach truck*. Unlike the moving-mast reach truck, on which the mast can be moved, this type of stacker moves its fork horizontally. If the outriggers are able to go underneath the rack, pantograph reach trucks, as opposed to moving-mast reach trucks, may perform storage tasks several loading units deep into the rack without the need for an extended fork.

The type of stacker that requires the narrowest aisle width is the *4-way reach truck*. It features the construction of a moving-mast reach truck but is additionally fitted with all-wheel steering. Since it is thus possible to drive sideways, this type of stacker is especially suitable for storing long loads.

Narrow aisle trucks are exclusively used for shelving operations. An aisle width of a mere 1.5 m or more is sufficient for them to be employed. Within the aisle, narrow aisle trucks run on rails alongside the shelves. Outside the aisle they change into unguided drive.

Owing to their lateral operability, narrow aisle trucks can move pallets both horizontally and vertically (diagonally), which leads to quicker cycle times. The cycle time is the time required for a single cycle or a double cycle. A single cycle comprises the pick-up and the storage of goods as well as the empty return run of the conveyor. Double-cycle runs also include the pick-up of another good from the same aisle and its transport to the point of departure. There should be a preference for double-cycle runs to ensure better time efficiency.

Narrow aisle trucks are capable of positioning heavy loads precisely at great heights. This necessitates solid rubber tires and thus a floor of good quality (see Fig. 6.15).

Narrow aisle trucks come in different forms. On a very basic level we can distinguish between *man up* trucks and *man down* trucks. With the man up

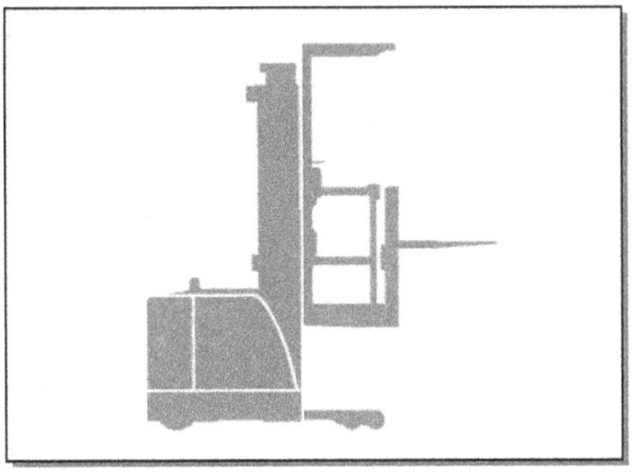

Fig. 6.15 Narrow aisle truck

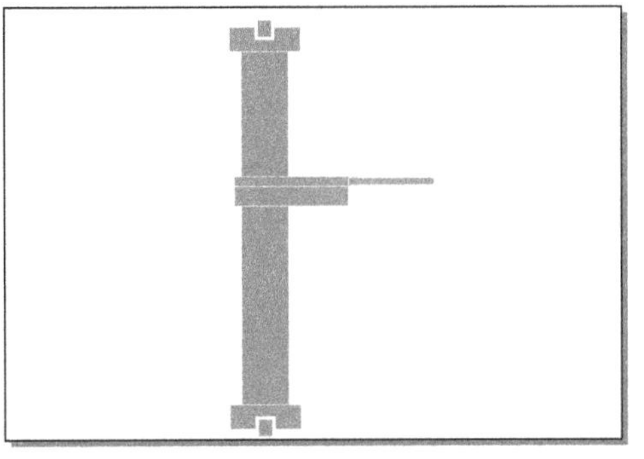

Fig. 6.16 Stacker crane

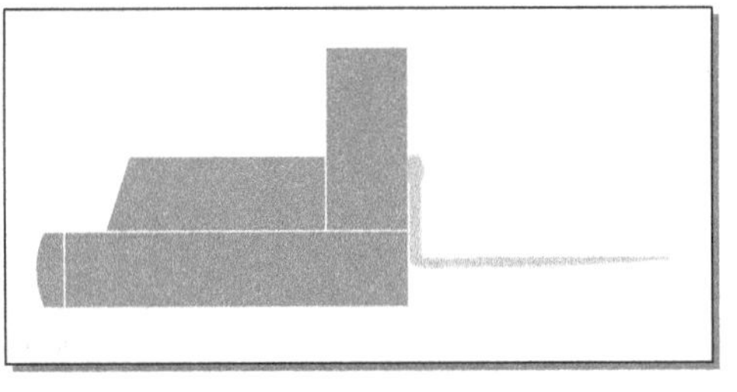

Fig. 6.17 Automated Guided Vehicle (AGV)

principle, the operator elevates with the load for increased fine-tuning of the fork and direct accessibility to the articles on the shelf. Man-down trucks are employed only if entire pallets are being handled. In this case the driver needs a positioning aid for precise pick-up of the pallet.

Stacker cranes have guide rails on the floor and for precise and safe at the top of the mast lateral operability. These conveyors serve to operate the bays of a storage rack either manually or automatically. They are predominantly floor-travelling and guided by rails. Only in exceptional cases, mostly with outdated installations, they

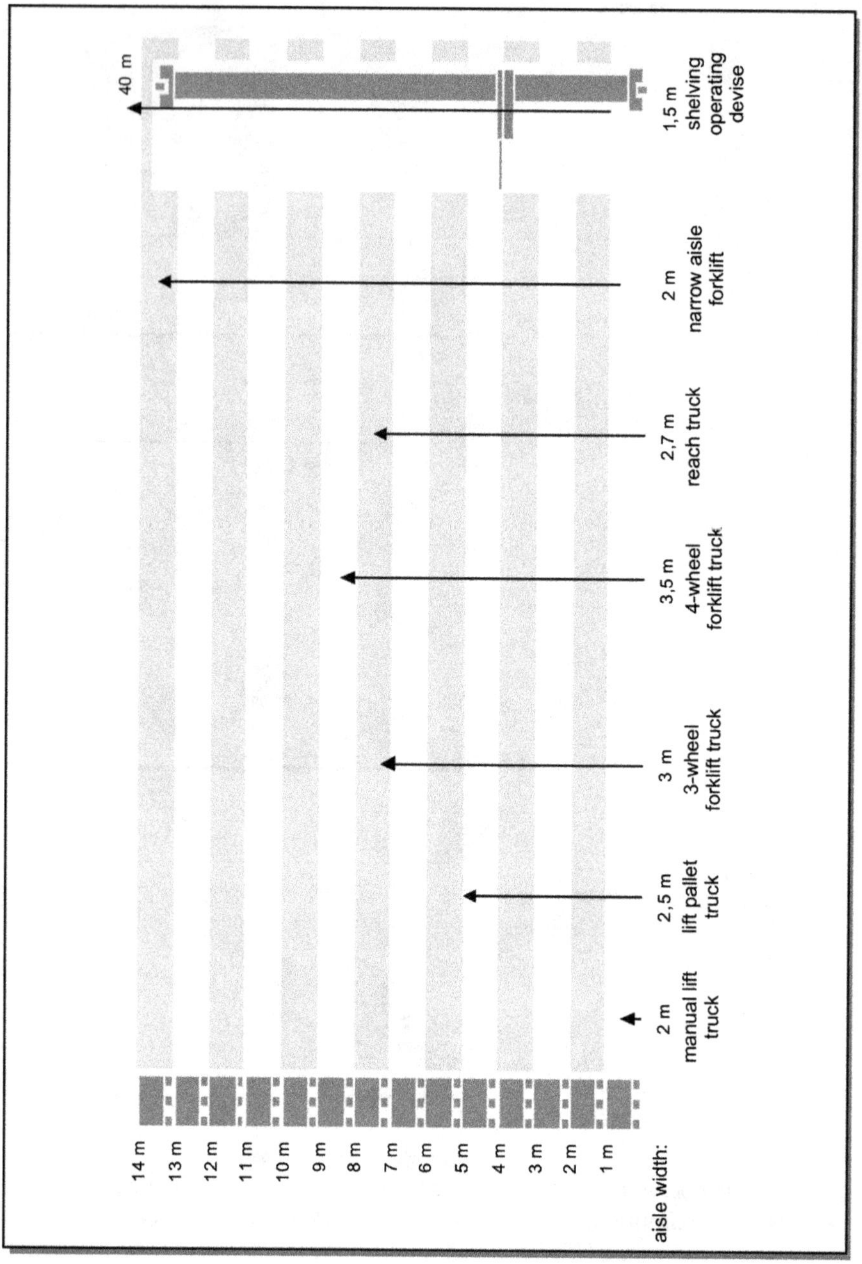

Fig. 6.18 Comparison of common conveyors

are suspended or attached to the racks.[8] In any case, however, they need to be guided from the ceiling or the upper edge of the rack. Stacker cranes cannot be driven outside the storage zone. The changing of the aisles is implemented by means of curves and switches or using a separate transfer carrier which picks up the entire unit, drives out of the alley, and changes aisles (see Fig. 6.16).

The additional guiding rail makes quick diagonal steering in high warehouses possible. Storage heights of 40 m can thus be reached without any problems. The stacker crane shown above features a telescopic fork to be used for pickup. Stacker cranes are often automated but also all types of conveyors covered so far can be operated automatically.

Automated Guided Vehicle systems (AGV) convey goods automatically throughout the entire interior area of the warehouse. Automated conveyors navigate through the warehouse autonomously using radio or infrared signals once they have received control data. To this end, orientation points are mostly used along which the AGV determines its position (see Fig. 6.17).

Depending on the controlling system, it may be difficult to change the routes (steered by induction loop), which renders the system rather inflexible. Compared to manually operated conveyors an AGV's conveying capacity is low. This is because its speed is limited to about 1 m/s so that in an emergency case it can come to a standstill as quickly as possible. AGVs used in closed-off indoor or outdoor areas reach much higher speeds. The fact that their operation is unmanned is of great advantage in cases where these conveyors are used for hazardous or high-precision tasks. Standardized load carriers and measurements for operating aisles are prerequisite for the use of Automated Guided Vehicles.

Figure 6.18 summarizes the most important differentiators of aisle width and stacking height.

Case Study 6.2: Conveying Capacity

Wilhelm Karmann GmbH – an automobile supplier from Osnabrück – specializes in the development and production of roofing systems. Additionally, the company is responsible for the production and assembly of complete series of models from various OEMs (Original Equipment Manufacturer). OEMs manufacture ready-for-sale products (automobiles in this case). A new goods receiving warehouse is being planned for the production of a new series of models. The supplied components (interior components) are unloaded from the truck using forklift trucks. How many forklift trucks are required if the following task is to be completed?

- Conveyance of 5.190 pallets per week over a distance of 311 m on average
- Conveying speed: 2 m/s
- Conveying capacity utilization: 40 %
- Working hours: 15 h/day; 5 day/week

[8] Cf. Jünemann and Schmidt (2000), p. 119.

6.3.2 Continuous Conveyors

Continuous conveyors can be classified as non-floor-bound, floor-mounted, and floor-bound devices. In most cases they are mechanized or automated, which allows them to be integrated in different conveyor systems. Continuous conveyors are mostly stationary, which means their layout is not very flexible, and they may pose an obstruction to other conveyors or work equipment.[9]

Roller conveyors are the type of continuous conveyor most commonly used in warehouses. They consist of many successively arranged, freely rotating rolls which are fitted between two steel profiles. Roller conveyors are available with or without an engine. Those equipped with an engine are moved by chains if heavy loads are conveyed, while belts are sufficient for the conveyance of smaller loads. Most roller conveyors are floor-mounted although they may also be non-floor-bound as a suspending conveyor from the warehouse ceiling (see Fig. 6.19).

Roller conveyors have proven to be a reliable, low-maintenance type of conveyor for indoor and outdoor use. They are, however, only suitable for packaged goods with at least one even surface. Changes in direction may be realized by curve structures or turntables. Using special roller track components the conveying goods can be accumulated to alter the distance between the goods or to singularize individual goods.

Chain conveyors are a variation of the roller conveyor. They convey the goods on engine-driven chains located on the side. Only goods with standardized measurements, such as pallets, can be moved on these kinds of conveyors.

Since box sizes can vary greatly, chain conveyors are more suited to transport pallets while roller conveyors lend themselves to the conveyance of boxes (see Fig. 6.20).

Belt conveyors are used to convey goods whose nature does not allow transport on rolls (e.g. bulk goods). Belt conveyors are comprised of a circulating conveyor belt made of rubber, synthetic, or India rubber. The belt runs on rolls or glides along a surface. At least one powered roll moves the conveyor, while the number of

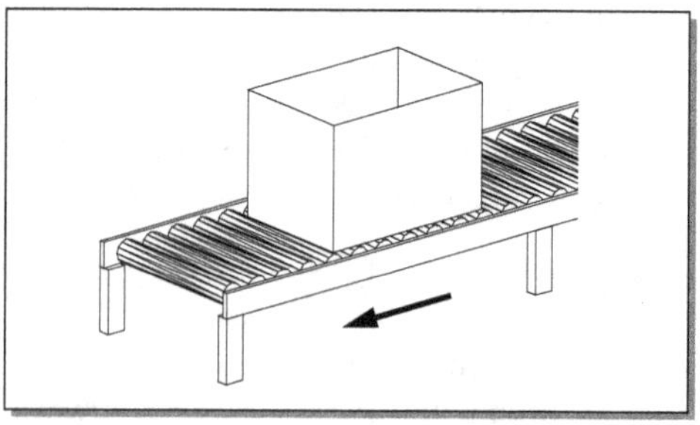

Fig. 6.19 Roller conveyor

[9] Cf. DIN 15201.

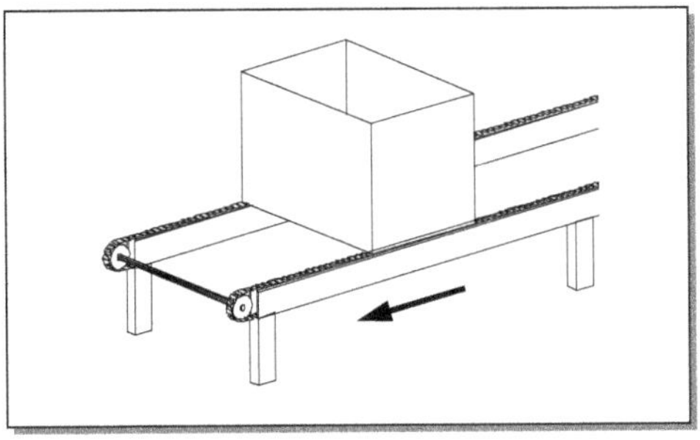

Fig. 6.20 Chain conveyor

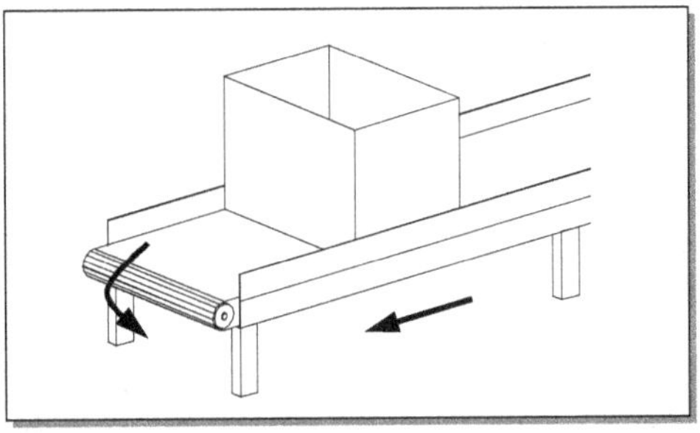

Fig. 6.21 Belt conveyor

powered rolls is dependent on the length and weight of the conveying good (see Fig. 6.21).

Belt conveyors generally reach higher speeds than roller conveyors. What is more, the dampening effect of the conveyor belt as well as the missing relative motion of the conveying good to the conveyor belt result in less noise being generated. Curves can be formed on the belt conveyor by using tapered rolls. Conveying routes can be generated by joining several separately powered belt conveyors together, which makes it possible to accumulate and sort out goods.

The advantages of roller and belt conveyors can be brought together in a combination of both these conveyor systems. Thus, the more inexpensive roller conveyors may be used for straight sections while belt conveyors are employed for curves and slopes since they provide better adhesion for the goods conveyed.

Circular conveyors transport general cargo by means of a suspension gear. The suspension devices are connected to a permanently running chain. Circular conveyors may run horizontally, vertically, or ascendingly. Consequently, this allows for an almost unrestricted layout in path configuration. The suspension gears can take different forms and may be designed for the particular load to be handled. Thus, they are available as platforms, trays, boxes, forks, or hooks, to name but a few.[10] At the points of pickup and deposit, the conveyor is pulled downwards, while for the rest of the conveying path it runs above the work area.

[10] Cf. Bode and Preuß (2004), p. 227.

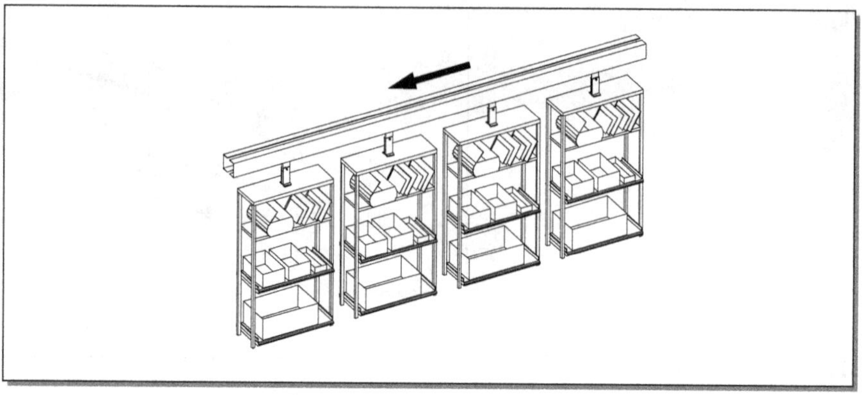

Fig. 6.22 Circular conveyor

This way, the conveyor does not take up any floor space and obstructions can be avoided (see Fig. 6.22).

Circular conveyors run continuously and they need to form a closed loop.

Power and Free conveyors are a variation of the circular conveyor. They feature separate tracks for drivers and suspension gears. If needed, the suspension gears are hooked into the drivers, which pull them along. This mechanism enables discontinuous conveyance and allows for switches to be integrated in the conveyor's path.

Electric telpher lines also belong to the category of circular conveyors. This conveyor need not run in a closed loop since each suspending device is self-propelled. On the basis of the same throughput, an electric telpher line is more lucrative over longer distances than a power and free conveyor. The latter is more suited for given distances and higher throughput. Throughput means the volume that is handled within a certain system over a certain period of time.

The *electric pallet ground conveyor* is a conveyor running on a system of tracks in which the individual elements are propelled by electric engines. Pallets are not only conveyed *straight* but crosswise transport also makes it possible to divert certain pallets. Gravity can be utilized on sloping sections along the path in order to forego the use of an electric engine. Changes in direction are realized by means of turntables or transfer platforms (see Fig. 6.23).

In comparison to discontinuous conveyors, the advantage of pallet conveyor systems is characterized by the possibility of transporting many pallets at the same time without the need for large numbers of ground conveyors. Goods can be transported to another floor more easily since a forklift truck, for example, would have to change floors as well, or the goods would have to be passed on to another truck. Changes in elevation can be overcome with the use of lifting tables. Inflexible conveying paths and obstruction of working area constitute a disadvantage.

Electric pallet conveyors are frequently used for storage in high-rack warehouses. In most cases, a discontinuous conveyor passes the pallets on to the rail system at a transfer point, from where the pallets are conveyed automatically.

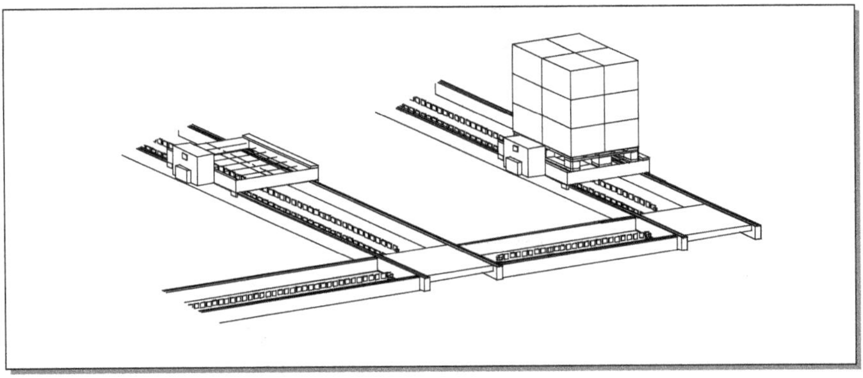

Fig. 6.23 Electric pallet ground conveyor

Floor-mounted drag chain conveyors are propelled from one central point. At certain intervals they are fitted with hooks and loops with which special carriers or pallet trucks can be attached and conveyed. The system may be made up of one or several main circles whose chains are propelled by several engines, depending on how much power is needed. Carriers can be diverted or led in automatically at the destination or departure point by using appropriate switch technologies.

This solid technology is especially common in transshipment terminals of forwarders since their conveying paths are rather straightforward and need not be changed. To preclude danger to crossing persons and to ensure easy attachment and disengagement of carriers, the conveying speed is low.

Case Study 6.3: Floor-Mounted Drag Chain Conveyors in Logistics Centers

The logistics service provider *Dachser GmbH & Co. KG.* uses floor-mounted drag chain conveyors in some of their logistics centers.

One of these conveying systems was implemented at the Dachser logistics center Berlin/Brandenburg in Schönefeld, where large volumes of orders are processed. Quick handling and high throughput rates are among the main objectives of the system. Furthermore, the operations can be carried out more steadily and smoothly than it would be possible using discontinuous conveyors (e.g. forklift trucks). As a consequence, the packages are not exposed to a high risk of damage and staff members enjoy low accident risk. Finally, the independence of staff members operating the conveyor system from the order volume leads to a degressive development of human resources with an increase in throughput volume.

The operations in the warehouse are divided into a goods-incoming and goods-outgoing process. The goods-incoming process comprises shipments which are intended for distribution at the site in Berlin/Brandenburg and which are usually received from other Dachser locations by 5 o'clock in the

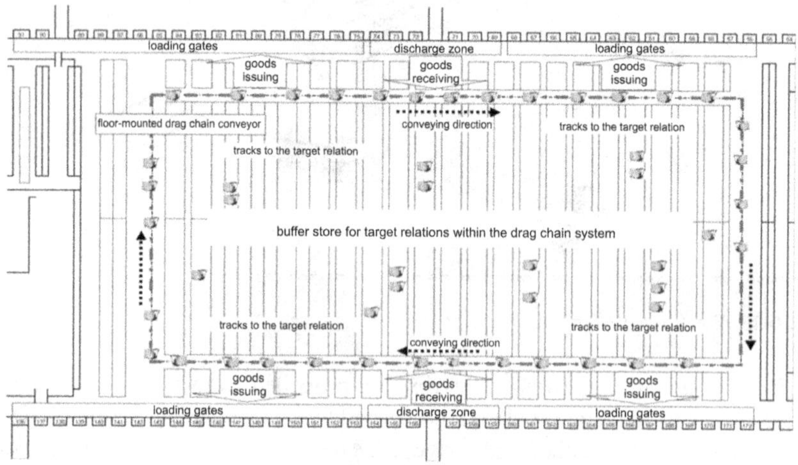

Fig. 6.24 Floor-mounted drag chain conveyor at the Dachser logistics center in Berlin/Brandenburg

morning. The goods-outgoing process deals with shipments which had been posted in the afternoon in the greater Berlin area and in the bordering state of Brandenburg by Dachser customers. These shipments have been taken to the logistics center where they are transshipped in the evening for on-carriage to national and international destinations. Transshipment in this process is carried out using a floor-mounted drag chain conveyor system (see Fig. 6.24).

The shipments to be transshipped are largely made up of standardized pallets. A pallet truck takes the pallets from the truck to the goods receiving zone, where they are scanned and hooked into the floor-mounted drag chain. Information about the pallet's destination was attached to each pallet during the scanning process. Once hooked into the conveyor system, the pallet is part of a continuously circulating system. As soon as the pallet reaches the loading gate of its target relation, the pallet truck is disengaged from the chain and deposited at the relation's buffer space. The parking spaces at the destination (rails) serve as a provisioning buffer until dispatch of the shipment on the same day. The empty pallet truck is hooked into the chain system again. The intervals between the pallet trucks as well as the conveying speed allow additional transshipment processes with different conveyors (e.g. forklifts) to be carried out simultaneously. This renders the entire transshipment process highly flexible.

6.3.3 Sorters

Sorters are used to separate and sort goods while using conveying technologies. Several sorter types are described in the following.

Push sorters are hydraulic push or pull arms which are installed above the conveyor belt. They push or pull the conveying goods towards chutes that are attached to the conveyor. The fact that they can also be added on to the conveyor at a later point and their compatibility with roller as well as belt conveyors is advantageous. The hydraulically operated pushers generate a high level of noise. Only sturdy goods can be sorted using this technique due to the pusher's high impact speed onto the goods, which also depends on the conveying speed.

Tilt-tray sorters have platforms that can be tilted to one side. These platforms are tilted so that the goods can slide down laterally under the force of gravity. Conveying goods may be fed in or discharged at any point of the conveyor. To this end, appropriate transfer or collecting equipment needs to be installed (see Fig. 6.25).

Tilt tray sorters do not require a separate conveyor, it is an integrated conveyor-sorter type. The goods are already separated on the conveyor or sorter before the handling process and can thus be deposited at their destination in a quick and accurate way. Problems may arise for goods with high friction since the tilting mechanism works in the same way for all goods. Tilt-tray sorters are very well suited for transporting robust, medium-weight goods. To achieve higher curve speeds, the trays may be tilted in the curves or the side walls could be molded into a bowl-shape.

A *crossbelt sorter* is a conveyor belt with traverse belt-trays, which are moved by an electric engine upon passing the discharge position, thus discharging the conveying good (see Fig. 6.26).

To avoid tipping of the goods, their bottom side should be flat due to the traction impacting on it. The conveyor's construction is not suitable for the transport of heavy goods. It is, however, supremely suited to be used in e.g. mail order companies, owing to its low noise generation and positioning accuracy if little traction is impacting on the goods.

Sliding sorters or slat sorters offer high-performance sorting of a variety of goods. This type of sorter consists of a slat conveyor – mostly made of extruded

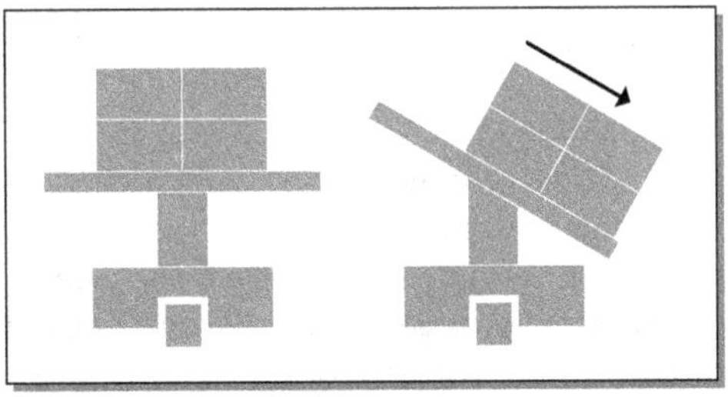

Fig. 6.25 Tilt-tray sorter

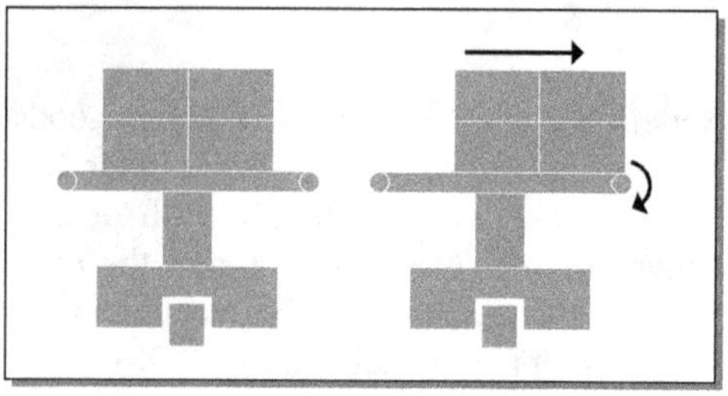

Fig. 6.26 Crossbelt sorter

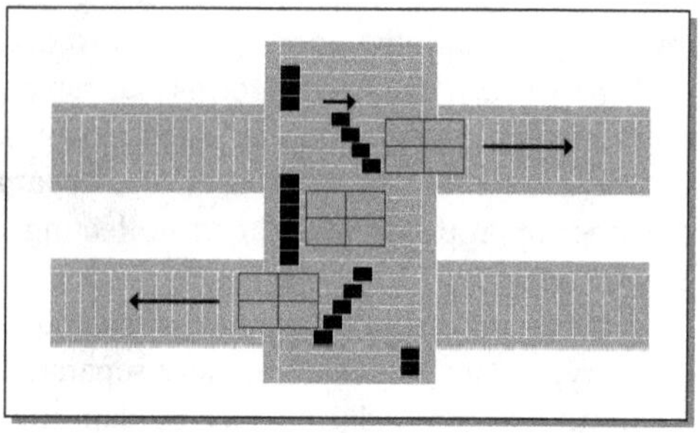

Fig. 6.27 Sliding sorter or slat sorter

aluminum carriers – on which the sliding blocks move. These push the goods towards the discharge exits (chutes) (see Fig. 6.27).

This system can be used to sort goods either to one side or to two sides. With this sorter, the number of the sliding blocks can be varied and thus be adapted to the requirements of the conveying good.

Case Study 6.4: Tilt-Tray Sorters in Parcel Centers

Deutsche Post AG uses tilt-tray sorters in their parcel centers to sort those parcels which are suitable for automated parcel handling, i.e. mostly parcels which do not exceed or fall below a certain maximum size or minimum size, respectively. Each carrier is equipped with two trays. If small parcels are being handled, each of the trays can carry one parcel. Bigger parcels take up the space on both of the carrier's trays.

Shipments turned in by Deutsche Post customers at, for example, post offices or post-office substations in the Leipzig area for delivery all over Germany, are collected by vans or trucks at these offices and transported to the parcel center in Leipzig. The shipments are unloaded there and individually put on a so-called feeding belt, as can be seen in Fig. 6.28. The feeding belt is located laterally to the sorter.

From there the parcels are fed consecutively onto the trays of the pre-sorter. In doing so, the system automatically recognizes free and occupied trays. Each pre-sorter is fitted with an address scanning device through which the parcel's ID and barcode as well as its address in plain text are identified. A picture is taken of each of the parcel's sides. The scanned barcodes are sent to the material flow

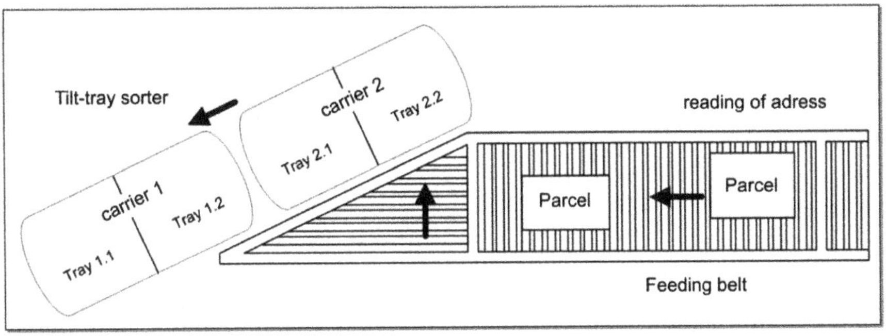

Fig. 6.28 Transition from feeding belt to tilt-tray sorter

computer (controller) while the delivery address, i.e. ZIP code, street, and house number are identified using the plain text. As soon as the material flow computer has identified the destination, it sends a feedback to the address scanner, which then terminates the process and deletes the picture. On the pre-sorter, the tray ID is then associated with the destination and discharge onto one of the connecting belts is determined.

After discharge onto one of the connecting belts, the parcels are conveyed onto a main sorter where they are scanned. The ID code and tray number are then sent to the material flow computer which confirms the destination. Subsequently, the computer decides which control system will trigger which tray to tilt the parcel into the exit chute. The tilting mechanism is triggered by an electronic contact prompting a discharge ramp to fold by which the tray is then tilted. The main sorters supply all terminal points in the parcel center. Shipments to other parcel centers are loaded onto swap bodies for on-carriage. Shipments for the center's own delivery district are loaded onto roller containers, which are then transported to the delivery bases.

Similar sorters are used at the destination centers with which the parcels are distributed to smaller delivery regions of the area.

6.4 Packages, Loading Units, and Load Carriers

The packaging of goods and the formation of packages serve more than purely logistical functions.[11] The packaging primarily has a *protective function* to preserve the goods in a saleable state and to prevent damage or spoilage. The *storage and transport function* enables the movement of goods within the logistics process.

The *identification and informational function* of the packaging supports the controlling of goods. This kind of data may be attached to the packaging in plain

[11] Cf. Jansen (1989), p. 79.

text, as barcodes, or using a RFID data carrier. Barcodes, which are used for identification, are made up of an opto-electronic script of lines that vary in thickness. RFID is a data transmission and storage technique based on radio technology and is also used for identification purposes. The data can, for example, contain information about the type of good, the content quantity, sender and receiver, weight, volume and handling instructions (fragility, cooling, information for installation etc.).

The *sales function* is especially relevant for consumer goods. Thus, it should contain buying stimulation to encourage customers to pick up the packaged item from the shelf. This includes advertisements, price, colorful packaging, or interesting shapes.

Finally, the packing sometimes also fulfills a *usage function*. This is mainly the case in the consumer goods sector, where e.g. food is consumed directly out of the packaging. The packaging also serves a usage function if goods are directly taken out of their packaging for assembly in production processes or if they are produced into the packaging.

In order to efficiently handle goods in general and packages or general cargo in particular, they are consolidated into *loading units*. *Load carriers*, such as pallets, wire-mesh boxes, or containers are generally used to form loading units.

Packaging falls into the broad categories of *disposable packaging* and *reusable packaging*. Disposable packaging is mostly made from cardboard and is designed for one-time protection during transport since the goods are only in some cases stored in it as well. Reusable packaging consists of more solid materials, such as plastics or metal. This type of packaging is characterized by the fact that the goods need not be taken out of the box during the entire supply chain to the installation location because the packaging fulfills all of the above mentioned packaging functions. Another advantage is the significant reduction in waste disposal costs compared to disposable packing and the environmental aspects associated with it. Return transport of empty packaging constitutes a disadvantage.

Packaging and *loading units* are further distinguished as follows.[12] *Packaging* is produced of packaging materials and serves to wrap up or hold together the packaged goods in order to render it able to be dispatched, storable and salable. *Loading units* are goods which are compounded for handling purposes by a load carrier. The formation of loading units requires additional effort. The advantages thus achievable, however, include efficient handling within the transport chain by forming greater pieces of cargo, cost-efficient usability of means of storage, conveyance, traffic and handling, as well as a general reduction of costs in the flow of materials and enhanced delivery service.

As mentioned before, one way of forming loading units is the use of *pallets*. Load-carrying pallets come in different forms, measurements and materials. Within Europe, the most common types of pallets are the *euro pallet* measuring 0.8×1.2 m and the *industrial pallet*, which measures 1×1.2 m (see Fig. 6.29).

[12] Cf. DIN 55405.

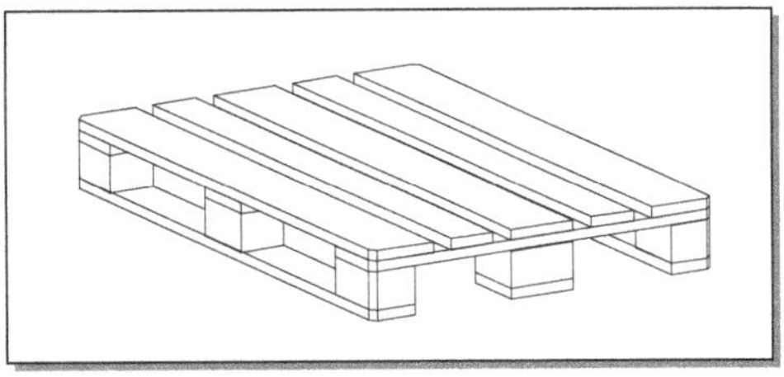

Fig. 6.29 Euro pallet

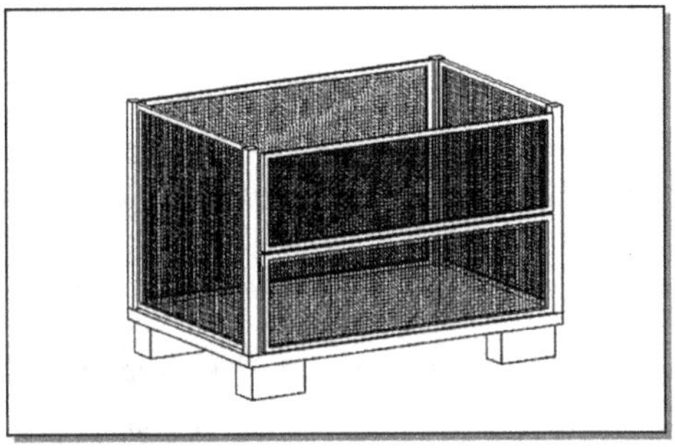

Fig. 6.30 Wire-mesh pallet

How many pallets with goods can be stacked on top of each other is dependent on the resilience of the good. Straps may be used to secure the load. Alternatively, a stretch or shrink-wrap procedure may be employed. With the stretch procedure, a flexible film is wrapped around the goods on the pallet so as to hold them together. If the goods are shrink-wrapped a film is pulled over the goods which is subsequently heated up. This makes the film shrink and hold the goods together.

Wire-mesh pallets have a higher stackability. At the same time, the need for re-packaging is reduced. Wire-mesh pallets are made up of three fixed wire-mesh walls and one two-part front wall, which is detachable. They are usually used for storage of small parts that cannot be stacked on pallets. They have the same measurements as euro pallets or industrial pallets. Collapsible wire-mesh boxes on which all walls can be collapsed are sometimes used as well in order to save space on the return transport of the empty pallet (see Fig. 6.30).

If the loading unit is supposed to be as small as possible, *small parts carriers* or boxes can be used to transport small parts.[13] These loading units are suitable for the formation of small batch sizes, which also increases flexibility. During transport the small parts carriers are consolidated into loading units on the pallets. Small parts carriers are available in different sizes which can all be combined on the pallet.

[13] Cf. DIN 30820.

Further examples of small parts carriers include:

- Stackable boxes (one side wall is slanted and only half-closed so that goods are accessible if boxes are stacked on top of each other)
- Rotary stacking container (stackable and can be turned 180° to nest inside one other)
- Folding container (foldable, mostly for light goods)

Closed load carriers or *containers* can be utilized to transport goods at sea, on the road, on rail, or in the air. Depending on capacity we distinguish between small, medium-sized, and large containers. Large containers can hold up to 14 euro pallets per layer. Inland containers have a rear door or several side doors and are standardized as 10-, 20-, 30-, and 40-foot[14] containers.[15] ISO-containers are also available in these four main versions.[16] They only have one rear door which makes them more difficult to load but delivers greater protection for the freight especially on overseas transport. ISO-containers are used world-wide, as opposed to inland containers, which are not used outside Europe (see Sect. 5.4.2).

Case Study 6.5: Usage Function of Packaging

In the consumer goods sector, packaging is adapted according to its use by the consumer. A famous example is the *5-Minuten-Terrine* by *Maggi GmbH*, which only needs to be filled with hot water so that the content can be directly eaten out of the packaging. Other examples include the *Joghurt mit der Ecke (Müller Corner range)* by *Alois Müller GmbH & Co. KG* (Müller Dairy Ltd.), yogurt in general, or drinking bottles.

It is also in the consumer goods sector that shipping units are formed out of the load carriers and the packaging containing the product. The packaging (e.g. stretch film) is removed and the product is made available for sale at the store on its load carrier (e.g. euro pallet).

In the manufacturing industry and especially in the automobile industry, the packaging fulfills a usage function in the production process. Manufacturers of components (supplying industry) which are needed during the car assembly process package the components at the end of production. The components are provided at the assembly in their packaging (largely reusable packaging) and are directly taken out of the packaging by an assembly line worker. The way the packaging is constructed must ensure an optimal position of the components for removal. This means, for example, that the workers should not be forced to bend down and that they should be able to remove the component without meeting any resistance.

[14] One foot corresponds to a length of 0.3048 m.

[15] Cf. DIN 15190.

[16] Cf. DIN ISO 668.

6.5 Picking and Handling

The process of taking goods out of the warehouse for compilation of a customer shipment is called picking. The receiver may be an external customer of the company or an internal accepting instance (e.g. production). The consumer information can be a sales-oriented order (customer order) or a production-oriented one (internal order). The process of unloading goods from a transport mode into the goods receiving area of a warehouse, from the goods issuing area of a warehouse into a transport mode, or from one transport mode into another is termed transshipment. Transshipment areas, which can be outdoors or indoors, are frequently used to effect transshipment between modes of transport. Additional picking processes may be carried out during transshipment if the shipments (packages or load carriers) need to be compiled into amended, customer-specific shipments. Such facilities are termed transshipment points or cross-docking points. If goods are also purposefully stored at these transshipment points over a certain period we speak of distribution warehouses, as described above.

6.5.1 Picking

The process of picking is divided into the sub-systems of *organization*, *materials flow*, and *data flow*. The *organization* of sections division can be *one-zone* or *multi-zone*. Articles with different properties, e.g. large and small or light and heavy, should be stored in separate sections to ensure optimal operation by the picking person. The order may be processed in a single-stage or multi-stage manner. Single-stage order processing leaves the entire picking process to be implemented by one picking person while with multi-stage order processing, only one stage is allocated to one person and the article is passed on for further sorting. Since multi-stage picking necessitates temporary storage and sorting, it is more suited to be used for order structures with long throughput times and which are not characterized by urgent or prioritized orders.

The materials flow is divided into provisioning, movement, withdrawal, and handing over. Provisioning is either static if the picking person moves towards the article (*man-to-goods*) or it is dynamic if the article is brought to the picking person at a picking workplace (*goods-to-man*). Movement is effected either on one level or on several levels using either manual or automated withdrawal techniques. The article is handed over either at one location (*centralized*) or at several locations (*decentralized*) in the warehouse.

The *data flow* is characterized by *data preparation* and *data transmission*. The data is prepared either according to order sequence (*Order Picking*) or in batches (*Batch Picking*). In the case of order picking the orders are processed synchronously to the order inflow. The processing may be carried out simultaneously or in series (consecutively). Batch Picking is a way of picking in which the orders are first collected. Individual items of several orders are subsequently compiled to one aggregated item. These orders are then allocated to the individual picking sections

Table 6.1 Basic principles of picking systems

<table>
<tr><th>Function</th><th colspan="4">Possibilities of Realization</th></tr>
<tr><td>Provisioning of articles</td><td>Static</td><td>Dynamic</td><td>Centralized</td><td>Decentralized</td></tr>
<tr><td rowspan="2">Movement of picking person</td><td>None</td><td>1-dimensional</td><td>2-dimensional</td><td>3-dimensional</td></tr>
<tr><td colspan="4">manually; mechanically; automatically</td></tr>
<tr><td rowspan="2">Withdrawal of articles</td><td colspan="2">Single item</td><td colspan="2">Aggregated item</td></tr>
<tr><td colspan="4">manually; mechanically; automatically</td></tr>
<tr><td>Handover of articles</td><td>Static</td><td>Dynamic</td><td>Centralized</td><td>Decentralized</td></tr>
</table>

and processed simultaneously or in series. As a last step, the aggregated orders are then distributed back to the original orders. This procedure is also called two-step picking.

Traditionally, we distinguish between *electronic data transmission* and *data transmission on paper*. The use of handheld devices, displays on picking vehicles, *Pick by Voice* or *Pick by Light* renders paperwork unnecessary. With Pick by Voice the order is transmitted to the picking person via headphones, while Pick by Light technologies activate small lights at the respective point of withdrawal. Confirmation of the picking order is done *actively* by the picking person or *automatically*, for example by means of an automatic withdrawal confirmation system on the shelf.

Using one picking system or another mainly depends on the range of articles and the volume of articles to be picked. In general, it seems recommendable to use single-stage picking systems for a small and homogeneous range of articles or in cases of little daily throughput. Multi-stage order picking or batch picking is more suitable for a more heterogeneous article range and higher throughput volumes.

The following outlines several common combinations of picking procedures resulting from Table 6.1.[17] Manual systems with static provisioning, one-dimensional movement, centralized handover, and manual withdrawal are suited for small numbers of picking items per day. Using a picking vehicle, the picking person drives through the aisles and brings the readily picked order back to the central picking station (see Fig. 6.31).

An extended range of articles would increase the transit times with this picking procedure. Growing throughput would soon lead to picking persons hindering each other in the aisles. In order to decrease transit times and to be able to distribute more picking persons to the aisles, it is more suitable to employ decentralized withdrawal in conjunction with a conveyor belt instead of picking vehicles (see Fig. 6.32).

The picking person puts the required number of articles into the passing carriers move e.g. by conveyors. The goods in the carriers are sorted at another location.

[17] Cf. Gudehus (2010), p. 679 et seq.

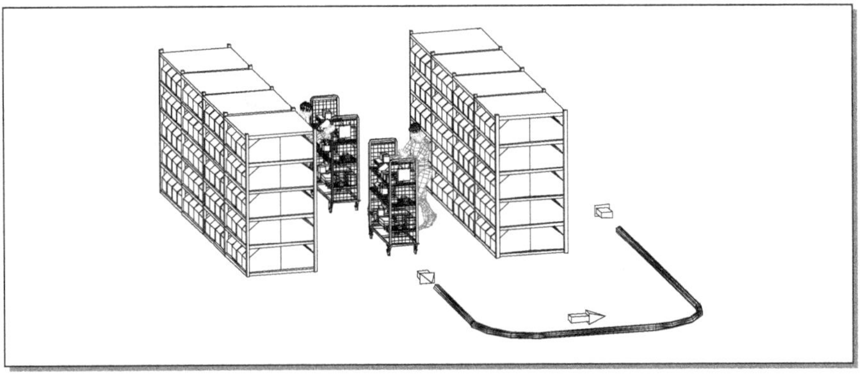

Fig. 6.31 Single-stage, one-dimensional picking with manual withdrawal and centralized handover

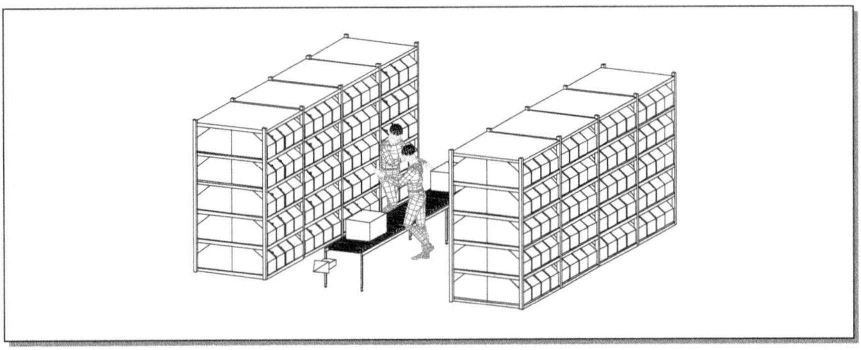

Fig. 6.32 Static provisioning, one-dimensional movement, manual withdrawal and decentralized handover

This two-stage commissioning procedure increases the throughput time of the order since the goods are on-carried and transshipped more often than with single-stage picking. This type of picking can also be implemented in a pallet rack warehouse. In this case, platforms are mounted along the aisles with approx. 2 m space from each other in which the picking persons are positioned. Loading with full pallets is implemented through special aisles by e.g. narrow aisle trucks. This type of storage is called *gallery storage*. Two-dimensional movement of the picking person is most suitably implemented if the range of articles makes the use of one-dimensional storage impossible due to its space requirements and if the number of picking items per day is low at the same time.

The two-dimensional movement better utilizes the height of the warehouse (see Fig. 6.33). In comparison to the use of one-dimensional picking with picking vehicles, using shelving operating devices or narrow aisles trucks reduces the aisle width.

The two-dimensional picking utilizes the same approach as one-dimensional picking. A multi-stage strategy is pursued to pick more orders within a certain timeframe from a larger range of articles. The picking persons are able to pick more

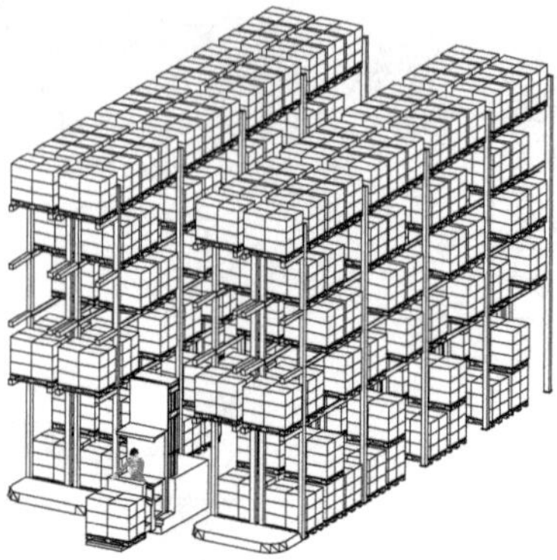

Fig. 6.33 Static provisioning, two-dimensional movement, manual withdrawal and centralized handover

articles from one shelving bay and they can operate more shelving bays than with a single-stage approach since the articles are only allocated to the orders in the next stage.

This approach may be automated with the use of automated storage and retrieval systems. The withdrawal device must be fitted with a collecting magazine so as to avoid return journeys to the transfer station after each withdrawal. This is why this type of withdrawal is more suitable for small goods.

If the picking and storage areas are separated – as is the case in high rack warehouses, for example – the loading units need to be retrieved and picked outside the warehouse before they are provisioned dynamically (see Fig. 6.34).

The goods to be commissioned are taken from the pallet, which is subsequently taken back to the warehouse, or open pallets are pooled. This is done in a decentralized way in the example presented and on-carriage is carried out using unmanned transport systems.

Other picking systems designed to better adapt to article characteristics and to achieve certain inventory and performance data can be formed by combining various types of stage systems, provisioning, withdrawal, and handover. The manifold possibilities of designing picking systems highlight the complexity of planning them. The picking system needs to meet service levels requirements of the processed order while at the same time keeping costs low.[18] The logistical service level is significantly influenced by an extended product range, decreased order sizes, and an increase in order frequency. This leads to an increase in the complexity and costs of picking. Improved picking performance while ensuring consistent

[18] Cf. Bode and Preuß (2004), p. 325.

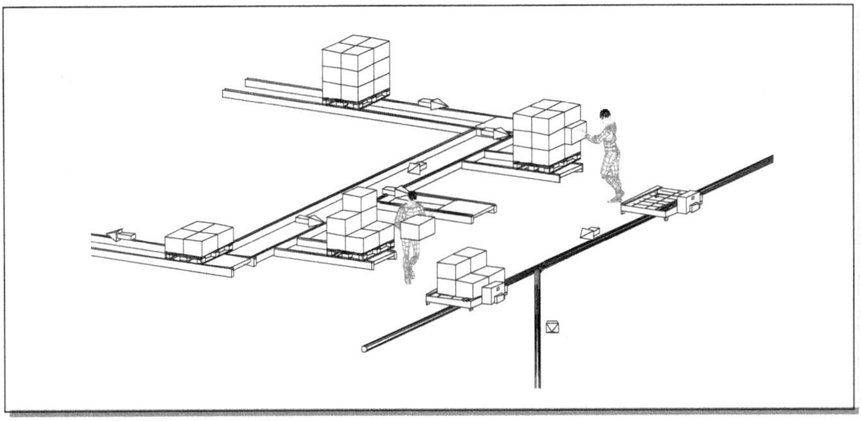

Fig. 6.34 Dynamic provisioning, one-dimensional movement, manual withdrawal and decentralized handover

quality standards can only be achieved by making adaptations to the picking system (more complex design).

6.5.2 Handling

Handling of goods means that a change of *load carrier*, *conveyor*, or *storage position* takes place. Since goods are *passive* objects, *active* handling techniques must be employed. This task is in the simplest case fulfilled by a warehouse worker. Warehouse workers use handling equipment if they cannot handle the goods under their own power or within the required timeframe. Even a fully-automated handling process without any warehouse personnel is viable. This type of handling equipment can be made up of *conveyors* and *load carriers*. Conveyors pick up the goods and overcome vertical and horizontal distances. Load carriers are used to consolidate goods and facilitate or enable their being picked up.

This section only covers handling between external and internal materials flow systems. The need for handling to take place between different means of transport is described in Chap. 5.

Handling between warehouses and external means of transport is most commonly carried out to establish a connection to *road traffic*. Which kind of handling equipment needs to be used is dependent on the way the goods are transported on the truck. In most cases the goods are on pallets and are not stacked. Loading and un-loading of the truck can be carried out using manual lift trucks if a loading bridge is available. If the pallets are stacked or have to be stacked, a conveyor with lift function, such as manual *lift pallet forklift trucks* or *forklift trucks*, is required.

The goods may also be loaded *loosely*, as is the case with parcel transport in the mail-order business or transport of tires in the automobile industry. To facilitate loading and un-loading, conveyor belts can be led into the truck. The loading person can drag the conveyor to the position of the current workplace and thus avoids moving in and out of the truck.[19]

The handling of goods between means of *rail transport or inland waterway transport* and internal materials flow systems is of minor importance for the supply of most warehouses.

Railroad wagons are always loaded and un-loaded from the side to avoid decoupling.

Goods that are being transported on *inland waterway vessels* are usually stored in containers. At the port, these containers are either loaded onto trucks first, which enables regular handling in the goods receiving and goods issuing areas, or they are taken to a loading zone using conveyors.

6.6 Warehouse Organization

Warehouse organization distinguishes between *operational warehouse organization* and *structural warehouse organization*. Operational warehouse organization defines the temporal and spatial organization of sequences which are also reflected in the order of warehouse sections that certain goods or parts of goods pass through. Structural warehouse organization defines work contents and competencies in a hierarchical structure.

The most important warehouse areas are the goods receiving area, the warehouse itself, and the goods issuing or outgoing area. The distances leading from the goods receiving area through the warehouse to the goods issuing area are called circulation area. The aisles between storage locations belong to the circulation area as well. It is also in this order that goods pass through the warehouse. Areas located in between these areas may fulfill additional functions. These include picking and packing areas, quality inspection zones, and provisioning areas in general. In addition, sanitary and administration areas for staff members must be taken into account.

The *goods receiving* and *goods outgoing* areas constitute an interface between internal and external traffic and are termed *loading zone*.

The *goods receiving area* mainly serves to accept the incoming goods and to forward them to the following warehouse areas (see Sect. 7.3.4). The goods are either stored or directly forwarded to the goods issuing area (transit area) to leave the warehouse shortly. Goods which do not meet the quality criteria after inspection in the goods receiving area are prepared for return shipment and also passed on to the goods issuing area. Stored articles first need to be compiled into a shipment (picking) upon receipt of a shipping order before they are taken to the goods issuing area.

[19] Cf. Jünemann and Schmidt (2000), p. 290.

The *goods outgoing area* provisions those goods that are ready for dispatch in their respective loading units and effects the loading process onto the vehicle (mainly trucks or trains).

Loading zones are designed according to *organizational* and *constructional* criteria. Routine tasks allocated to staff members on a daily basis have to be considered in the organization. Planning is further based on the availability of technological facilities (conveyors) and on space requirements. The size of the loading zone is determined by different performance variables, such as arrival times of vehicles, their loading capacity, and load factors throughout the day and over periods of seasonal fluctuations. The peak load describes a percentage margin with which temporary peak performance is expressed. In addition to the performance characteristics, the properties of the goods to be loaded have to be taken into account when designing the facilities for handling between loading zone and truck.

The way in which the goods are handed over should be adapted to the vehicle. Today, height-adjustable ramps that are attached to the building and which the conveyor crosses to directly reach the vehicle's loading platform are most commonly used. As the following figure shows, the ramps (also termed loading bridges) are designed for rear discharge of the trucks (head ramp). The relatively high space requirements for maneuvering constitute a disadvantage (see Sect. 4.2.6). The so-called saw-tooth ramp, which aligns the trucks in a zigzag pattern, is a good alternative to this. If no ramps are used, the height difference between vehicle and warehouse must be overcome using a lift truck. In cases where the conveyors carry out loading operations outdoors, the suitability of the goods and conveyors for certain weather conditions must be taken into account. Conveyors need not access the loading platform of the trucks, aside from the fact that this is not even always possible (see Fig. 6.35).

The right choice of *loading gates* depends on the intended closing and opening times – which are important, for example, when loading chilled goods – and on the space required. Hinged and folding doors open relatively slowly and take up space inside or outside the building. Since sliding doors are slid to the right or left, they

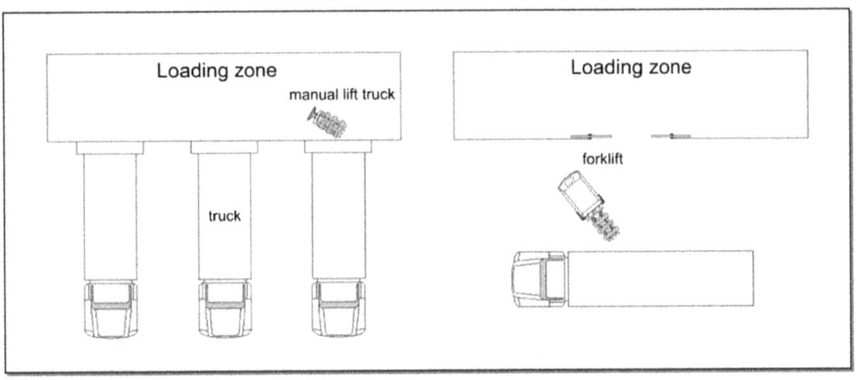

Fig. 6.35 Loading and un-loading of trucks with and without the use of ramps

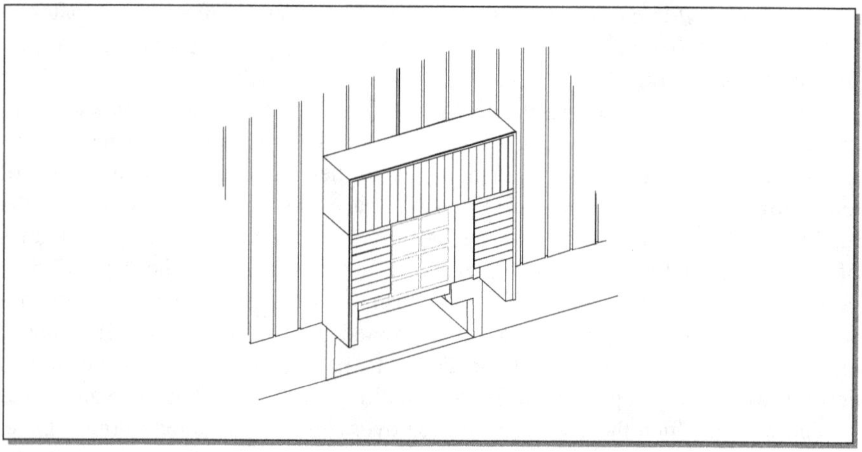

Fig. 6.36 Gate system

preclude a dense arrangement of several doors next to each other. Shorter distances can be realized with roll-up doors or sectional doors as they are pulled overhead when opened. These door types are used preferentially because they open and close quickly and do not take up much space.

A complete gate system consists of a loading bridge, a shelter (optional), and of the door itself. The main purpose of a gate is to enable loading and un-loading of trucks. Additionally, improved conditions for the protection of the goods can be achieved with outdoor shelters or temperature and air locks inside the building. What is more, drafts can be avoided this way, which leads to an improvement of the working conditions for the loading personnel (see Fig. 6.36).

Gate houses offer the best insulation. They are built in front of the actual gate opening and contain a loading bridge and a second gate, which creates a temperature lock.

The most common layouts for goods receiving and goods issuing areas are shown in the figure below (see Fig. 6.37).

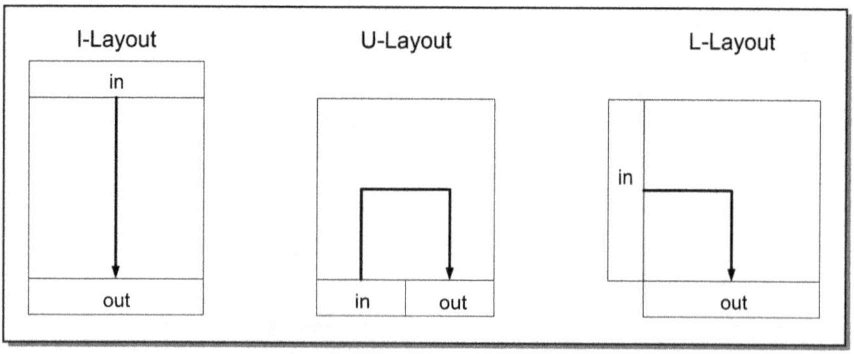

Fig. 6.37 Possible goods receiving (ingoing) – goods issuing (outgoing) layouts

By virtue of their shape, the letters *I*, *U* and *L* imitate the flow of goods. The *I-layout* provides for all goods to travel through the entire warehouse. This is also the layout's disadvantage since the frequently handled goods have to travel the same distance as the less frequently handled ones. The straightforward flow of goods can be seen as an advantage. Both the *U-layout* and the *L-layout* enable frequently handled goods to be stored close the goods receiving and goods outgoing areas to avoid travels through the entire warehouse. Goods that are supposed to be moved directly to the goods issuing area without intermediate storage do not have to pass through the entire warehouse as is the case with an I-layout. The U-layout is greatly advantageous in that all gates are situated on the same side of the warehouse. Thus, the space needed for the maneuvering of the trucks is located on one side and smaller than with the other layouts. By-passes for trucks are therefore not necessary and the warehouse premises can be kept smaller. Since the traffic takes place on one side, however, obstructions and confusions of goods receiving and goods outgoing areas are more likely to occur.

The following describes the organizational design of the actual area where the goods are stored – the so-called *storage zone*. Storage zones can be classified according to various characteristics (also see Sect. 6.2):

- Storage temperature (temperature-controlled – heated or chilled – or not temperature-controlled)
- Weight and volume of articles
- Turnover rate (fast or slow-moving consumer goods, ABC articles)
- Property relations of articles (customer, consignment, or customs warehouse)
- Degree of hazardousness of article (hazardous goods classification)
- Value and fading frequency of articles

Storage zones should not be planned too small to take advantage of synergies. Neither should they be too large so that the area can be taken in at a glance.

Storage zone management aims at minimizing distances, producing even utilization of existing storage capacity and at avoiding an overaging of the stored goods. To this end, various strategies for storage space allocation and storage and withdrawal are available.

With *Fixed storage space allocation,* each article is allocated a fixed storage space while the space remains empty if the article is out of stock. Fixed storage space allocation can theoretically be practiced without the use of IT systems.

Free storage space allocation within fixed areas or storage zones allocates storage spaces to the article according to its turnover frequency. Articles with a high turnover rate are located within short reach. In *chaotic storage systems*, each article is allocated the nearest available storage space or the nearest free storage space which is best suited for the article.

These strategies may be combined with permission or prohibition of additional storing in storage spaces that are already partly occupied. If additional storing is permitted, the age of the articles may be of no relevance. Quantity adjustments upon withdrawal may be permitted or prohibited. This means that loading units are opened to comply with the order quantity.

The practicability of *withdrawal strategies* depends on the storage type used and on the degree of complexity of the storage management. Withdrawal of those articles that had been stored first is called *First In – First Out* (FiFo). Live storage shelves, for example, automatically realize a FiFo strategy. Using appropriate control devices or marking, however, a FiFo strategy can be implemented for any storage technique. If the most recently stored article is taken out first, we speak of a *Last In – First Out* (LiFo) strategy. This strategy is utilized if goods are stored whose age is irrelevant, e.g. goods that are not subject to a sell-by date.

Apart from FiFo and LiFo, other strategies can be adopted and combined. Thus, withdrawal may be organized according to the shortest distance to the point of withdrawal and then be combined into orders with optimized distance coverage. Furthermore, withdrawal can follow a certain order or is implemented according to quantity adjustments. For example, withdrawal may be organized according to size or weight of the articles. In the case of perishable goods it is best to combine the FiFo principle with the *FeFo principle*. The FeFo (First Expired – First Out) principle provides for goods whose sell-by date will be reached next to be taken out first. Quantity adjustment means that only full or open loading units are taken out.

Case Study 6.6: Storage Space Requirements

At the goods distribution center of *Amazon.de GmbH* in Leipzig, articles are re-ordered according to the following principle. Whenever the stock reaches the reorder level of three loading units, six loading units are ordered. The loading unit of one article requires two storage spaces. One loading unit always contains ten articles. This means that the loading units have to be depleted one after the other so that 30 articles are left if three loading units are in stock. The following question needs to be answered in order to assess how the storage space must be allocated in order to take up the least amount of storage capacity:

How many storage spaces must be made available if

(a) Fixed storage space allocation is implemented and
(b) A chaotic storage strategy has been chosen?

Case Study 6.7: Warehousing Processes

The trading company *Gebrüder Heinemann* KG a distributor of international brand articles on the travel market and, amongst other things, operates duty-free shops on airports. Both consolidation and flexibilization of their logistical activities to international market requirements and their own positive business development have led to a turnkey-ready logistics center being built in Hamburg Allermöhe with *SSI Schäfer* as the general contractor. The logistics center is one of the most modern and high-performing ones world-wide. It keeps 35,000 articles in stock with an outgoing tonnage of about 1,000 t per day, which is equivalent to 85,000 picked boxes (see Fig. 6.38).

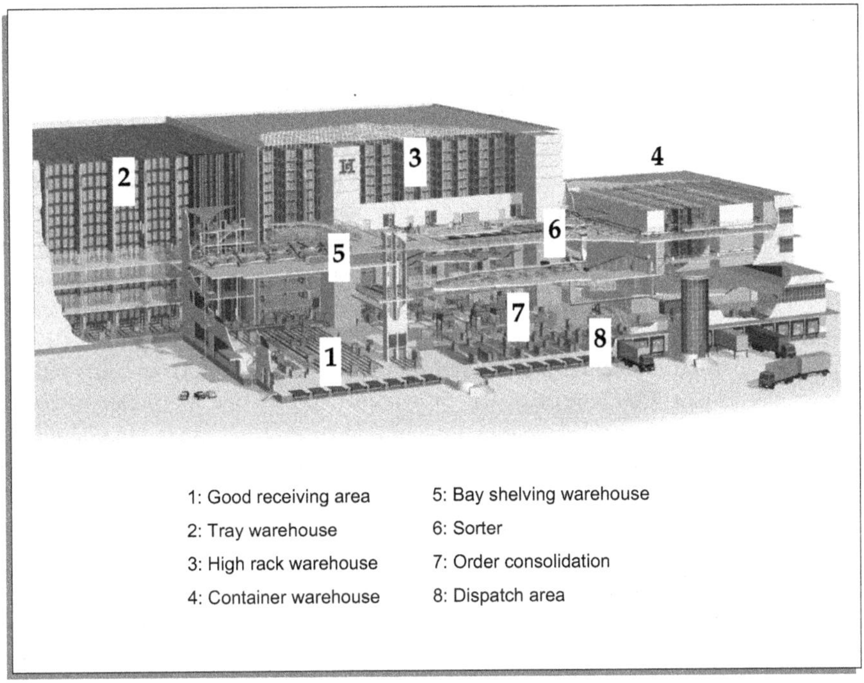

Fig. 6.38 Warehouse layout Gebr. Heinemann, KG (SSI Schäfer (2007))

The warehouse is divided into a gallery storage warehouse for pallets, a container storage warehouse with picking zone, a bay shelving warehouse, and a tray warehouse.

In the *goods receiving area* the goods are completely transshipped from the truck onto conveyors and separated for the different warehouse sections.

Cigarettes, coffee, candy, and spirits are stored on whole pallets in the *gallery storage warehouse*. IT-supported quality inspection and labeling of the pallets are carried out in the goods receiving area. Accurate weighing is especially important since all inspections of the outgoing goods will be based on comparisons of weight later on.

The gallery storage warehouse consists of eight aisles, each furnished with one shelving operating device which stores and retrieves the pallets in the location allocated to them by the warehouse management system. To achieve the highest possible storage capacity, the pallets are stored double-deep. The gallery storage warehouse supplies the various storage and picking areas with goods. The *tray storage warehouse* serves to provide whole cartons for direct withdrawal by the dispatch area or to re-stock the picking area.

High-priced articles, such as perfume, cosmetics, cigars, or articles for individual picking are transported from the goods receiving area to the second warehouse level. The articles are separated into system containers at 16 repacking stations. The packed containers undergo a shape control and are electronically recorded using a barcode. Conveyors turn the containers in the right direction and transport them to one of the storage locations allocated by the warehouse management system within one of the 36 picking zones of the *container storage warehouse*. The containers are stored double-deep as well.

Accessories, such as suitcase sets, toys, and designer articles are likewise separated in the goods receiving area and – analogous to the picking articles – taken to the second warehouse floor using conveyor technologies and vertical conveyors. The articles are then distributed manually to the storage locations of the *storage shelves*. Goods and storage location are associated with each other using a barcode reader to enable the warehouse management system to retrieve the manually stored articles at any time.

The distribution center is designed for approximately 1,600 outgoing pallets per day. Apart from the dispatch of whole pallets from the gallery storage warehouse, goods are picked in the individual storage areas, analogous to storage procedures. The routes of the picking vehicles in the manual storage areas are optimized by the warehouse management system.

Picking systems have been integrated into four levels of the gallery storage warehouse to carry out retrieval. Goods pallets are provided for whole-box picking. The shelving operating devices automatically supply the picking station according to the stipulated minimum stock levels. These picking stations are partly *dynamic picking stations for* slow-moving articles in order to realize a greater range of articles. The warehouse management system provides the respective pallets for picking. Once the goods have been picked, the pallets are re-stored into the gallery store. Double deep pallets with fast-moving articles are continuously provided at the static picking stations. Whole boxes are picked manually in the gallery store on the pallets located on the picking vehicles. Staff members receive all data via a radio data transmission terminal which is attached to the picking vehicle. Thus, the picking staff can be directed through the warehouse on optimized routes which take article weights into account. By means of a control weighing and for the purpose of accurate documentation of withdrawals and inventory, the pallets undergo a quantity check by weight comparison both at the beginning and at the end of picking. The picked pallets are temporarily stored in buffer zones and are combined with the rest of the order in the dispatch area.

Articles – especially perfume and cosmetics – are picked purely by individual picking in the *bin storage warehouse*. The articles are supplied from the 36 silos of the warehouse by a dynamic picking system. In order to pick the articles, the warehouse management system requests a box that is adapted to the picking volume. An order label is attached to the box, which associates the box with the transport tray through a barcode. The tray with the box on it then travels through the picking warehouse fully automatically. Depending on the goods to be picked the tray is diverted at the respective picking station where the articles are stored and provisioned in a time-efficient way. The tray's diversion prompts the *shelving operating device* to trigger the immediate transport of the storage container with the requested articles from the storage zone to the provisioning zone. Static and dynamic picking stations provide the respective articles for picking. Picking can now take place at 77 picking points. The box is sorted out by means of a radio scanner. Subsequently, all picking items and picking quantities are indicated by a pick-to-light installation at the point of picking. Using these indications and the data from the radio terminal, the picking person can now compose the articles. The picking process is confirmed and triggers a control weight mechanism along with a booking in the system. By means of a variance analysis, the system determines the contents of the box. After picking is finished, the staff member pushes the tray with the box back onto the conveyor. The tray is then transported to the next picking station where picking continues. If the picking order is completed, the box is discharged from the container storage warehouse. A continuous scale checks the order once again during this process. All picked orders are transported to a control terminal. If any discrepancies are detected, the staff member at the control station is informed and performs a manual check. Incorrect pickings are fed back into picking cycle through the conveyor for completion. The conveyor system carries the inspected boxes on to the tray collection point. The trays are collected and led back into the system. The picked boxes go through a strapping machine, after which the conveyor transports them to the sorter.

Picking from the assortment of accessories and slow-moving articles takes place in the *bay shelving warehouse*. By means of multi-order picking, up to eight orders can be processed at the same time. The picking list to do so is given to the staff members through mobile hand terminals. Depending on the article and order, whole boxes may be retrieved in the tray warehouse. The goods are then repacked from the boxes into cartons. The content is recorded and checked by scanning the EAN code. After attachment of a dispatch label, the cartons are transported to the sorter.

Cartons from the automated picking warehouse and from the bay shelving warehouse are conveyed to a *sorting machine* for initial order combination in the *dispatch area*. They are sorted there according to orders and customers or destinations and then discharged through the respective chute. Staff members at the chutes compile the cartons on a pallet. Upon completion of the palletization, the warehouse management system checks whether the pallet weight is correct, in another control weighing. Pallets from the perfumes and accessories area are then transported to the goods issuing area. The warehouse management system determines whether a pallet goes directly through the shrinking machine to the allocated loading gate or if it should be collected at a loading station in the loading area, together with the pallets from the gallery storage warehouse. Staff members from the loading zone decide how they would like to load the goods onto the matching pallets and communicate their decision to the warehouse management system via radio scanner. Subsequently, another control weighing takes place to avoid loading mistakes. On the way to the dispatch area the pallets are secured by an automated pallet shrinking station. Within a live storage system, four transfer carriages then distribute the pallets to gravity roller conveyors for dispatch. From there, the shipments are loaded onto the trucks. This way, up to 100 trucks are sent off with goods on a daily basis.

The extremely complex movements of goods are managed and comprehensively monitored by SSI Schäfer's *warehouse management system* named ant®. ant® controls all processes from the *goods receiving area* to the *goods issuing area* and manages the data transfer between shelving operating devices, picking persons and mobile radio terminals. Its main objective is exact trip management of the trucks to the gates as well as just-in-time actuation of retrieval processes so that loading times can strictly be adhered by and the loading zones do not become blocked as a consequence of early order provisioning. Ongoing checks and inventory management through numerous weighings and the resulting reduction of shortages to under 1 % only become possible through the use of an advanced management system. Additionally, the use of state-of-the-art visualization technologies makes it possible to clearly visualize the processes and movements of materials.

Review Questions

1. Explain the basic functions of a warehouse.
2. Name the areas in which continuous and discontinuous conveyors can each be applied.
3. What steps would you take in designing a system for reusable packing and what must be taken into consideration?
4. Assuming a throughput of 1,000 items per day, is single-stage manual picking or multi-stage automated picking more suitably implemented?
5. Allowances for a possible expansion of storage capacities should always be made. Which provisions can be taken in the choice of the warehouse location and storage technology?
6. Name possible ways to secure palletized loading units.
7. Name the advantages and disadvantages of unmanned transport systems and the resulting task areas.
8. According to what criteria would you determine the use of stackers?
9. What needs to be considered when opting for a type of sorter?
10. What possibilities do we have to design and equip a warehouse as flexibly as possible?

Additional Literature

Ackerman, K. B. (1997). *Practical handbook of warehousing* (4th ed.). Alphen aan den Rijn: Kluwer Academic.

Arnold, D., Kuhn, A., Isermann, H., & Tempelmeier, H. (2002). *Logistik*. Heidelberg: Springer.

Bode, W., & Preuß, R. (2004). *Intralogistik in der Praxis: Komplettanbieter der Intralogistik*. Suhl. Wirtschaftsverlag.

Fischer, W., & Dittrich, L. (2003). *Materialfluss und Logistik. Optimierungspotenziale im Transport- und Lagerwesen*. Berlin: Springer.

Frazelle, E. (2002). *World-class warehousing and materials handling*. New York: McGraw-Hill.

Gudehus, T. (2010). *Logistik* (3. Edition). Berlin: Springer.

Jünemann, R., & Schmidt, T. (2000). *Materialflusssysteme: Systemtechnische Grundlagen* (2. Edition). Heidelberg: Springer.

7 Inventory, Stock and Provisioning Management

Learning Objectives

Inventory management, stock management, and provisioning management play an important role in logistics. On the one hand, they directly influence the delivery service. On the other hand, inventory costs are a significant cost driver in the supply chains.

This chapter aims to present different approaches to inventory and order planning as well as to the application of established warehousing strategies, of fundamental principles of provisioning, and of procurement-logistical concepts. Furthermore, different approaches in developing cooperative inventory management strategies within the supply chains will be discussed. From this the reader will get comprehensive references in order to meet the challenge between high product availability and low inventory (costs).

Keywords

- Function of inventory
- Demand-oriented (pull) or action-oriented (pull) supply of goods
- Availability versus destocking
- Warehousing strategies
- ABC analysis
- Just in time/Just in sequence
- ECR, VMI, CPFR
- Regional freight carrier concept
- Cooperation concepts

H. Gleissner and J.C. Femerling, *Logistics*, Springer Texts in Business and Economics, DOI 10.1007/978-3-319-01769-3_7,

7.1 Inventory Basics

7.1.1 Contents and Objectives

Inventory management, *stock management,* and *provisioning management* serve logistical purposes and purposes related to warehousing. The objective of *warehousing theory* is to minimize the total inventory costs, order costs, and delivery costs. The basic aim is the reduction of stock and the depletion of safety stock. To this end, decisions regarding order sizes, order dates, and delivery dates have to be made; usually by the material planning department. Other points of consideration include *stock-keeping* of inventory and *delivery.*

The *logistical aim of inventory management* is to optimize the entire supply chain across the different stages to adequately serve the demand. Problems may arise as a result of this multi-stage structure and due to the isolated and often uncoordinated flows of goods at the different stages of the supply and distribution chain. This is owing to the fact that stocks are kept at all these stages to meet the demand of the subsequent stage.

Depending on its position in the value chain, inventory can comprise material goods (raw, auxiliary, and operating materials), intermediate, or semi-finished products (parts, components, modules), or finished products (end products). Players in the supply chain are producers who manufacture ready-for-sale end products. Suppliers produce parts, components, and modules and can therefore be categorized as parts suppliers, component suppliers, module suppliers, or system suppliers. The end products are offered to the consumers through different distribution and sales channels, such as trading enterprises, wholesale stores, and retail stores. Thus, procurement organs and *procurement logistics* as well as distribution organs and the associated *distribution logistics* play a part in the value chain.

7.1.2 Determination of Requirements

There are three traditional methods available for the *determination of material demand*:

- *Deterministic* methods, where the requirements are calculated based on the production program
- *Stochastic* methods, which make use of consumption statistics and forecasts
- *Heuristic* methods, which allow to make inferences as to the requirements by means of comparisons with similar products (analogous estimation) or intuitive estimations without relying on numerical data

The kind of raw materials and parts required and the point in time for their procurement can be derived from the determination of requirements based on the *types of requirements* shown in Fig. 7.1.

Primary demand is dependent upon external factors, such as fluctuations in demand, the state of the economy, or seasonal demand and can therefore hardly

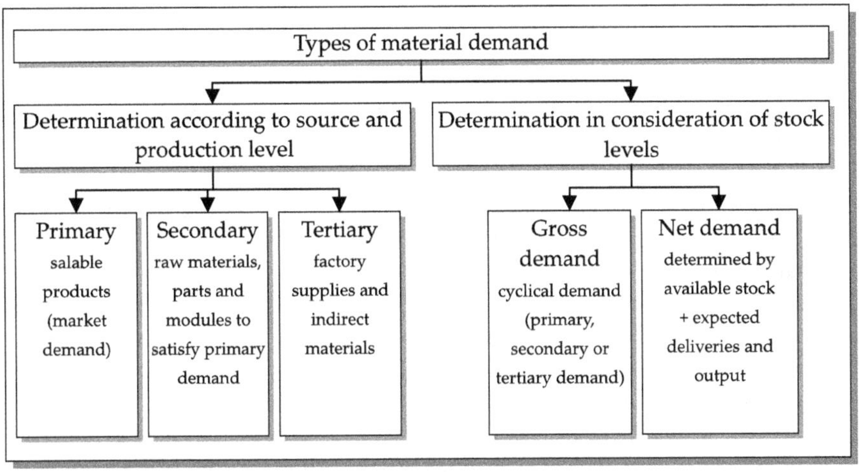

Fig. 7.1 Distinction of types of material demand (Schulte (2001), p. 113)

be influenced. *Secondary demand* derives from primary demand and is calculated on the basis of stock lists. *Tertiary demand* is of minor importance in determining demand since it mainly determines quickly and easily obtainable materials which are not included in the planning stage but are procured as per to consumption.

The distinction between *gross demand* and *net demand* is relevant in relation to the point at which demand occurs. The stock of inventory is taken into account while determining these types of demand.

7.1.3 Functions of Inventory

Inventory can assume different *functions*, which are independent of materials demand.[1]

The *cycle stock* or *adjustment function* of inventory offsets the temporal imbalance between supply and demand within an economy. According to production theory, it is advisable to produce larger quantities even though the products may not be in demand or sold either before or immediately after production. The period between when the end-product becomes available (supply) and when a need for the product arises (demand) can be minimized by means of inventory management. This is to cut costs for storage space and capital commitment.

The availability of raw materials and goods needed for the production of a specific product helps the manufacturer achieve shorter delivery times and higher

[1] Cf. Heiserich et al. (2011), p. 116 et seq.

flexibility to meet different customer requirements. This is referred to as the *acquisition function* of inventory.

Inventory allows *Economies of scale* to be collected if the purchasing department manages to obtain quantity discounts and favorable terms of transport through consolidation.

Another function of inventory is the *safety stock* function. In this case, stock is kept to ensure the availability of goods at all times. This is important, for example, in the procurement process so that production need not be discontinued if suppliers are unable to provide raw materials or semi-finished goods in time or if delivery is delayed due to busy transport networks or inclement weather conditions. On the other hand, sudden fluctuations in demand may be offset using safety stocks. Without influencing the production process, unsold goods may be kept as inventory in warehouses. Conversely, if sales are above average, demand can be met using inventory stocks and sales shortfalls may thus be avoided.

Furthermore, inventory has a *function of speculation.* Through stock-keeping and the availability of storage capacities, it becomes possible to purchase goods today even if they will only be needed later for production or resale. Such decisions may be economically sensible if market prices are low (e.g. with promotional offers) or if prices are likely to increase in the future (e.g. if prices for raw materials go up or if demand increases).

It is only through the *refinement function* of inventory that certain products become salable after production. Examples of this are typical ripening processes of food and drinks.

Finally, inventories may serve to ensure consistent quality for products that hail from different provenances (place of origin, production series etc.). In these cases, inventory takes on an *assortment function.*[2]

7.1.4 Inventory Cost

Keeping inventory stocks entails *inventory costs* in the form of storage costs (building, technology, staff) und inventory carrying costs (capital commitment, shrinkage, obsolescence). Unavailability of out-of-stock goods leads to production downtime and drops in sales, which can be classified as out-of-stock costs, sales shortfalls, and shortfalls in profit margin. The portion of costs and the share in results that inventory has in the respective reference and target figures can be determined by *material and inventory classification.* Probably the most famous method of inventory classification is the ABC and XYZ analysis. It is the aim of this

[2] Cf. Ihde (2001), p. 44 et seq.

type of analysis to assess a specific inventory's share of value in e.g. turnover and to determine regularities in consumption. Case study 7.1 will demonstrate how an ABC and XYZ analysis can be applied.

Case Study 7.1: ABC-XYZ-Analysis

SoLog, *Solutions for Logistics* PLC is a logistics service provider specializing in logistical services for the entertainment sector. In order to harmonize the internal storage and picking processes with the demand patterns for articles, an ABC and XYZ analysis is carried out. 9 articles have been selected here to demonstrate the procedure of the analysis.

The products to be analyzed are listed in tabular form and are given a rank according to the value of the yearly requirement for each item number (see Table 7.1).

Calculation: value = quantity issued/consumed x item cost.

As a next step, the products are categorized into class A, B, and C based on the original ranking (See Table 7.2).

All inventory including the strongest-selling products up to the position where the accumulated percentage first exceeds 70 % are grouped *A class* inventory. The remaining positions up to around 90 % are *B class* products. The residual 10 % of turnover are taken up by *C class* products.

The differentiation between the individual groups may vary. Action plans for each group of inventory can be deduced from the analysis.

In particular, these are:

A-Articles

- Complex, exact procedures for disposition
- Procurement in very short intervals
- Minimized procurement times
- Precise inventory control

Table 7.1 Ranking according to yearly requirement

Item No.	Item Quantity	Price per unit in €	Value of yearly requirement	Rank
1001	10,000	3.25	32,500	5
1002	12,800	1.10	14,080	8
1003	8,000	1.75	14,000	9
1004	6,000	17.50	105,000	3
1005	12,400	1.60	19,840	6
1006	8,000	6.25	50,000	4
1007	4,000	136.20	544,800	1
1008	6,800	2.10	14,280	7
1009	12,000	37.85	454,200	2
			1,248,700	

Table 7.2 ABC class list

Rank	Item No.	Value of yearly demand in €	Proportion of the overall value in %	% Accumulated	Class
1	1007	544,800	43.6 %	43.6 %	A
2	1009	454,200	36.4 %	80.0 %	A
3	1004	105,000	8.4 %	88.4 %	B
4	1006	50,000	4.0 %	92.4 %	B
5	1001	32,500	2.6 %	95.0 %	B
6	1005	19,840	1.6 %	96.6 %	C
7	1008	14,280	1.2 %	97.8 %	C
8	1002	14,080	1.1 %	98.9 %	C
9	1003	14,000	1.1 %	100.0 %	C
		1,248,700			

- Extensive research into procurement markets
- Careful choice of suppliers
- Very exact calculation of order quantity and time of order placement

C-Articles

- Determination of requirements based on consumption
- Simplified methods of order processing
- Long order intervals and large order quantities
- Large safety stock inventory

B-Articles

- In the middle, to be treated as either A or C articles on a case-by-case basis

However, analyzing the goods solely by their proportion in the overall value is not sufficient. Apart from their value, consistency in their consumption is also a decisive factor. The XYZ analysis determines these consistencies:

- X-items: consistent consumption and therefore high prediction accuracy
- Y-items: fluctuating or seasonally increasing or decreasing consumption
- Z-items: highly volatile consumption and therefore low prediction accuracy

This analysis also allows for action plans to be deduced in conjunction with the ABC analysis. Thus, groups of AX, AY or BX items are highly suitable for just-in-time delivery while BY and CX items are suited to a limited extent (see Sect. 6.3). Additional analyses need to be carried out with regard to replacement times, frequency of item changes, out-of-stock costs, as well as maximum inventory, minimum inventory, and safety stock.

7.2 Inventory Planning and Management

7.2.1 Disposition Procedures

In a general sense, disposition is defined as the quantitative arrangement of orders with their individual service requirements and the scheduled allocation of internal orders to available resources.[3]

Disposition is carried out in the procurement of charge materials for production and in the provision of end-products for trade. A conventional *lot size calculation* may be employed to optimize the procurement of materials for production. This determines the ideal order quantity while taking order and storage costs into account. Scale of discount, price hikes, and promotional offers etc. may also be factored in.[4]

In the trade sector dispositioning takes place on certain levels. Central warehouses may dispose with suppliers, regional warehouses with central warehouses and branch stores with regional warehouses In so doing, different procedures are implemented which can generally be broken down into demand-led procedures and program-led procedures.

Applying *Consumption-led disposition*, procurement becomes dependent on demand. This demand is determined by outward materials movements within the framework of inventory control. Order quantities and time of order placement are established on the basis of these movements and existing inventory. The producer then places the order with the supplier. Subsequently, the supplier processes the order and identifies the order quantity and the delivery date.

Program-led disposition is characterized by its dependency on the production program. Based on an initial forecast about the future demand of a specific product (primary demand), the parts and components needed for the production of the end product are identified (secondary demand). These requirements are then reconciled with the existing stock inventory (net requirements) so that procurement requirements (determination of order quantity) and provisioning dates (order scheduling) can be established.

Another consumption-led disposition procedure is called *stochastic disposition*. It utilizes forecasting models that are based on past figures. Exponential smoothing, for instance, is used as a forecasting procedure to visualize *trend functions* (constant functions, linear functions, quadratic functions). This is a recursive calculation of exponentially smoothed averages for which only the actual demand of the latest period and the smoothed average of the preceding period are required.[5] The effect of past averages on the new forecast can be altered by means of smoothing parameters.

---

[3] Cf. Gudehus (2010), p. 43.

[4] Cf. Stölzle et al. (2004), p. 84 et seq.

[5] Cf. Stölzle et al. (2004), p. 63 et seq.

This procedure (exponential smoothing of the first order) is suitable for articles with stable sales and demand structures. Extraordinary fluctuations in demand triggered by e.g. seasonality or marketing activities can be taken into account using seasonal factors (exponential smoothing of the second and third order).[6]

7.2.2 Warehousing Strategies

Individual *demand-led disposition procedures* differ with regard to the parameters used by the manufacturer to determine order quantities and the time of order placement. These procedures, also termed warehousing strategies, include[7]:

- Order rhythm – lot size (t,q)
- Order rhythm – order level (t,S)
- Order point – lot size (s,q)
- Order point – order level (s,S)

Applying the *order rhythm – lot size procedure* (also called t-q policy), orders are placed in regular intervals, which is represented by the parameter t (see Fig. 7.2).

The inventory level is always restocked by the same amount q. Uneven outward stock movements may lead to highly fluctuating inventory stock levels S. Thus, a high volume of outward stock movements within a given interval t bears the risk of inventory shortfall.

Using an *order rhythm – order level procedure* (also referred to as t-S policy), orders are placed in set intervals, represented by the parameter t. The inventory is always restocked up to a pre-defined level S. The capacity limit can therefore never be exceeded. Due to varying order volumes, suppliers may experience difficulties in

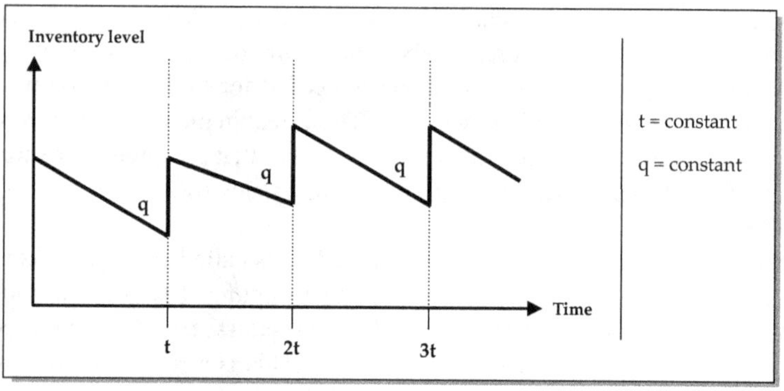

Fig. 7.2 Warehousing strategies and order policies of the t-q policy

[6] Cf. Kernler (2003), p. 65 et seq.

[7] Cf. Stölzle et al. (2004), p. 91 et seq.

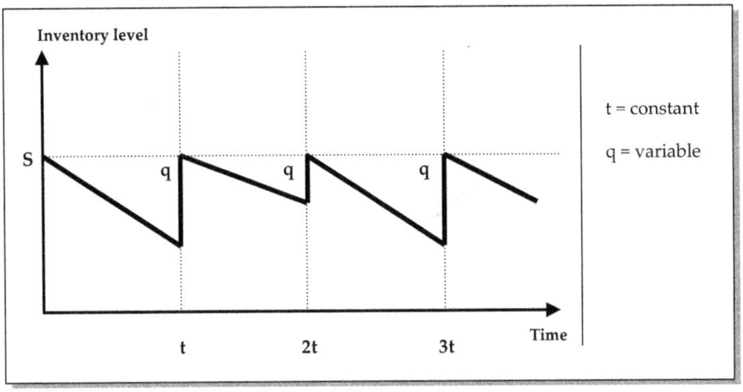

Fig. 7.3 Warehousing strategies and order policies of the t-S policy

defining realistic capacity levels, which lead to shortfalls or delays in delivery in cases of capacity bottlenecks. Downtime consequences can be interruption of production (see Fig. 7.3).

With the *order point – lot size procedure* (also termed s-q policy), a fixed quantity q is ordered as soon as the inventory level reaches or falls below a certain order point s. Orders are therefore placed in irregular intervals since outward stock movements are dependent on demand and thus may differ from one period to another. The risk of inventory shortfalls is relatively low. They are only experienced if the inventory level has dropped near the order point s, no orders have been placed yet, and if a larger quantity than s is needed for the next withdrawal from stock (see Fig. 7.4).

Applying the *order point – order level procedure* (also called s-S policy), an order is placed as soon as the inventory level reaches or falls below a certain order point s. The inventory is restocked with the difference to a pre-defined order level S. Orders are placed in irregular intervals since outward stock movements are dependent on demand and thus may differ from one period to another. The risk of inventory shortfalls is relatively low since the warehouse is filled up to the

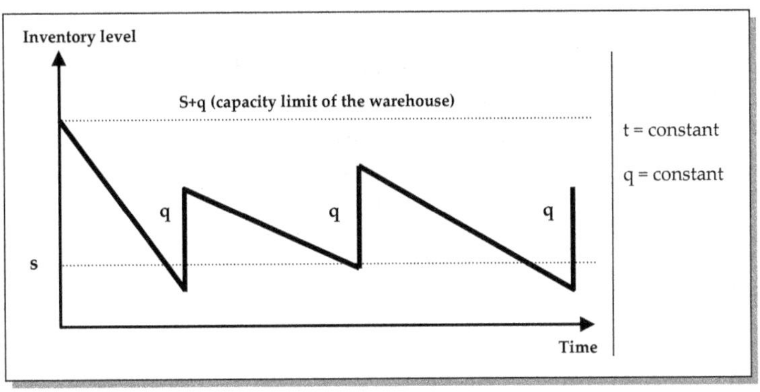

Fig. 7.4 Warehousing strategies and order policies of the s-q policy

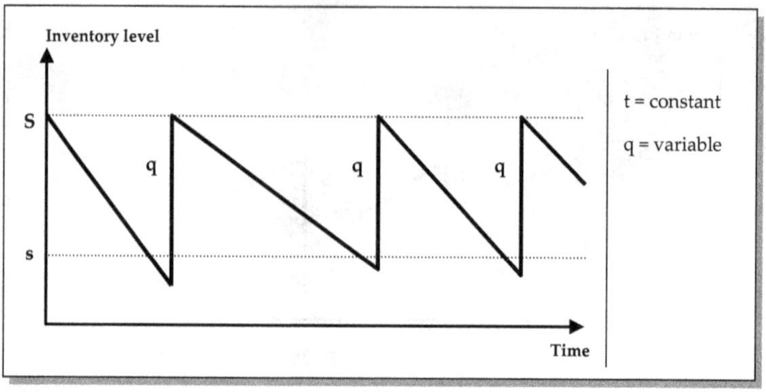

Fig. 7.5 Warehousing strategies and order policies of the s-S policy

maximum level, i.e. high storage costs are deliberately incurred. Analogous to the order point – lot size procedure, shortfalls may occur if the inventory level has dropped near to the order point s (see Fig. 7.5).

7.2.3 Safety Stocks

Regardless of the actual demand, each inventory-carrying stage within a supply chain is planned on the basis of requirements forecasts which, in turn, are based on previous downstream demand structures. Orders are always triggered if the inventory level drops below a pre-defined order point (see order point procedures) which should still allow for demand to be met during the replenishment lead time. Since demand and replenishment lead times are subject to fluctuations, it is necessary to keep *safety stocks*.

Safety stocks are established based on the service level and on the expected probability distribution for demand. The precise amount of safety stocks depends on the extent of forecast errors, expected demand, and on the probability with which this demand occurs. Safety stocks can be calculated by means of the Gaussian normal distribution, which standardizes the frequency of deviations from a mean value. By specifying a certain multiple of the standard deviation, a probability can be established with which a specific demand can be met by keeping a certain level of stock. With a service level of 84 %, a safety stock of one standard deviation of the forecast demand is required. If statistical safety of 97 % is aimed for, a safety stock of two standard deviations is needed. These statistical safety measures are only applicable with normally distributed frequency distributions.[8]

[8] Cf. Alicke (2005), p. 63.

7.3 Provisioning of Goods

7.3.1 Individual Procurement and Bulk Procurement

Individual procurement on demand, in-stock provisioning, and just-in-time provisioning are basic concepts in the provisioning of goods. *Individual procurement on demand* is characterized by low stock levels and therefore incurs minimal storage costs. Risks and disadvantages include delays in production, loss of customers or turnover, and potentially high purchase prices. In the case of *in-stock provisioning or bulk provisioning*, goods are kept in stock. Keeping inventory secures the production process or the supply of customers but also incurs high storage and capital commitment costs. On the other hand, purchase discounts may be applicable.

7.3.2 Production Synchronized Provisioning

Using *production synchronized provisioning*, only the materials needed are procured and provisioned. This type of provisioning may be employed between two stages of production, between production and assembly, or between supplier and assembly.[9] A self-managed control circuit governs the internal production and orders are processed according to current demand and existing inventory levels. This procedure termed KANBAN (card) is organized de-centrally and utilizes a production concept based on the pull principle.

The pull principle is a control concept whereby replenishment orders are triggered after demand occurs. With the push principle, however, replenishment orders are placed according to plan. Applying the KANBAN principle, the self-managed production or procurement at each upstream stage is set off by the demand of the ensuing downstream stage. Materials must not be ordered in too large quantities, too early, or in reserve by the production stage (drain). To realize the potential of the KANBAN principle for the suppliers (source), they must not produce buffer stock and they should only initiate production upon receipt of an order while guaranteeing the flawless quality of the products. The short-term management of production is passed on to the staff members. Cards are used as part-specific information carriers between the provisioning unit (source) and the consuming unit (drain).

7.3.3 Just-In-Time Concepts

On the basis of the KANBAN principle – which Toyota developed into an efficient procedure for production – just-in-time concepts between suppliers and buyers have emerged.[10] High predictability of the demand of the goods is a crucial prerequisite

[9] Cf. Schulte (1996), p. 301.

[10] Cf. Stölzle et al. (2004), p. 133 et seq.

for this. Ideally, no inventory stocks are kept, which also precludes storage costs. However, high transport costs are likely to arise due to small transport lots and high transport frequency. In addition, there is a risk of production downtime and delays in production. To reduce these risks it is vital that buyers work closely together with suppliers or with one of the supplying branches in the proximity.[11]

Just-in-time provisioning through suppliers is based on a three-stage process:

- *Framework agreement*: forecasts regarding capacities and demand for the next 1 or 2 years, rolling
- *Forecast delivery schedule*: Stipulation of terms of delivery and period-specific obligation to take delivery
- *Just-in-time delivery/dispatch call-off*: Specification of exact quantities, delivery dates, and points of delivery

A stable exchange of information between supplier and consignee is prerequisite for just-in-time delivery. This can be secured, for instance, through a common inventory control or by granting the purchasers access to the supplier's order processing systems and production planning systems (PPS). Furthermore, the supplier needs to offer an extremely high service level as well as exceptional quality assurance so that the purchasers need not conduct quality checks.

Amongst other things, the selection of goods suitable for just-in-time delivery is dependent on the predictability of demand. AX, AY or BX parts are predestined for this while BY and CX parts are suitable to a certain extent (see case study 7.1). Further analyses must be carried out with regard to, for example, the composition of the parts, whether parts are used for a serial product or one variety and whether parts are to be assembled in pre-fabrication (component configuration) or in the assembly of the end product. Additionally, frequency of item changes, out-of-stock costs, as well as maximum inventory, minimum inventory, and safety stock should be taken into consideration.

Just-in-time delivery concepts can be broken down into block delivery and just-in-sequence delivery. With *block delivery*, the supplier receives call-offs several times per day. Following a buffering at the supplier's site, the parts are delivered directly to the buyer's production site. The parts are unsorted since the composition of the parts is not dependent on a specific sequence. *Just-in-sequence delivery*, however, is a direct delivery of small quantities transporting parts to the buyers production site several times per day and synchronously to production. Apart from the timely provisioning of parts at the point of assembly, a sequential delivery according to the sequence of assembly is taken into consideration (just-in-sequence).[12]

[11] Cf. Ihde (2001), p. 274 et seq.

[12] Cf. Vahrenkamp (2007), p. 343.

Case Study 7.2: JIT and JIS in the Automobile Industry

Based on volume and value, the VW plant Mosel procures more than 50 % of its purchased parts through JIT and JIS procedures. The responsibility of the respective suppliers comprises disposition, container management, and transport, including provisioning of the components at the point of assembly. The JIS suppliers are located at a distance of about 30 km from the plant. This makes it possible to process an order and provision the parts just-in-sequence at the assembly line within a few hours after the order was placed by the assembly department at the OEM's site (VW).

Using this principle, the OEM reduces production costs due to short assembly and delivery times and as a result of freed-up production areas and production capacities. This type of provisioning is especially effective with modules suppliers, such as chair producers who deliver pre-finished systems (see Chap. 12). The responsibility of coordinating sub-suppliers is thus passed on to the modules supplier, which reduces the number of direct contacts. All in all, this can be regarded as a win-win situation between the OEM and the modules supplier. Long-term partnerships are forged as a result of a close cooperation. The supplier benefits by taking over certain proportions of the value added, by gaining know-how, and therefore by sharpening the competitive edge.

7.3.4 Goods Receiving Functions

Irrespective of individually applied provisioning principles, goods need to be received physically. This takes place at the *goods receiving department* where all operational actions as to receipt, quantity control, and quality control of the delivered goods take place. In particular, this includes the functions of[13]:

- Receipt of delivery
- Comparison of order and delivery data against type of goods delivered, quantity of goods delivered, and time of delivery
- Approval to unload at a specific unloading point
- Checking goods during unloading with regard to shortfall quantities and damages by means of visual inspection
- Reporting defects and, if necessary, refusing acceptance and returning goods
- Precise quantity control through counting, measuring, weighing, and recording the results
- Comparison of results against delivery documents
- Labeling of goods and approval to store

[13] Cf. Fortmann/Kallweit (2007), p. 83.

From a procurement-logistical viewpoint, a great amount of information is necessary to ensure smooth goods receipt. This includes[14]:

- Categorization of the delivering means of transport according to type, size, and loading height
- Delivery volume and quantity, such as average and maximum volume of delivery, delivery time, number of deliveries per day
- Form of delivery, such as pallets, containers, boxes, or other types of loading aids

Special importance is placed on the *incoming goods inspection* since it has to be verified whether the functionality and quality features agreed on with the supplier have been adhered to. Depending on their extent and thoroughness, quality controls may take a considerable amount of time, which delays the provisioning of the incoming goods for production or dispatch. For this reason, the task of carrying out quality controls is more and more frequently assigned to suppliers. They conduct these controls using stipulated control parameters, control methods, testing equipment, and rejection criteria.

7.4 Transport, Warehouse, and Location Concepts

7.4.1 Regional Freight Carrier Concept

In addition to the procedures of goods provisioning, numerous *transport, warehouse, and location concepts* have been established.[15] A *regional freight carrier* is a logistics service provider who organizes the composition and bundling of shipments from suppliers on behalf of a buyer within a specific area and implements transport of the shipments to the production site of the consignee.

The buyer determines:

- The allocation of the regional freight carrier to the location of the supplier
- The processes the supplier should follow at the point of contact with the regional freight carrier, e.g. scheduling of readiness to dispatch, specification of pickup notification from supplier to regional freight carrier
- Specifications to be observed by the regional freight carrier as to means of transport, containers, load carriers, timeframe for unloading, and delivery times

The regional freight carrier is responsible for:

- Implementation of collection rounds (timetable)
- Consolidation of individual shipments at central transit points

[14] Cf. Fortmann/Kallweit (2007), p. 85.

[15] Cf. Stölzle/Gareis (2002), p. 402 et seq.

- Transport to consignee in complete, *point-to-point shipments*

The prerequisites for the regional freight carrier concept are:

- Delivery conditions *ex works*, i.e. the regional freight carrier delivers by order of the consignee
- Suitability of goods for consolidated transport
- Limited need for speediness
- Negotiable delivery dates and loading times between suppliers
- Sufficient spacial concentration of the suppliers

Concentrating on one carrier per collection area facilitates scheduling through central disposition. This increases the reliability of incoming shipments. Cost advantages result mainly from the consolidation of goods flows. Bundling a number of individual shipments into consolidated shipments results in a reduction of traffic and thus in a decrease both in environmental pollution and transport costs. In addition, bottlenecks at the buyer's goods receipt may be avoided and the coordination of delivery dates can be simplified.[16]

7.4.2 External Provisioning Warehouse

An *external provisioning warehouse* serves the purpose of merging input materials from several suppliers at a jointly managed warehouse.[17]

The prerequisites for this are:

- Horizontal cooperation between suppliers
- Consolidation during pre-carriage to the warehouse
- Consolidation during on-carriage to the consignee
- Consolidation for picking and loading carried out jointly by suppliers
- Synchronous delivery

Splitting warehouse costs and (jointly) remunerating the employed logistics service providers are difficulties in the realization of external just-in-time warehouses. The advantages are a reduction of traffic, increased reliability due to geographical proximity, as well as greater suitability for just-in-time or *just-in-sequence* procedures. *Consignment warehouses* can be combined with just-in-time warehouses but are different from them in that they are managed by one supplier. The supplier stores contractually stipulated inventory stocks in them, from which the consignee removes the goods in demand. The storage area is normally provided by the consignee. The goods remain the property of the supplier until removal from stock. The resulting capital commitment costs are borne by the supplier.

---

[16] Cf. Vahrenkamp (2005), S. 233.

[17] Cf. Stölzle/Gareis (2002), p. 410.

7.4.3 Supplier Settlements and Supplier Parks

External provisioning warehouses are different from *supplier settlements* and *supplier parks*. The latter ones are designed to provide settlement space for several suppliers of mostly one buyer and/or of other service providers. Mostly, these settlements are not devised as production sites but as assembly sites. The reason for this is the fact that assembly is in most cases more customer-specific than production, which renders a re-location close to the assembles site of the OEM worthwhile. Among the objectives are cost-cutting by making use of synergies (park/facility management, IT networks, conference rooms, canteen, public transport etc.) and improved service through increased flexibility, reliability, and sped-up processes (see Sect. 4.2.6).

Case Study 7.3: Supplier Park Smartville

Supplier and industrial parks are very common in the automotive industry, which is demonstrated by the *Smartville* project in Hambach, France. The different models of *smart* cars have been produced there since 1997. The supplier park is not located directly next to the production site, but the five system suppliers and two logistics service providers are contained within the premises, forming a *cooperative factory* together with smart France, a subsidiary of Daimler. The layout of the production plant resembles a plus sign as can be seen in Fig. 7.6 and is carried out by smart France. The system suppliers are located around this center and supply prefabricated modules, such as car bodies, cockpits, rear axles with

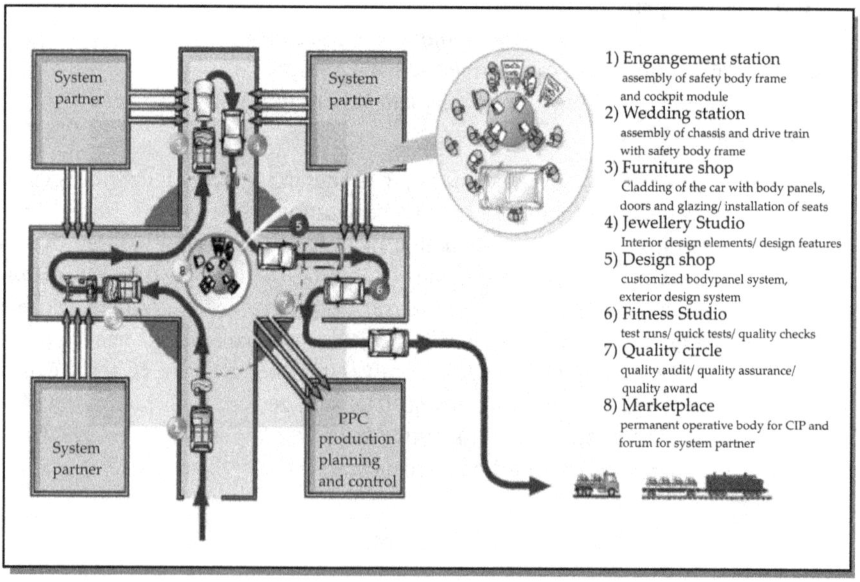

Fig. 7.6 Factory premises at smart France (Smart GmbH 2007)

drive, front-end modules, doors, and plastic covers directly to the assembly line. The advantages of this layout are the short distance between supplier and final assembly as well as the ease of extending the entire structure. Just as on a common production site, the suppliers located in immediate proximity are integrated in the production and develop a high sense of responsibility for the processes. Furthermore, by contractual agreement, smart France motivates its partners by encouraging them to make suggestions for improvement of the end product throughout the entire product life cycle. About 70 % of the materials delivered to the assembly lines are provisioned by the system suppliers located in Hambach. 20 % of the remaining materials are delivered by JIT suppliers which facilitates production with lean inventory. One of these suppliers is the engine factory MDC in Kölleda (Thuringia). The engines are delivered in swap bodies which dock to the respective assembly station. Thus, the distance between material removal and point of assembly hardly exceeds 10 m. The production plant in Berlin is informed at a 3 days notice prior to assembly start in Hambach as to which delivery is required. This becomes possible by specifying so-called *pearl chains,* which signal that from that point on the sequence of production is not changed anymore (see Sect. 12.2.1). Therefore, Daimler is able to schedule its tours well ahead of time and utilize its truck capacities to the fullest.

The residual 10 % of the production materials are made up of standard parts and small parts. These are delivered to a supplier warehouse (consignment warehouse) where they are buffered for 3–10 days on average. While still in the warehouse, these parts remain the property of the producer until the logistics service provider feeds them into the assembly processes.

All modules and bought-in parts generally remain the property of the suppliers up to a defined counting point at the site in Hambach. Billing is then carried out based on the requirements and according to the parts list. This renders incoming goods inspection unnecessary since the supplier carries the responsibility for the delivery date and the quality of the goods up to the point of assembly. Hence, smart France produces without keeping any inventory. This takes inventory risks off smart France and direct capital commitment costs can be avoided.

Having the five main suppliers on site reduces the risk of supply bottlenecks due to delivery problems such as traffic jams and so forth.

7.5 Cooperation Concepts of Inventory and Provisioning Management

7.5.1 Efficient Consumer Response

Apart from requirements planning and management, different forms of cooperation between suppliers and producers or between producers and the trade sector are becoming more and more important for the optimization of inventory along logistics chains.

Efficient Consumer Response (ECR) is a cross-company approach to optimize flows of goods, information, and cash with the aim of an overall optimization of the system including suppliers, producers, the trade sector, and end consumers. The focus is on increasing the efficiency and productivity of the entire value chain and not that of individual links in the chain. To this end, cooperative partnerships are forged between producers and the trade sector for efficient replenishment, efficient promotion, efficient assortment and efficient product introduction.[18] In particular, this calls for cooperation concepts in logistics (Supply Chain Management, *Supply Side*) and for cooperation concepts in marketing (Category Management, *Demand Side*).[19] For our purposes we shall focus on the logistical approaches. These approaches primarily seek to overcome uncoordinated and isolated procedures along the supply chain, to build mutual trust and understanding, and to disclose necessary information.

Efficient Replenishment aims at an efficient management of supply. This pull-system which displays characteristics of just-in-time delivery of industry production depends on the actual sales. Its goal is to synchronize the production of the manufacturers and their suppliers with customer demand by linking all stages in the supply chain together (customers, trade, headquarters, warehouse, branches, producers, logistics service providers) within one integrated system. For the implementation thereof, the instruments of Supply Chain Planning, Supply Chain Execution, and Supply Chain Event Management are used (see Fig. 7.7):

- *Supply Chain Planning* deals with the planning of logistical resources in procurement and distribution, such as inventory, delivery, and transport capacities
- *Supply Chain Execution* supports the operational processes (management and control), such as order processing, stock management, and transport
- *Supply Chain Event Management* monitors all activities and generates notifications or warnings in cases of deviations (anomalies in inventory, provisioning process)

Continuous Replenishment Programs (CRP) are geared towards securing direct and automated replenishment by means of immediate transmission of stock and inventory data. *EDI systems (Electronic Data Interchange)* are used for information exchange since sped-up or partly automated replenishment can be realized by avoiding media disruptions. Manual entry of data from faxes or emails into ERP systems would be an example of media disruption.

---

[18] Cf. Heiserich et al. (2011), p. 256 et seq.

[19] Cf. Hertel et al. (2005), p. 173 et seq.

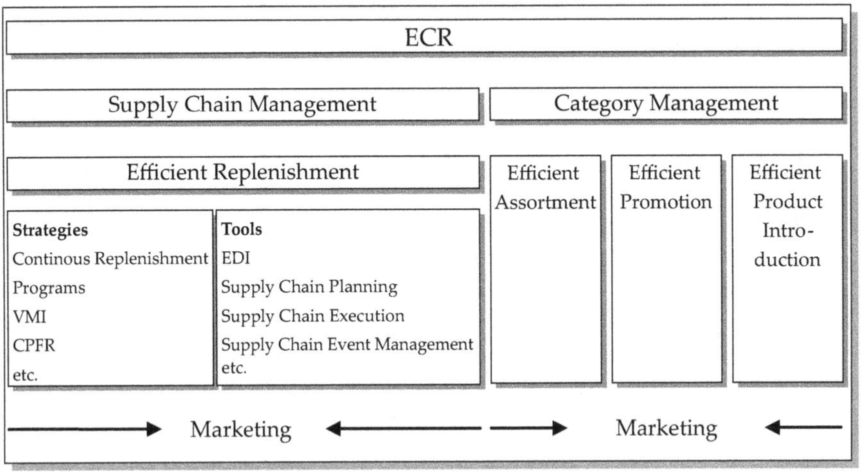

Fig. 7.7 ECR overview (Cf. Gleißner (2000), p. 101)

7.5.2 Vendor Managed Inventory

Along with CRP, *Vendor Managed Inventory (VMI)* is an ECR strategy which optimizes replenishment. While planning and management of goods flows has so far been at the discretion of the trade sector respectively the trade company, we can now observe that the responsibilities of procurement logistics are increasingly being assigned to the producers and suppliers. This means that only the producer makes decisions about delivery rhythms and quantities delivered. The trade sector and buyers, on the other hand, forego the establishment of delivery quantities and delivery dates.

The process of Vendor Managed Inventory involves a continuous monitoring of inventory at the customer's site (warehouse, branch) and – citing the example of a producer-vendor relationship – is based on the processing of sales data (scanning cash register) in retail stores and on the transmission of sales figures and inventory data to the producer. The sales forecast resulting from this is used as the planning guideline by the producer. Thus, the producer determines the delivery quantity based on quotas, order time, and available stocks. An order confirmation is then sent to the trade company. Order processing and delivery is carried out by the producer while the inventory level and the inventory range are jointly determined by the vendor and producer.

A mild form of VMI is *Co Managed Inventory (CMI)*. Here, the trade company is still responsible for processing the order. The supplier issues suggestions as to the order, which only have to be confirmed or adapted by the consignee.[20] If a VMI

[20] Cf. Stölzle et al. (2004), p. 143.

strategy is being introduced, CMI may initially be used on an interim basis to alleviate the reservations of both parties and ultimately make a transition to VMI.

7.5.3 Collaborative Planning, Forecasting and Replenishment

Collaborative Planning, Forecasting and Replenishment (CPFR) is an advancement of ECR in the field of sales planning.

It is advanced in the sense that forecasting and planning of sales are modified and disposition is re-structured. It is especially important to take fluctuations in demand and uneven order frequencies into account and not to amplify their effects through wrong forecasts at the individual production stages and inventory-carrying points in the supply chain.[21] Above all, peaks in demand need to be analyzed accordingly.

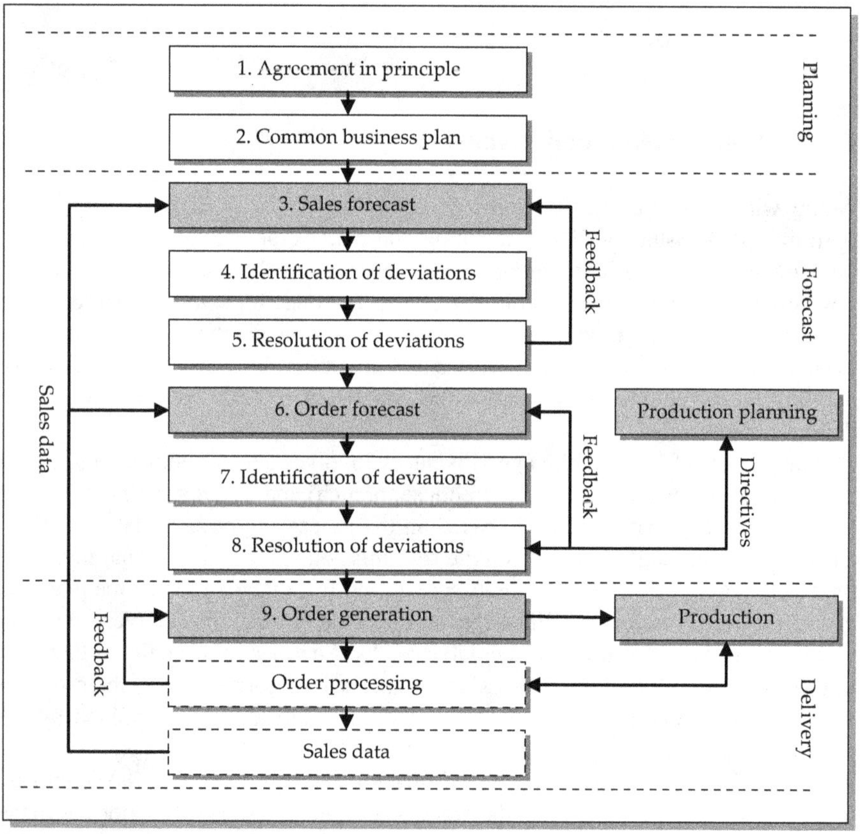

Fig. 7.8 CPFR process model (Cf. Kuhn/Hellingrath (2002), p. 112)

[21] Cf. Ihde (2001), p. 321.

This requires a detailed analysis of past purchase patterns (seasonal influences throughout the year, weekly trends). Special emphasis therefore lies on the collaboration in the forecasting phase, in which the joint handling of critical deviations turns into a learning process through giving feedback.[22] This integrated planning approach is shown in Fig. 7.8.

Additionally, marketing activities need to be taken into consideration (campaigns, advertisements). Electronic marketing, electronic market research, and data from Customer Relationship Management (CRM) systems may help to enrich sales planning with valuable information. Integrated and improved methods for planning and disposition as well as shortened cycles facilitate greater availability and thus more reliable deliveries. In addition, more flexible planning becomes possible since changes in plan at one stage entail consistent changes at all upstream and downstream stages.

The following effects can be achieved through improved transparency:

- Greater product availability with low stocks
- Timely disposition of updated production plans in which changes in plan on the buyer and supplier side can be taken into account
- Quicker reaction to changed customer demand
- Higher capacity utilization of transport means through coordination of distribution plans

Successful implementation of CPFR increases the effectiveness of ECR and/or VMI. However, both ECR and VMI can be employed independently and do not require the use of CPFR.

[22] Cf. Hertel et al. (2005), p. 197 et seq.

Review Questions

1. What are the goals of inventory management?
2. What are the functions of inventory?
3. Name the most important types of inventory costs.
4. What is the difference between order-point procedures and order-rhythm procedures?
5. What are safety stocks?
6. Define the different principles of goods provisioning?
7. Name the most important approaches of Efficient Consumer Response (ECR).
8. Who employs regional freight carrier concepts and runs external just-in-time warehouses?
9. Describe the procedures involved in just-in-time concepts.
10. What statements can be made on the basis of ABC and XYZ analyses?

Additional Literature

Alicke, K. (2005). *Planung und Betrieb von Logistiknetzwerken: Unternehmensübergreifendes Supply Chain Management*. Berlin: Springer.

Arminger, G. (2004). Absatz- und Bestellprognosen im CPFR-Prozess für den Handel. *Supply Chain Management*, o. Jg., 2004, Nr. 3, S. 45–52.

Burt, D. N., Petcavage, S., Pinkerton, R. (2009). *Supply management*. New York: McGraw-Hill.

Cavinato, J. L., Flynn, A. E., & Kauffman, R. G. (2006). *The supply management handbook* (7th ed.). New York: McGraw-Hill.

Gudehus, T. (2004). *Logistik* (3rd ed.). Berlin: Springer.

Hahn, D., & Kaufmann, L. (2002). *Handbuch Industrielles Beschaffungsmanagement* (2nd ed.). Wiesbaden: Gabler.

Kernler, H. (2007). *Logistiknetze: Mit Supply Chain Management erfolgreich kooperieren*. Heidelberg: Hüthig.

Monczka, R., Handfield, R., Giunipero, L. C., & Patterson, J. L. (2011). *Purchasing and supply chain management* (5th ed.). Mason: South-Western College Publishing.

Sydow, J., & Möllering, G. (2009). *Produktion in Netzwerken: Make, buy & cooperate*. München: Vahlen.

Zipkin, P. J. (2000). *Foundations of inventory management*. New York: McGraw-Hill.

Logistics Network Planning

8

Learning Objectives

Designing logistics systems means dealing with complex planning and decision problems. The following chapter will examine details and strategies in logistics planning with a focus on planning logistics networks, warehouse networks, and transport networks.

Besides basic structures and configurations of logistics networks, the reader will be introduced to contemporary concepts of designing transport and storage networks. These are mainly employed in distribution. As logistics networks are increasingly being operated by the logistics service providers, this chapter will highlight aspects of outsourcing, tendering, and selecting transport and logistics service providers.

Keywords

- Planning tasks and details
- Logistical goal conflict
- Logistics networks and structures
- Warehouse network structure
- Transport network structure
- Cross docking
- Capacity planning
- Warehouse layout planning
- Outsourcing

DOI 10.1007/978-3-319-01769-3_8,

8.1 Strategic Network Planning

The strategic logistics goals which network planning is trying to reach are improvements of the logistics performance. The goal conflict between market-driven logistics services and minimal logistics costs poses a distinct challenge (see Fig. 8.1).

The image of the product and the resulting customer expectations define which service is in line with the market. The definition of low logistics costs varies from company to company. Minimum costs, however, are always the goal.

For the basic strategic logistics network planning, the synoptic or the incremental planning approach can be employed. The synoptic approach is based on long-term goals; the processes necessary for reaching these goals are comprehensively, gradually, and systematically introduced. The incremental approach, on the other hand, finds solutions to subproblems without explicitly defining goals in advance. This is why the latter strategy is often called a strategy of incoherent steps or muddling through.[1] A general statement about which planning approach is to be employed for the respective problem cannot be made. The synoptic approach stresses a structured working method while the incremental approach provides more room for creativity.

The general method of strategic solution finding – which is also valid for logistics planning processes – is divided into three steps:

- Definition of guidelines
- Strategy formulation
- Strategy implementation

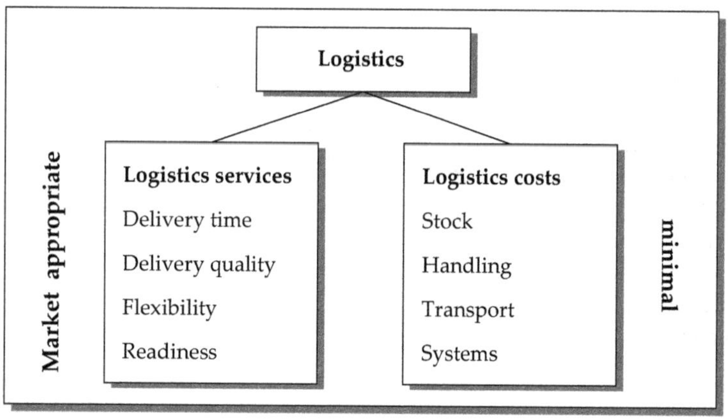

Fig. 8.1 Goal conflict in logistics performance (Cf. Schulte (2009), p. 8 et seq)

[1] Cf. Bea/Haas (2009), p. 221.

Guidelines are based on general company goals and values. They are the starting points for logistical *strategy formulation* (see Sect. 3.1). The first step is to examine the company environment to find chances and risks, and the company itself for strengths and weak points. The results are used to develop different strategic options from which the preferred logistics strategy is chosen. During the *strategy implementation*, strategic programs are created which help to employ the strategy. The results from implementing the new logistics strategy are compared to the original goals. If the goal and the result deviate from each other, changes to the strategy will be made at the beginning of the decision-making process.

The following will give an overview about common logistics network models and their advantages and disadvantages. They are the basis for strategic planning processes.

8.2 Overview Logistics Networks

8.2.1 Representation of Logistics Networks

The planning task of logistics and supply chain management is to create the preconditions for entering new markets, in order to introduce new or existing products into these markets as successfully as possible. This necessitates the creation and maintenance of a corresponding logistical infrastructure – i.e. logistics networks – for procurement, production, and distribution logistics. The following chapter focuses on the planning of warehouse and transport networks for distribution logistics.

Logistical networks can be represented using a graph-theoretical representation of vertices and edges. Vertices represent the realization of physical processes, e.g. warehousing, handling of goods, or data processing. Edges represent the installations for transport processes, flow of goods, and information transmission. Such a net then consists of sources of goods, e.g. manufacturers' plants, points of sale of merchant intermediaries, and the locations of consumer demand.

A detailed description of logistical nets is possible by further division into relevant levels:[2]

- *Functional level*: Description of tasks for the distribution of goods (warehousing and transport)
- *Process level*: Details about the number of stages in the process (degree of fragmentation of the supply chain)
- *Spatial level*: Localization of the logistical process actualization
- *Institutional level*: Aspects of the organizational realization (creation and tasks of functions and processes)

[2] Cf. Stieglitz (1999), p. 91.

8.2.2 Domains of Decision Making in the Network Design

Domains of decision making in the network design can be divided – according to their time horizon and their contents – into strategic, tactical, and operative aspects. The following planning contents may serve as an example:[3]

- *Strategic (long-term)*
  - Problem of gradation: Gradation/function of the vertices
  - Problem of locations: Location of the vertices
  - Problem of numbers: Number and dimension of the vertices
  - Problem of allocation: Allocation of sources of goods and points of consumer demand to the vertices
- *Tactical (medium-term)*
  - Transport organization
  - Allocation of stock and representation of stock size within the network
  - Definition of the purchase and supply frequencies
  - Definition of the inventory management structure
- *Operative (day-to-day)*
  - Vehicle employment and route planning
  - Definition of the ordered quantities and order times

8.2.3 Structures of Logistics Networks

Regarding the complexity of logistic networks, distinctions can be made between one-step direct nets, two-step indirect nets, multi-step indirect nets and hybrid, and combined nets. Figure 8.2 shows examples from consumer goods logistics with respect to the distribution. These basic structures can also be applied to other fields (see Sect. 12.2).

[3] Cf. Stieglitz (1999), p. 97; Corsten and Gössinger (2007), p. 155 et seq.

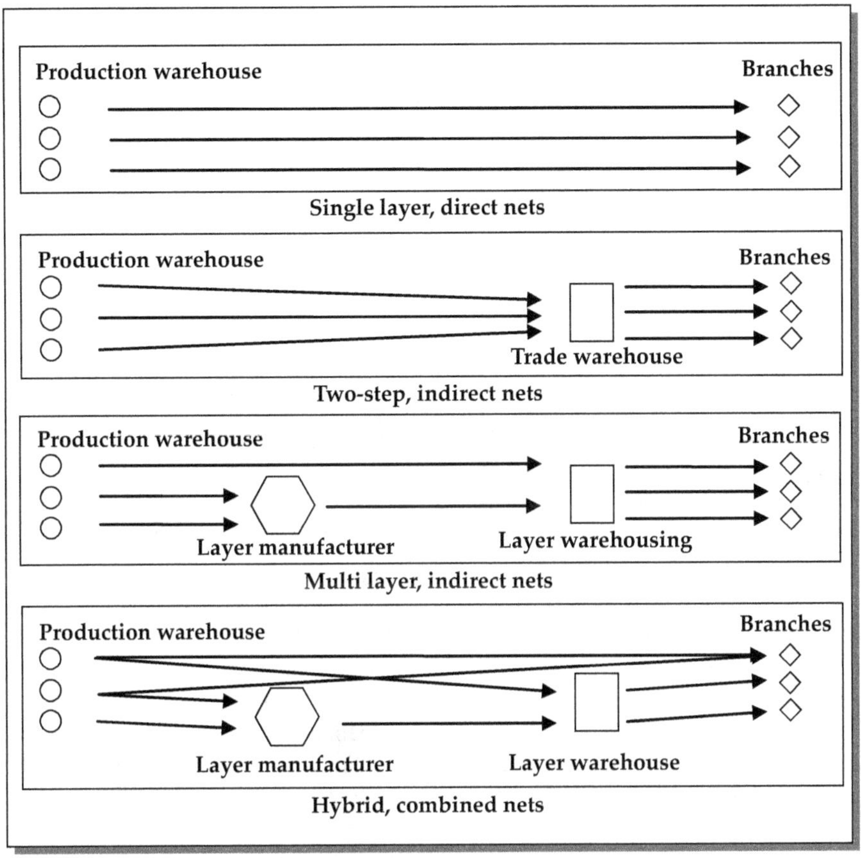

Fig. 8.2 Representation of a network for the flow of goods (Cf. Stieglitz (1999), p. 95)

8.3 Warehouse Networks

8.3.1 Warehouse Network Planning

The locations of network nodes are defined during the process of *location planning*. Changes in the strategic framework conditions are often the reasons for location problems. Possible dimensions of decision-making are:

- Production versus distribution locations
- Static versus dynamic time horizons
- One- versus multi-product strategy

Location factors are divided into quantitative (e.g. transport costs, property prices, staff costs, support measures, taxes) and qualitative factors (e.g. geographical location, condition, infrastructure, possibilities for recruitment, legal aspects). These factors inform location strategies aimed at the creation of capacity: adding capacity

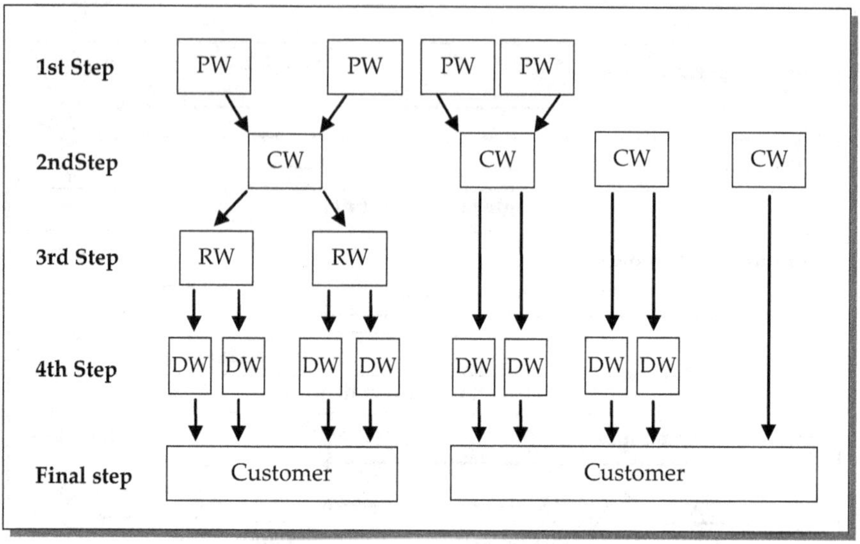

Fig. 8.3 Number and degree of centralization of distribution systems (Cf. Schulte (2009), p. 460)

in new regions or increasing capacity in existing locations; capacities may also be concentrated in existing locations by (partially) shutting down other locations.

For any location, the structure of a warehousing system is analyzed as a vertical one and a horizontal one. The horizontal structure is defined by the number, capacity, and geographical location of the warehousing and handling processes on every warehousing stage; it defines the degree of centralization of a distribution system. The vertical structure of local warehousing and handling processes defines the hierarchy and the number of steps of a warehousing system. Figure 8.3 shows these correlations.

The definitions of the resulting warehousing types are as follows:[4]

- *Production warehouse (PW):* Storing of finished goods from a production site for short-term balancing of stock volume
- *Central warehouse (CW):* Storing of a full product range, supply for lower warehouse levels or, in a centralized structure, for distribution to deliver a customer order
- *Regional warehouse (RW):* Buffer for a specific sales area to relieve earlier or later warehousing stages
- Distribution warehouse (DW): Separation function to compile the ordered quantity in a specific sales area (spot delivery)

8.3.2 Degree of Centralization

With the existing goal conflicts between centralized and decentralized structures, the decision about the *degree of centralization* of a distribution warehouse system is an

[4] Cf. Delfmann (1999), p. 193.

important one. There are two basic advantages to centralized warehouse concepts: First, the possible reduction in the number of connections between the points of delivery and the points of receipt; second, the possibility to considerably reduce the stock, especially the safety stock. Additional, more efficient warehouse technologies, e.g. automated, which are only profitable with a large handling volume, can be employed. This is why for some time there has been a trend to reduce the number of warehousing stages. Centralization and direct supply to companies is more and more taking the place of multi-stage decentralized structures.[5]

Centralization of the warehouse can be achieved by eliminating a full warehouse level or by reducing the number of warehouses on one level. These measures lead to a decrease in the *density of the warehouse network* which means the total volume of goods is stored in fewer warehouses. This, however, creates a higher potential for bundling in delivery to warehouses. The catchment areas of the warehouses – represented e.g. by the average number of branches supplied to by one warehouse – grow. This results in the average *transport distance* growing. On the contrary, the creation of a decentralized regional warehouse system is opted for if the delivery time is the most important factor, i.e. in cases where the supply from a central warehouse to individual regions could only be realized with delivery times or at delivery costs not tolerated by the market.

A concept often practically applied is *selective warehousing*. This is a two-level warehouse structure in which goods with a low prognosis risks, e.g. fast-moving goods, are stored in decentralized regional and spot delivery warehouses that do not carry the full product range. Inventory and B and C articles (slow-moving goods) are centrally stored to economize the stock (see Fig. 8.4).

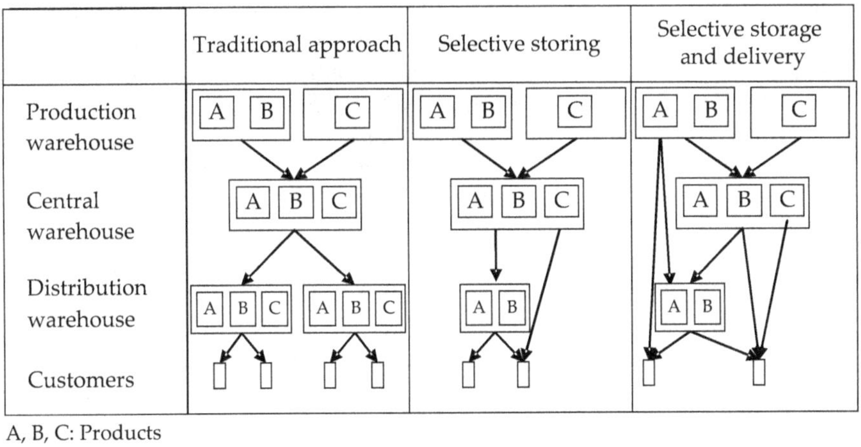

Fig. 8.4 Concept of selective warehousing (Cf. Schulte (2009), p. 460)

[5] Cf. Baumgarten and Thoms (2002), p. 53.

In addition to selective warehousing, further optimization is possible if goods are delivered from the points on which they are stored without touching warehouses on earlier or later stages. This is called selective delivery. Both concepts are often employed in spare parts logistics.

Case Study 8.1: Spare Parts Logistics

The company *Bosch und Siemens Hausgeräte GmbH (BSHG)* manufactures home appliances. It was founded in 1967 as a joint venture between *Robert Bosch GmbH* and *Siemens AG*. The company now has 43 plants in 15 countries in Europe, the USA, Latin America, and Asia.

The basic requirement of fast troubleshooting is an effective spare parts logistics. It directly becomes a part of customer experience. Therefore it is also temporarily much more coupled to the time of the resale. This insight has inspired the slogan *The first product is sold by the sales department; all further products are sold by the service*. In 2000, the project Total Customer Logistics was started. The individual steps were gradually implemented by 2005. The foremost goal was an increase in customer satisfaction while reducing the costs at the same time. The logistics concept that was created will be shortly introduced.

The global spare parts logistics network is based on the global manufacturing networks of the product areas. Globally, there are seven *central warehouses* on four continents. To prevent uneconomical small transport activities, intercontinental moving of goods is consolidated on the central warehouse level. The flows of goods are clearly defined. The BSHG plants deliver goods only to the central warehouse assigned to them. One central warehouse is assigned to the subsidiaries in every country. The BSHG's largest *central spare parts warehouse* is located in Fürth, Southern Germany. With the introduction of direct delivery from a central warehouse, it was possible to shut down eight *regional warehouses* in seven European countries and 55 *customer service warehouses* in Germany and Austria (see Fig. 8.5).

By now, 800 customer service technicians in Germany, Austria, Belgium, and the Netherlands receive spare parts overnight directly from the Fürth central warehouse. Orders from German technicians are taken every day until 3 p.m. After the order-related commissioning, the parts are handed to the transport company in reusable boxes for delivery by 7 a.m. the next day – directly to the technician's vehicle or to another point previously agreed on. Parts not needed on the previous day are returned. Secondary processes like sorting, disposing, and new packaging are integrated. Prior to each order, a clear statement is made which spare part is needed. For this, the stock lists and exploded drawings are available on the brand homepages, besides a number of general information.

The size of the appliance which is to be repaired determines the logistics strategy: *repair at the customer's* or *repair in a workshop*. In Europe, more than 2,000 company technicians are responsible for repairing stoves, washing machines, dish washers, and cooling devices directly in households. Small appliances, on the other hand, are repaired in central repair workshops to use the economies of scale of centralization.

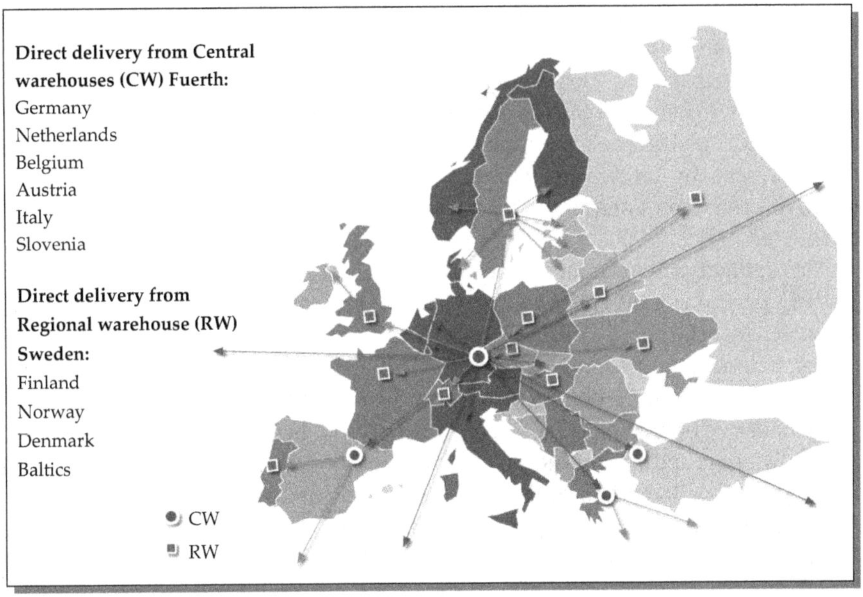

Fig. 8.5 Logistics network BSHG (BSH GmbH (2005))

With this concept, including its effects on the repair and core business, customer satisfaction could be measurably increased during the last 5 years. Further market share in the core business could be won and the costs for spare parts logistics could be lowered by 26 %.

8.3.3 Capacity Planning

The configuration of the warehouse network substantially influences the required floor space, the spatial and staff capacities, as well as the layout of a warehouse location. Furthermore, the technical equipment on storage, handling, and sorting installations and the resulting process organizations must be considered.

The *capacities* can be divided into qualitative and quantitative aspects. The qualitative achievement potential of a warehouse location must be in line with the respective customer and order requirements regarding flexibility and productivity. Quantitative capacity dimensioning defines which capacities are actually available for a given period of time. It has to consider the current order volume – e.g. represented by the number of storage places, the number of orders within a given period of time, and the chronological demand history – with respect to the possible monthly, weekly, daily, or shift performance.

Anticipatory capacity planning must also take in account that the spatial and handling potentials as well as the staff capacity of a warehouse location must cover different workload volumes. Expected dimensions, ranges, and times of demand fluctuations must be part of the planning. Most of all when planning commissioning capacities, solutions which are balanced between the danger of too limited capacities – especially during (seasonal) peaks – and the problem of too small workloads must be found. Staff adjustments to demand fluctuations are possible with flexible work time models using working time accounts (e.g. for a full year) and by employing part time and temporary staff. However, possible negative effects on commissioning quality, productivity, and reliability must be considered.

The defined processes necessary to realize the desired result are planned in *process organization*. The necessary qualitative capacities are largely defined based on the class and degree of automation of the equipment as well as the corresponding qualification requirements of the staff. It must be noted that a possibly desirable high degree of automation, described as the relation between equipment performance and staff performance, is considerably influenced by the types of storage and commissioning tasks, the homogeneity of the goods, as well as the degree of palletization of the packaging. This has effects on the quantitative capacity characteristics like possible order processing times and performance features that are, in turn, to be considered for the qualitative capacity planning.

8.3.4 Layout Planning

Besides the projected number of orders or the number of goods to be stored, *layout planning* demands a rough idea of the flow of goods which is influenced, among others, by the warehouse technology that is employed and by the spatial arrangement of the equipment. Layout planning can be divided into basic and fine layout planning. In basic planning, the property and building layout is planned based on the local circumstances. Besides the dimensioning of the full floor space requirement, the structuring of the floor space for the warehouse and for the yard (e.g. parking, waiting, and maneuvering spaces for vehicles and load carriers) must be planned. Furthermore, enlargement options of the location must be taken into account. Steps in the *building layout* planning are: the definition of the number and the arrangement of gates and ramps; space and room requirements for discharging, conveying, storing, commissioning, temporary storing; providing and other functional areas; the spatial arrangement and the shape of the building which represents the fine layout planning. (see Sect. 4.2.6).

With respect to the floor space capacity of the mentioned functional areas, the *technical equipment* of the storing, sorting, and commissioning areas (shelves, packaging machines, palletizers, lifting trucks, stackers) must be specified, as well as the *staff capacities* according to their numbers and necessary qualifications. The focus of the quantitative dimensioning of e.g. the commissioning capacities would be planning the staff requirements and deployment. Based on the expected volume of orders – as seen from the number of orders, number of order items, and number of pieces per order item within a given period – the number and working times of the staff, determined by the number and length of the shifts, is defined.

8.4 Transport Networks

8.4.1 Transport Network Planning

For logistics networks, the planning of the *transport networks* must occur analogue to the planning of the warehouse structures. Tasks of transport comprise supplying goods to plants, warehouses, and selling places as well as the disposal of materials and goods which are no longer needed. The transport services necessary for this are realized in transport nets. Transport nets consist of vertices, joined together by edges. *Vertices* of transport nets are sources, selling, and handling points. The first and the second point are locations where transport objects enter or leave this transport net; loading and discharging processes take place here. *Edges* connect vertices by transporting. The appearance of the edges is partly determined externally, e.g. by the traffic route. Two relations lie on one edge. One relation is a source-shopping point relationship (sender-receiver).

The task of *transport network planning* is to design the process organization of the transport of goods between the sources and the selling points. The goal is to create a net with the most effective transport connections between the individual points by using the given infrastructure. Transport net planning defines whether the edge is realized by regular or irregular transport intervals. Regular transport intervals are scheduled based on demand projections and framework contracts. Irregular transport intervals are planned based on existing transport orders (see Chap. 5). Table 8.1 gives an overview of the basic elements of transport nets.

Table 8.1 Basic elements of transport nets (Adapted from Janz (2003), p. 21)

	Vertices	Edges
Function	▪ Source, selling point ▪ Handling point (type)	▪ Short distance ▪ Long distance
Institution	▪ Transport contractor ▪ Sub contractor ▪ Cooperation partner	▪ Transport contractor ▪ Sub contractor ▪ Cooperation partner
Characteristics	▪ Road accessibility ▪ Loading and discharging areas ▪ Floor space ▪ Staff ▪ Tools ▪ Handling technology ▪ Warehouse equipment	▪ Traffic infrastructure ▪ Number and types of vehicles ▪ Employed vehicles ▪ Geographical characteristics ▪ Distances ▪ Altitude differences
Organization/ processes	▪ Sorting processes	▪ Regular transport intervals ▪ Irregular transport intervals

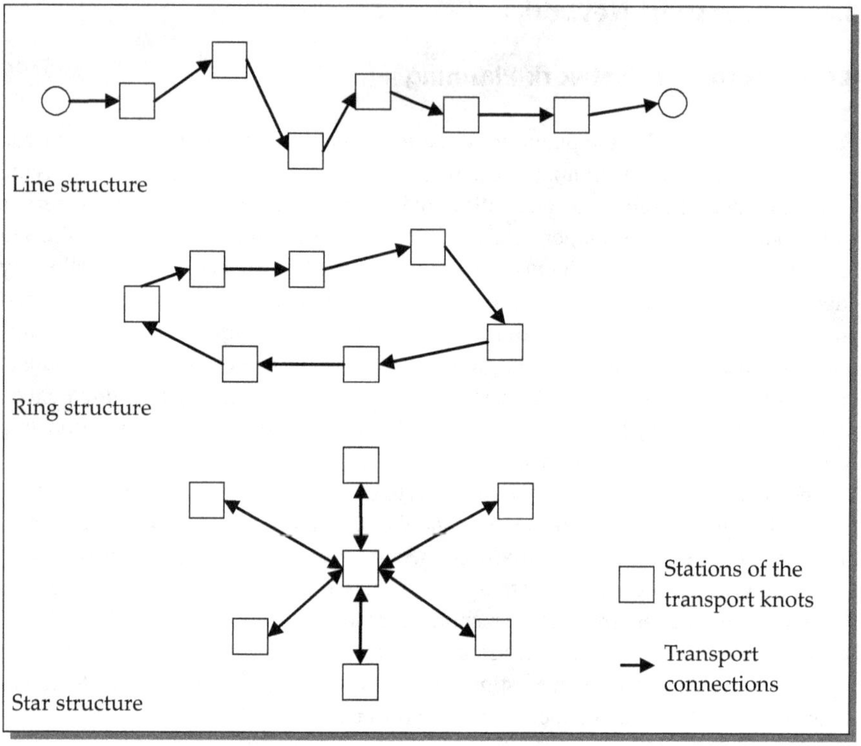

Fig. 8.6 Forms of net structures (Cf. Gudehus (2010), p. 778)

8.4.2 Transport Network Configurations

Configurations of transport nets can be divided into one- or multi-stage forms of line, ring or star structures; see Fig. 8.6.

Consecutively arranged vertices and edges characterize line structures. In this structure, transports have a high number of empty runs. Those tend to be prevented with ring traffic. The average vehicle utilization is increased and the cycling of the means of transport is optimized. From these basic network structures, further forms can be derived which are employed mainly in practical transporting: grids, hub and spoke nets, and hybrid net structures (mixed forms).

In *grids nets,* all shipping and receiving depots are connected by direct traffic (complete network). There is a direct exchanging of goods. There is no need for an additional exchange of goods. The individual depots have a collecting and a dispatching function as well as a sorting and providing function for the main run. The number of relations in such a direct transport net increases squarely with the number of depots (see Fig. 8.7).[6]

[6] Cf. Vahrenkamp (2007), p. 262.

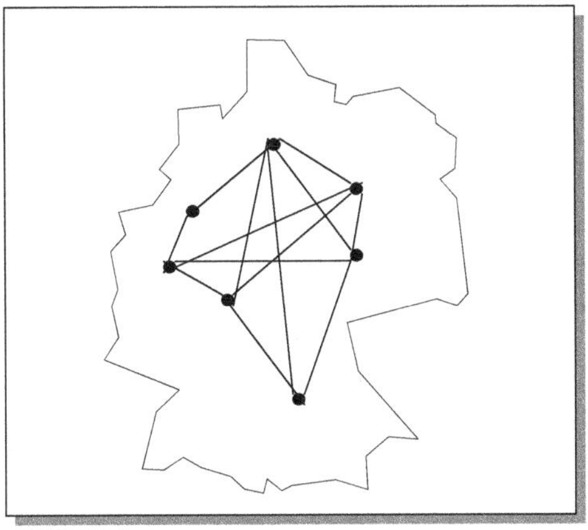

Fig. 8.7 Net configuration: grid

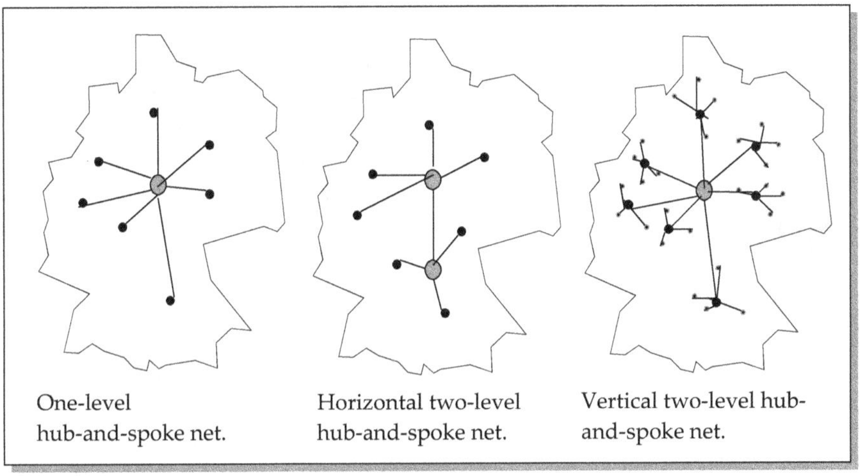

Fig. 8.8 Net configuration: hub-and-spoke

Hub-and-spoke nets are characterized by the traffic between the dispatching and the receiving depot being channeled to and from a central handling depot (hub). There are no direct ways between two depots, i.e. all transports are made via the hub(s) (see Fig. 8.8).

Functions of depots are sorting, consolidating, and forwarding. They are distributed across the area that is to be supplied; they have a regional collecting and delivering function. The number of relations for a hub is exactly twice as high as the number of depots.

The number of connections for hub-and-spoke traffic is calculated as follows:

$$n \text{ connections} \, (n = \text{number of depots}).$$

For exclusive point-to-point traffic in grids, the number of connections is calculated as follows:

$$(n-1)+(n-2)+\ldots+1 \text{ connections or}: [n^{*}(n-1)]/2.$$

Case Study 8.2: Hub and Depot Planning

A logistics service provider is planning a distribution network in a new location. The focus is on the reduction of connections because of a calculation showing that many small flows of goods and many ramp contacts are pushing costs. Ten depots are needed to cover the area.

Calculate the number of connections for a hub-and-spoke system and the point-to-point traffic.

With the result, what do you recommend and what must be considered when comparing the costs?

Hub-and-spoke systems realize bundling advantages. By transporting shipments with different destinations when feeding into the hub or shipment from different sources when delivering from the hub, vehicles can be utilized better (economies of density) or larger vehicles can be employed (economies of size). At the same time, the sorting efforts in the depots (spokes) are reduced because the shipments must be *sorted* to one relation only, the hub. Thanks to the consolidation by the hub, larger quantities may be transported as an unsorted complete shipment from a customer directly to the hub. Without the hub-and-spoke concept, transports on routes with little traffic could only be offered with a high time bundling (see Fig. 8.9).

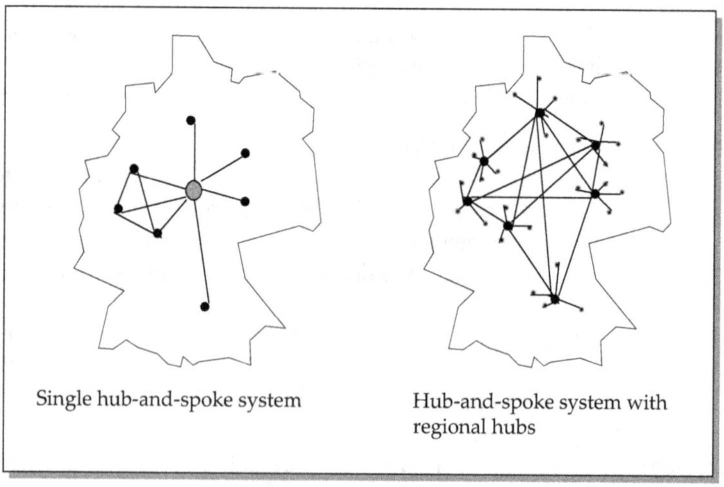

Fig. 8.9 Combinations of grid and hub-and-spoke nets

Indirect routes and additional handling processes are disadvantages of hub-and-spoke systems. Decision criteria for implementing a hub-and-spoke system are the volume of goods to be transported within the net, the time reserves available for the consolidation steps, and the costs necessary for installing consolidation stations.

Besides pure grid and hub-and-spoke systems, there are a number of *combinations of these two configurations*. In such a net, relations with little traffic are supplied via a hub; relations with a high traffic level are supplied directly. Additional options are *single hub-and-spoke systems* and *hub-and-spoke systems with regional hubs*.[7]

Case Study 8.3: Distribution Nets

Sites where goods are stored, handled, or compiled according to customer or order requirements are called distribution centers. From experience, different forms of distribution centers have developed. With respect to their functions, they are divided into regional and trans-regional distribution centers. A 2006 study by a logistics initiative from the German state of North Rhine-Westphalia shows different modes of use for such distribution nets in different forms in this state.

Regional distribution centers are mostly pure delivery sites operated by retailers for supplying goods to their branches. On average, e.g. the regional logistics centers operated by the retailer *Aldi Sued* supply consumer goods to 50 branches in a region. An Aldi Sued distribution center has an average size of 2.5 ha and an average property size of around 20 ha. In one center, 150–180 people are employed. The investment volume for every site is 40 million Euros. A different example for regional distribution centers is the food warehouse in Oberhausen, Germany, operated by the company *Lekkerland* which, among other activities, supplies convenience goods to gas stations. On a 10 ha property, a center in one configuration stage has been built. It has a floor space of 1.8 ha and employs 500 people.

Trans-regional distribution centers are, among others, used for central storing and delivering goods to regional distribution centers. In Europe-wide distribution structures, trans-regional centers also have the function of national distribution. Such centers are mostly located in peripheral regions with important trans-regional traffic connections (interchanges, terminals for intermodal transport) and strategically planned near markets and production sites. The distribution center operated by the drug store chain *dm Drogeriemarkt* in Meckenheim near Cologne is an example of a trans-regional distribution center. The goods are supplied by the manufacturers, stored, and compiled for the branches as needed. The distribution center has a size of 2 ha on a property of 6.5 ha. Around 50 million Euros have been invested, and 500 jobs are planned. Besides the center in Meckenheim, dm owns four more in Germany. An example

[7] Cf. Arnold et al. (2008), p. 784 et seq.

for the distribution using only one national distribution center is the e-commerce company *QVC*. On a 6 ha property, the company invested 100 million Euros in Hueckelhoven, Germany. About 1,000 employees work there.

Besides regional and trans-regional distribution centers, more and more *European distribution centers (EDC)* are built as a byproduct of market expansion. One advantage is that the goods can be stored in one location, regardless of their origin. Orders can be compiled fast and efficiently and delivered to large European customers. The distribution center operated by *3M* in Juechen is an example of an EDC. Besides Germany, 3M products are delivered to the Netherlands, Denmark, Sweden, Norway, Finland, Poland, and Russia from there. The European distribution center is 4.2 ha and a total investment volume of more than 50 million Euros. At the moment, 200 people are employed by the center. Another example of an EDC is the site operated by the cosmetics manufacturer *Shiseido Co, Ltd.* in the Duisburg harbor. The flows of goods from the production sites are bundled there and distributed across Europe to around 6,000 trade partners. The distribution center which is located on a 4 ha property is not operated by Shiseido itself but by the logistics service provider *GEODIS*. As contract logistics service providers, more and more specialized companies operate full distribution sites of customers from industry and trade; as a part of industry logistics concept, they may also build distribution centers for their customers (see Chap. 12). An example is the logistics provider *Fiege*, which is building a 4.2 ha logistics center for the tire manufacturers Bridgestone/Firestone and Yokohama in Dortmund. Another example is the outsourcing of production logistics of the confectionery producer to the logistics service provider *tts Global Logistics*, which in the meantime has been taken over by the Fiege Group. Since then, the logistics have been handled in a 3 ha central warehouse in the logistics park Cologne-Eifeltor with around 90 employees.[8]

8.4.3 Transport Consolidation

Another design variable of transport nets is *transport consolidation,* which has to be a part of the planning. Consolidation is the combination of shipments by more than one loading and/or receiving points during collection and/or distribution runs. Such a route-related consolidation follows the consignment and/or consignee acceptance[9] approach, as shown in Fig. 8.10.

[8] Cf. NRW Landesinitiative Logistik 2006, p. 48 ff.

[9] Cf. Ihde (2001), p. 225 ff.

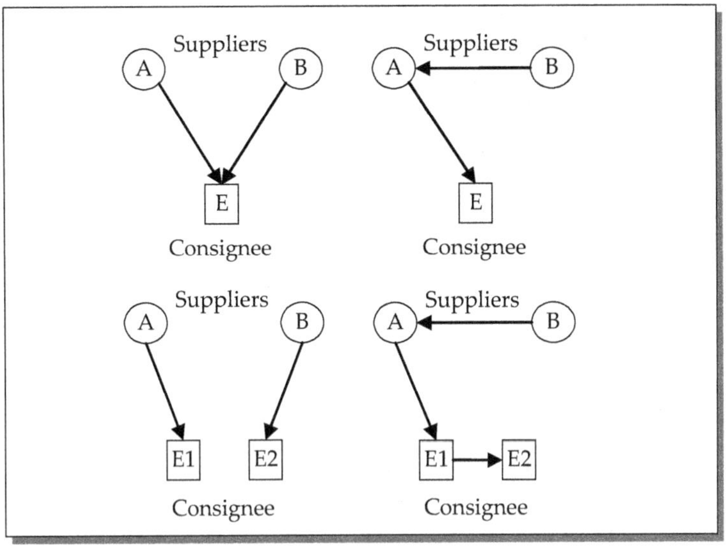

Fig. 8.10 Load and consignee acquisition (Cf. Ihde (2001), p. 226)

Consolidation can also be achieved by including a handling point without stock (transit point): Shipments are regrouped according to relation and receiver. Effects of consolidation are, among others:

- Creation of polynomial transport chains
- More effort compared to direct delivery
- Additional handling processes
- Growing scheduling efforts for the flows of goods to be consolidated
- Better utilization of the means of transport because of a higher relation-related quantity
- Creation of constant connections in cases of long-term trading relations

In procurement and distribution logistics of the consumer goods industry, numerous *consolidation concepts* are employed.[10] Table 8.2 gives an overview about the most important ones.

The *multi-pick concept* says that the ordered quantities of a retailer and its branch for a defined replacement period are collected during a defined tour of manufacturers and sent bundled to the trading warehouse (see Fig. 8.11).

This requires giving the collecting and delivering job to one logistics service provider. Effects are the increase of the average utilization of means of transport and of the discharged quantity per delivery.

The *by-pass concept* includes picking up shipments that are already commissioned for the branch at the manufacturer site. The goods are not stored

[10] Cf. Stieglitz (1999), p. 150 ff.

Table 8.2 Consolidation concepts (Cf. Stieglitz (1999), p. 150)

Point of acceptance \ Location of the branch-related commissioning	Transit point	Manufacturer/supplier
Trade warehouse	*Multi-pick concept*	*By-pass concept*
Branch	*Cross docking*	*Direct store delivery*

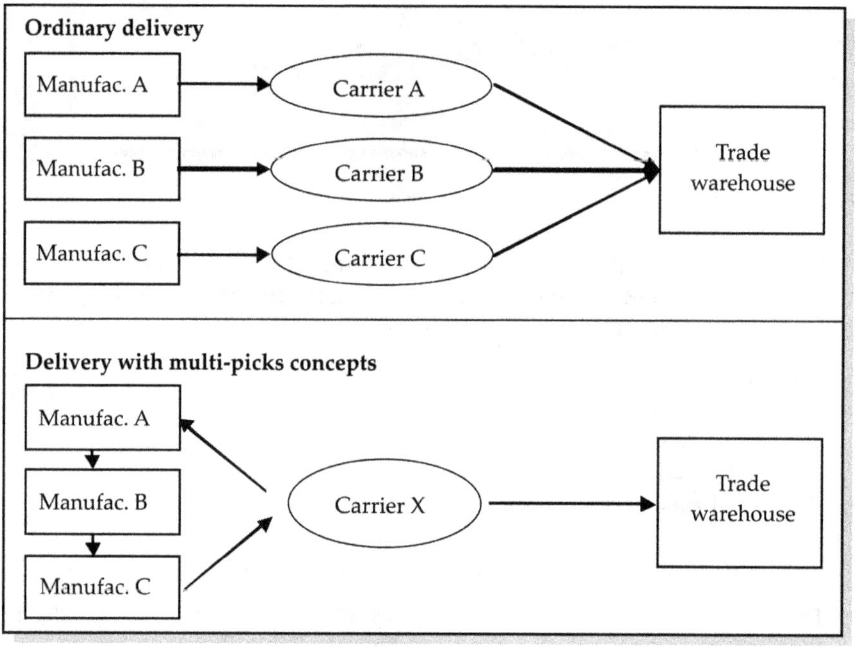

Fig. 8.11 Ordinary delivery versus delivery with multi-pick concepts (Cf. Stieglitz (1999), p. 150)

but passed through. During this process, an order joining takes place with branch shipments from the warehouse of the retailer. With the branch orders being divided into order parts (warehouse and manufacturer), not only the provision of the shipment from the trade warehouse must be realized in the order cycle but also the branch commissioning by the manufacturers. Advantages are the reduction of handling efforts as the loading and discharging processes in the warehouse are no longer necessary.

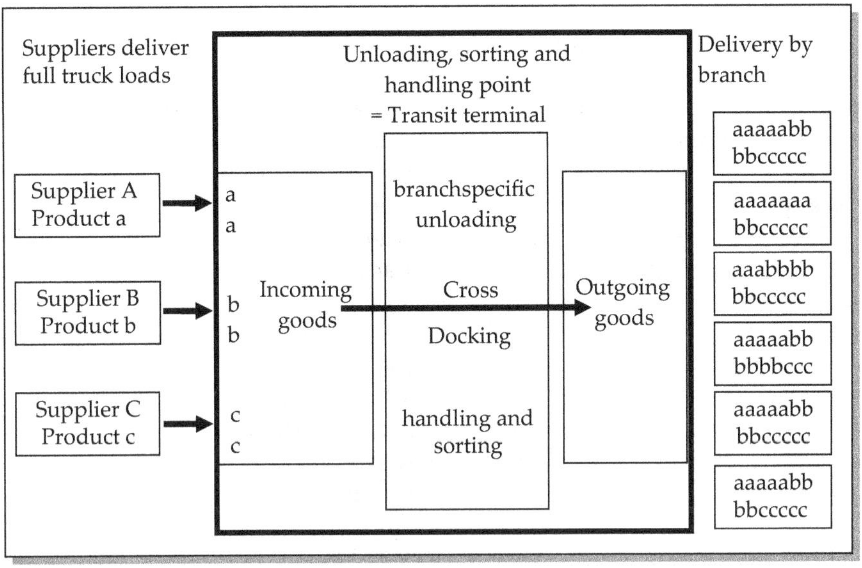

Fig. 8.12 The cross-docking principle (Cf. Kotzab (1997), p. 159)

The *cross-docking* and *merge-in-transit concepts* are closely related to the reduction of warehouse levels.

For cross docking, the handling process means that goods are cross-exchanged (cross) between delivering trucks from different carriers that dock to the goods receipt of the transit terminal at the same time (docking) and empty delivery vehicles which are ready for loading in the goods outwards. The basic requirement for this concept is advance information, in order to control the operation factors; i.e. there must be a clear coordination regarding minimum and maximum order volumes with the suppliers. High reliability is required here, with no stock buffer planned between the manufacturer and the branch. The cross-docking platform must be able to handle a high volume in little time. Some effects are a clear reduction in the order lead time, from the placement of the order to the goods arriving at the branch, a reduction in stock in the branch, the increase of the goods handled, and the storing and take-out processes no longer being necessary (see Fig. 8.12).

With respect to its steps, cross docking (CD) can be divided into:

- 1-step CD: The manufacturer commissions for the branch in advance; then the shipments of more suppliers are joined in the cross-stocking point for the delivery to single branches without manipulation to each shipment (e.g. pallet) itself.
- 2-step CD: The manufacturer sends pallets by sort that are recompiled at the cross docking point for final shipments to single branches.

The *merge-in-transit* concept must be seen in the context of production sites located all over the world which makes it more and more necessary to join products from a customer order not in a warehouse but to merge the individual items in handling terminals (in transit) to complete orders before the actual delivery.

The *direct-store delivery* concept makes it possible to compile the goods for a branch already at the manufacturer. The latter must make the commissioning efforts. Depending on the size of the shipment (vehicle utilization) combined runs can be made directly to the branch or compiled in a transit point after handling of the shipments with respect to the relation to define delivery runs.

The basic possibilities for using consolidation potentials usually depend on the terms of delivery agreed on by the supplier and the customer. They are important instruments, especially in international movement of goods. The interpretation of standard contract terms is defined in the International Commercial Terms (INCOTERMS), which regulate the obligations of buyer and seller regarding costs and risks. The terms EX Works (EXW), Free On Board (FOB), Cost, Insurance, Freight (CIF), and Delivered Duty Paid (DDP)[11] are common. With this, the terms of delivery determine acquisition price.

The following will examine the choice between DDP and EXW more closely.

In case of *DDP*, the suppliers bear the costs and corresponding duties, e.g. for delivery time and condition. Sometimes they will assume further duties like e.g. the discharging of the goods. The buyer will assume the risk at the point of receipt, i.e. the seller is responsible for complete and punctual delivery.

In case of *ex works*, the buyer bears the transport costs and takes responsibility for delivering the goods to their destination. The supplier does not need an own distribution structure. The buyer must organize the planning and controlling of the collection and delivery at the points of receipt and order a logistics service provider. Changes in the terms of delivery from DDP to EXW are becoming more and more common. Basic requirement for changing to EXW is the bundling of volume from individual or cooperating manufacturers.

With the freight costs a part of the pricing for DDP, it must be ensured that the correct freight cost is calculated for EXW. The changes in the terms-of-delivery strategy influence the distribution of the system leadership for goods movements. This may result in strong conflicts of interests between the supplier and the buyer (see Sect. 3.3.2), which is especially true for the cooperative division of the savings resulting from the consolidation.

Case Study 8.4: Net Design of METRO Group Logistics

MGL METRO Group Logistics is the logistics service provider for all companies of the METRO Group. As logistical service and competence center, METRO Group Logistics controls the flows of goods of the METRO Group sales brands. With founding of METRO Group Logistics in 1995 and the implementation and further development of the innovative concept of procurement logistics, the METRO Group was able to strongly increase efficiency. The core points of this concept will be briefly examined in the following.

[11] Cf. Pfohl (2000), p. 189 et seq.

The network structure was changed to a mix of relation loading, direct loading, and classic goods transport. The cross-docking terminals are located in a way that there are always handling points close to branches and suppliers. The disadvantages of ordinary cross-docking structures which often feature some less central terminals for the complete branch supply, causing unnecessarily long ways, are no problem. Each one of the three distribution systems is employed efficiently. The relation loading to terminals near the branch is suitable for suppliers with a large share in the product range; their trucks can cover longer distances without handling. Secondly, direct delivery from cross-docking terminals close to the manufacturer to large branches is suitable as soon as the trucks are fully utilized. Thirdly, in all cases in between, i.e. for all ways between suppliers with smaller quantities and branches with a smaller daily requirement, the system provides for a transport with two interruptions. For this, the collected shipments from different suppliers in source regions are consolidated into main run relations. In the destination regions, goods from different suppliers and source regions are delivered to different branches, which can be described as classic transport of goods. The network structure can be seen in Fig. 8.13.

By using many possible vertices in the source or destination regions, the METRO Group Logistics system uses the advantage of shorter distances, which prevents many unnecessarily long transports. The large number of cross-docking terminals would be, were they owned by METRO Group, uneconomic. The logical – but until then uncommon – consequence was using the existing nets of established service providers.

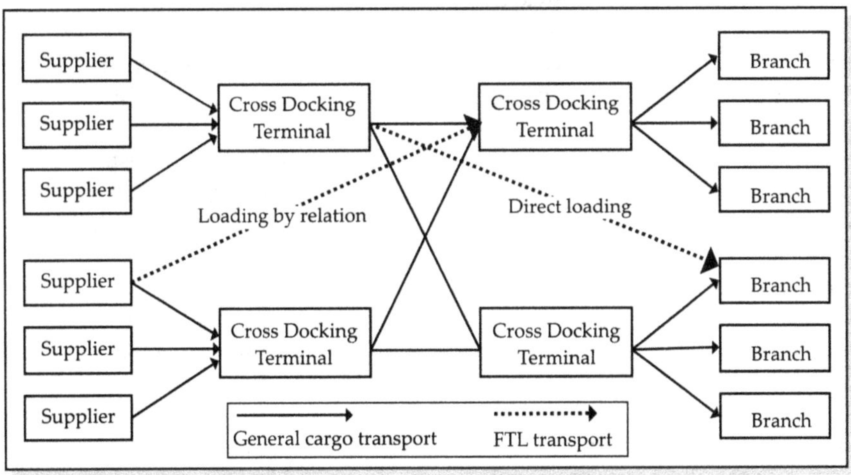

Fig. 8.13 Cross-docking variants in the METRO Group Logistics system (MGL METRO Group Logistics 2002)

The controlling and coordination of the METRO Group procurement logistics system is accompanied by a change in the system leadership in the supply chain from the supplier to the retailer.

However, METRO Group does not always demand the control over logistics. The supply chain partner with the structurally most favorable qualifications – i.e. long-term – should bear the logistics efforts or find a logistics service provider. To ensure this rule is obeyed, METRO Group expects being paid for logistics services exactly the sum of the process costs taken off the suppliers by the METRO Group service (cost equivalent).

In case of system leadership change, the following advantages and disadvantages for METRO Group can be named: The suppliers do not have to plan runs and concentrate on the core business, the manufacturing of products. However, they are forced to exactly meet the production deadlines defined by METRO Group.

The main advantage for METRO Group is the reduction of ramp contacts in the branches because the trucks are better utilized and more precisely timed. Though production fluctuations from the suppliers now have a stronger effect on the whole transport organization.

The procurement logistics system introduced has by now been implemented in many countries where METRO Group does business. It has, however, always been adapted to the country-specific economic-geographical and logistical characteristics. Besides procurement logistics, METRO Group Logistics offers further services regarding sales brands and product ranges as well as cross-border services.

8.4.4 Transport Logistics Concepts

Numerous transport logistics concepts have developed from the different structures of transport nets. In this context, the freight village and city logistics will be introduced.

Freight villages are transport logistics vertices on which logistics service providers are present. At least carriers (road/rail/waterway/airport-multi modality), an interchange in the form of handling points for short and long distance runs (trans-regionality), as well as diverse logistics service providers and other pre- or post-transport service providers (multifunctionality) should be parts of a freight village. Aims are transport-logistical advantages to realize rationalization potentials, e.g. by bundling transports and by shared use of information and communication systems as well as databases. Ecological goals are also possible, e.g. the prevention of high traffic volume by transport bundling and preventing empty runs. Furthermore, it is an advantage for regional economic policy, e.g. securing jobs, supporting small and medium-sized businesses, and a balanced spatial structure.

Location requirements for freight village are, among others:[12]

- Position within the region
- Frequency/main direction of goods suitable for freight villages
- Space requirements of 60–200 ha
- Infrastructural connection/features
- Highway connection
- Main train line
- Inland waterway connection (optional)
- Terminal for combined transport
- Train terminal for combined transport
- Shipping port

The service providers in the freight village should at least offer the functions inspection of goods, storing, order handling, commissioning, confectioning, mounting, packing, route planning, disposition of transports, and shipment tracking.

Despite the advantages, there are large difficulties in establishing freight villages. Although around 40 freight villages were planned in the 1990s in Germany, only a few have actually been realized (e.g. Bremen).[13] The reasons for this are:

- Carriers/forwarders/service providers are not willing to move after investing in a larger site
- Carriers fear modal split changes from road to rail
- Large space requirements make it difficult to find a suitable property
- Required road connections lead to high infrastructure costs
- Communities, railway and freight village operators experience financing restrictions

City logistics concepts try to solve the problems that come up with delivering to and collecting from inner city areas.[14] The city development is characterized by a strong imbalance of the traffic flow, i.e. a usually much larger flow of goods into the city than from it. The storage spaces in business have been reduced to a minimum what, in turn, necessitates daily deliveries, or even more. The structure of inner city traffic is mainly characterized by carrier and road freight transport as well as by cars and utility vehicles for quick delivery. Pedestrian zones and determined delivery times additionally restrict deliveries. The times during which retailers accept deliveries are highly restrictive which leads to unproductive waiting times of the vehicles. Increasing packaging disposal raises the need for transports in urban areas.

To solve these problems, numerous city logistics have been developed and partly realized in some cities. The central approach is the partial relief of the inner areas from transport traffic by bundling the delivery and collection runs and by reducing

[12] Cf. Aberle (2009), p. 560 ff.

[13] Cf. Berg (2003), p. 418.

[14] Cf. Vahrenkamp (2005), p. 413 et seq.

the number of daily runs to the inner city. The utilization of the remaining runs is therefore increased and the inner city congestion intensity is reduced. Further considerations are the use of public transport (trains and subways) for inner-city transport of goods and the realization of city terminals for carriers, directly located on the edge of the city.

8.5 Outsourcing, Tendering and Placing of Transport and Logistics Services

Cost effects are the most common argument for logistics outsourcing. Costs are reduced by optimizing – after the logistics service provider has reached a high average utilization, e.g. sinking costs per item with a growing transport volume or economies of scale in warehousing. High fixed costs caused by capacities targeted towards high peak loads can be made variable by outsourcing. Labor costs can be reduced by sector arbitrage, e.g. because of more favorable wage agreements for service providers. Furthermore, the reduction of opportunity costs should be realized so that companies can concentrate on their core business and not on logistics. In the latter case, it is possible to calculate a lost profit in the form of opportunity costs.

Besides positive effects on costs, *performance increases* can also be a part of logistics outsourcing.[15] These result from the increase in efficiency by specialization because of the service provider's know-how and the employment of possibly more advanced technology. However, higher degrees of outsourcing do not automatically lead to more success but rely on *planning* and *realization*. The process of successful outsourcing of logistics services can be divided into six phases:[16]

1. Creation of a full company logistics concept
2. Definition and quantification of the required services
3. Developing and adopting the placing policy
4. Realization of the tendering
5. Evaluation of the offers
6. Assessment of the services and costs

In the full company logistics conception, the nature and the quantity of the required services, the logistics structures, and the limits of the own logistics networks have to be defined. Rules have to be implemented in the form of benchmarks e.g. for transport costs, costs and prices of internal logistics services, estimated investments and operating costs for logistics centers. For the *definition and quantification of the service requirements*, the required service quantities, rules for the realization of the services and payments related to the areas in which the service was realized, must be defined. The goals of the outsourcing must be defined with the *developing and adopting of a placement policy*. The core competences of

[15] Cf. Wildemann (2007), p. 134 et seq.

[16] Cf. Gudehus (2010), p. 987.

the logistics services must be defined and, subsequently, which services must still be realized internally. Furthermore, the limits of the outsourcing must be defined.

The *realization of the tendering* should consist of the following: A short description of the service requirements, a table with the required service quantities, a price table to be filled by the service provider, and the general purchasing conditions for logistics services. Tendering documents for warehouse and system services should include the following:[17]

- Specification of the required functions and services
- Definition and identification of clear process and service chains
- Definition of service and quality requirements
- Statement about the required service quantity
- Requirements for information and communications systems
- Performance recording and payment

The *assessment of the offers* can be made according to formal, functional, and commercial criteria:

- Formal criteria are, among others:
  - Quality of the offer (Completeness, presentation)
  - Price tables completely filled
  - Offer documents signed
- Functional criteria rate e.g. the
  - Concept solution: The offered solution fulfills all functions and requirements
  - Functionality: Method of the realization (technology, equipment etc.), capa bility of the IT competence
  - Service realization: Availability of sufficient capacities or resources, qualifications and experience of the staff
  - Relevant and suitable references

Commercial criteria help to rate the following:

- Prices for the different services, including reference objects, fixed costs, vari able cost rates etc.
- Yearly costs resulting from the planned quantities and services costs
- Payment modalities like payment deadlines and escalator clauses
- Liability/warranty: Regulating the amount and the length of the warranty agreements and defining the liability sum and the penalty for faulty services

For performance recording and payment, agreements must be made about the planning period and the payment periods as well as quality deficits and penalties. In detail, this means agreements about the utilization risk, e.g. by fixed payments, agreements about price adjustments (escalator clauses) and regulations about

[17] Cf. Gudehus (2010), p. 1005 et seq.

discounts in the form of logistics discounts (reference values, e.g. full packaging units, full pallets with one kind of article, long lead times) and quantity discounts (depending on the number of loading units, transports etc.). The tendering is completed by including *performance measurement* by performance monitoring using fault statistics, customer complaints and quality reports as well as an active outsourcing controlling to monitor the services offered and the price development on the market for logistics services (see Chap. 10).

Besides the planning and realization, the ability and willingness for cooperation between the outsourcing company and the logistics service provider are becoming more and more important as a *success factor*. This may become visible in open communication, mutual trust, and collaborative planning.[18]

Case Study 8.5: Distribution Warehouse Structure Planning

The goal of this case study is to apply the information from Chaps. 6, 7, and 8 to a simplified planning task.

The basic questions for a network planner are the following:

- How many warehouses?
- In which locations?
- What types of warehouses?
- With which capacities?

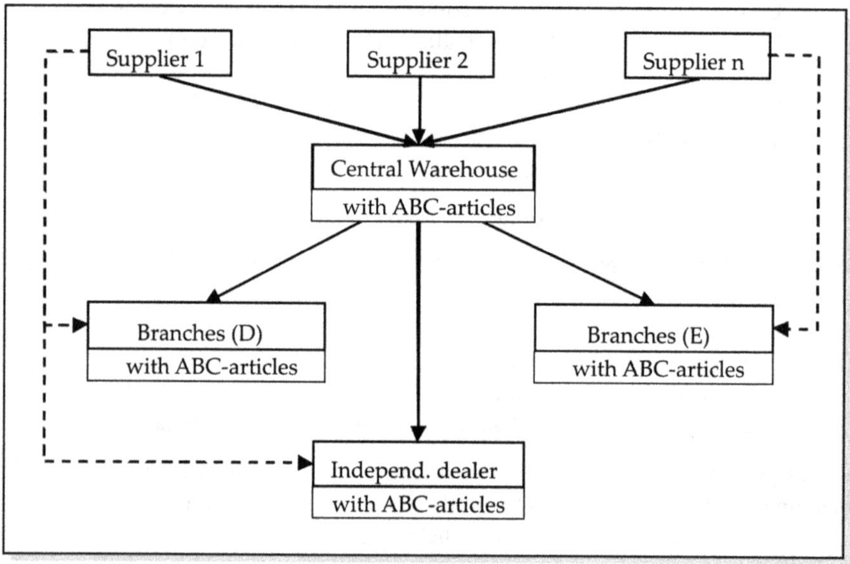

Fig. 8.14 Existing distribution structure

[18] Cf. Bretzke (2007), p. 176 et seq.

- Which production sites deliver to which warehouses?
- Which customers are supplied to by which warehouses?
- Which means of transport are employed?
- Which product quantities are produced in which production sites?

The answering of these questions is aimed at one basic goal: To ensure a delivery service which is in line with the market and, at the same time, by minimum logistics costs.

A furniture store chain is reviewing its existing distribution structure. At the moment, 143 suppliers deliver their goods to the central warehouse. The central warehouse near Kassel, Germany, carries the full product range (A-, B-, and C-class items); the same is true for the branches and the connected independent dealers. Direct delivery to the German (D) branches is realized by four suppliers and with around 30 products. Direct delivery in Spain (E) is realized by four suppliers, with around 520 products (see Fig. 8.14).

Now you are asked to outline alternative network structures!

Review Questions

1. What is a logistics net and what structures may it have?
2. What is meant by *degree of centralization*?
3. Give concepts of transport consolidation.
4. What transport-logistical concepts do you know?
5. Explain cross docking.
6. What forms of distribution sites do you know?
7. Describe changes in the terms of delivery?
8. Name advantages of hub-and-spoke nets?
9. Design a net for the flow of goods for the food consumer goods sector.
10. What are important factors in the capacity planning of a distribution center?

Additional Literature

Aberle, G. (2009). *Transportwirtschaft: Einzelwirtschaftliche und gesamtwirtschaftliche Grundlagen* (5th ed.). München/Wien: Oldenbourg. 2009.

Aden, D. (2001). Neue Chancen für Logistik-Dienstleister durch die Globalisierung. In: Baumgarten, H. (Ed.) *Logistik im E-Zeitalter. Die Welt der globalen Logistiknetzwerke*. Frankfurt.

Ballou, R. H. (1995). Logistics network design: Modeling and informational considerations. *International Journal of Logistics Management, 6*(2), 39–54.

Baumgarten, H. (Ed.). (2001). *Logistik im E-Zeitalter. Die Welt der globalen Logistiknetzwerke*. Frankfurt.

Baumgarten, H., & Thoms, F. (2002). *Trends und Strategien in der Logistik: Supply Chains im Wandel*. Berlin: Verbum.

Baumgarten, H. et al. (Ed.). (2003). *Springer experten system logistik-management*, 1. Berlin.

Bell, M. G. H., & Iida, Y. (1997). *Transportation network analysis*. Chichester: Wiley.

Blom, F., & Harlander, N. A. (2003). *Logistik-Management: Der Aufbau ganzheitlicher Logistikketten in Theorie und Praxis*. Renningen: Expert-Verl.

Francis, R. L., McGinnis, L. F., & White, J. A. (1992). *Facility layout and location: An analytical approach* (2nd ed.). Englewood Cliffs: Prentice-Hall.

Gudehus, T. (2010). *Logistik*. Berlin: Springer.

Ihde, G. B. (2001). *Transport, Verkehr, Logistik*, Teil 3. München: Vahlen.

Janz, O. (2003). *Integriertes Transportnetzmanagement: Angebots- und nachfrageorientierte Planung und Steuerung komplexer Transportnetze*. Köln: Eul Verlag.

Maltz, A. B., & Ellram, L. M. (1997). Total cost relationship: An analytical framework for the logistics outsourcing decision. *Journal of Business Logistics, 18*(1), 45–66.

Pfohl, H.-C. (2004). *Netzkompetenz in Supply Chains. Grundlagen und Umsetzung*. Wiesbaden: Gabler.

Powers, R. F. (1989). Optimization models for logistics decisions. *Journal of Business Logistics, 10*(1), 106–121.

Schulte, Chr. (2009). *Logistik*, 5. Edition, Kapitel 8. München: Vahlen.

Stieglitz, A. (1999). *Die Reorganisation handelslogistischer Versorgungsketten*. München: Huss.

Vahrenkamp, R. (2007). *Logistik: Management und Strategien*. München: Oldenbourg.

IT in Logistics 9

Learning Objectives

For the realization of cross-company supply chains, the informational linking of departments within the company and of cooperating companies is an important means. Furthermore, the internal logistics processes must be supported by adequate information and communications systems. There are strong relationships amongst these systems and the logistics functions. The goal of this chapter is to introduce the many areas of application for identification, information and communications systems, and their potential use in logistics systems. Additionally, the respective effects on logistical performance, quality, and costs will be explained.

Keywords

- Information requirements in logistics and supply chain
- IT standards in business
- Flows of information and their effects
- Electronis Data Interchange (EDI) principle
- Identification systems – barcode and RFID
- Stock management systems
- Planning of transports and runs
- Enterprise Resource Planning (ERP)
- E-logistics

H. Gleissner and J.C. Femerling, *Logistics*, Springer Texts in Business and Economics,
DOI 10.1007/978-3-319-01769-3_9,

9.1 Electronic Data Transmission

9.1.1 Communication Standards

The complexity of logistical processes and the high number of parties involved require strong efforts for information and communication as well as for controlling and documentation. To control the logistics processes, diverse information with mostly large data volumes are necessary. As an example for a shipment these are: information about the addresses of sender, receiver, shipping company, and carrier; information about the goods like weight, volume, size, special shipping and handling requirements regarding cooling, fragility, preservability etc.; and information about the price, the date of the shipment, and the time of the delivery. Traditionally, this information can be communicated orally or in written form and be attached to the shipment; it can also be sent prior to the shipment by fax.

However, a manual or paper-based process alone is not sufficient anymore for a fast information supply that is ahead of physical events. To realize logistics processes in the best way possible, it is necessary that e.g. information about the delivery of a shipment is available way before the actual delivery for the recipient and other parties like e.g. the carrier. This is made possible by electronic data transmission technology. The same is true for realizing concepts with streamlined process and organizational structures in logistics and in the company as a whole.

For this, the data should be available in standardized formats which allows for all involved parties in the process chain to read and process the data without further transmission, so-called media disruptions. This means a change of transmission medium or data carrier in the information chain, e.g. transferring information from an inventory card (paper) or a delivery note (paper) into an IT system or vice versa. The problem with such media disruptions is the possibility of errors in the process. Conversion in this context means e.g. the harmonization of the data-technological sentence structure (syntax) with respect to the denomination, numerical characteristic, sequence and length of fields in data sets. If media disruptions or conversion processes are avoided during data transmission, information is not only available faster for status requests or for adding controlling commands but the overall transaction costs can be reduced in the process.

Transaction costs are defined as the costs necessary to initiate, coordinate and realize the transfer of goods from the original seller to the consumer.[1]

Caused by the requirements of worldwide communication and the necessity to rationalize internal processes, efficient concepts for managing the data intensity and data volume have been developed – also, and especially, in logistics.

This is often called information logistics. It must fulfill the following requirements:

- Avoidance of media disruptions (orally – written – electronic data format)
- Reduction of conversion efforts (intervention-free data transmission)

[1] Cf. Gleißner (2000), p. 21 ff.

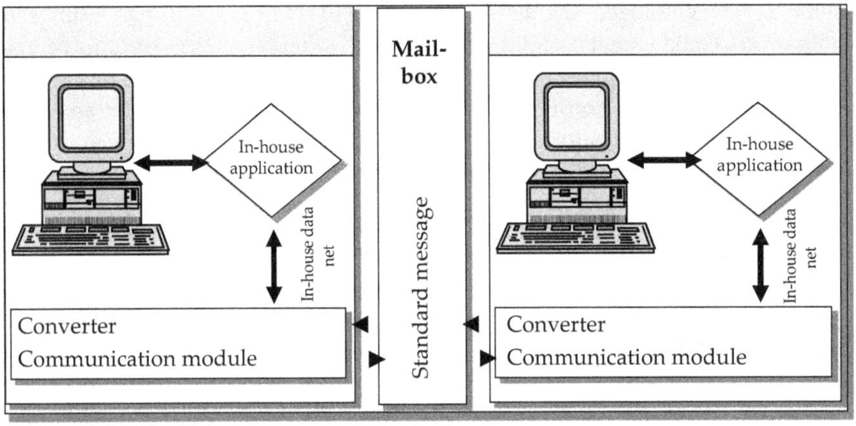

Fig. 9.1 The EDI principle (Cf. GS1 Germany (2006a), p. 5)

- Avoidance of data redundancies
- Data availability before, during, and after the physical logistics process
- Availability of information when needed, with respect to time and demand
- Linking to existing internal information and communications systems as well as interfaces to the Internet
- Trans-company and trans-national communication standards.

Electronic Data Interchange (EDI) is the automated and intervention-free exchange of structured data between business partners according to uniform standards. For this, the data is transmitted from a data processing application of the sender to one of the receiver without personal interferences or modifications (see Fig. 9.1). Structured data transmitted via EDI is characterized by a precise definition of their composition. The information to be exchanged must be determined with respect to the syntax and to the semantics to make possible the communication between the partners. The syntax describes the order of the signs and sign-compositions in a message. The semantics, in this instance, describes the meaning and the content of a sequence of signs.

With this, the preconditions for automatic processing of the exchanged data without further delay are created. From a technical point of view, EDI is a controlled, so-called file transfer between data processing systems.

Different standards are available to create syntax and semantics, i.e. a unified *language* between the communication partners. These standards can generally be divided into open and proprietary standards. Open standards can – with the exception of limitations by the sector or country they are used in – basically be used in every company. Proprietary standards, on the contrary, are only for the data exchange between a company and its business partners.

The advantage of proprietary standards to be tailored exactly to the requirements of the partners, therefore being effective and independent compared with standard solutions, is neutralized by the disadvantage of their limited possibilities of application. Trade companies are facing special difficulties; they have to be able to work with and adapt to different manufacturers' standards. For this reason, open standards can be considered far more important. They are more and more facilitating the growing information network of all levels of the economy.

The following outlines the most important developments in open EDI standards:

ANSI X.12: Based on standardization efforts begun in the USA in the 1960s, this trans-sector and –function standard was developed from 1978 on, coordinated by the American National Standards Institute ANSI. Today, ANSI X.12 is the most important American EDI protocol standard. It is also available as an EDIFACT subset.[2]

VDA: Since 1978, there had been recommendations by the German Automobile Industry Association VDA for the data exchange between automobile manufacturers and the supplying industry which have by now been extended by the data exchange with logistics service providers. Developed by a VDA working group, these recommendations are a sector-specific standard. The documentation is available directly from the VDA as the central coordinator. With this, the automotive industry was a pioneer of the systematical development and application of standards for electronic data exchange.

ODETTE: An international sector-specific solution for the automobile industry and their suppliers, ODETTE (Organization for Data Exchange by Tele Transmission in Europe) is a European further development of earlier VDA standards. It has by now been made into an EDIFACT subset as well.

SEDAS: SEDAS (*Standardregelungen einheitlicher Datenaustauschsysteme* – Standard Regulations of Uniform Data Exchange Systems) was developed by a German association for rationalization, GS1.[3] It was the first national protocol standard for data exchange in consumer goods trading, including the manufacturers. Orders, invoices, market research data, as well as reference data like product and customer information are to be transmitted using the SEDAS format. However, national sector-specific standards are limitations, so despite its widespread use, SEDAS will successively be replaced by EANCOM®.

EDIFACT: To eliminate basic limitations by national or sector-specific standards, the United Nations' Economic Commission for Europe (UN/ECE) began to

[2] EDIFACT subsets are sector-specific specifications based on the general framework of EDIFACT.

[3] GS1 (formerly CCG) is a German rationalization association in Cologne, provided equally by institutions of trade and industry. The best known regulations created under coordination of GS1 are agreements concerning international numeral systems (e.g. EAN and corresponding barcodes), electronic data exchange as described above (e.g. SEDAS, EANCOM) or, in logistics, multi-way transport systems and pallets. For some time now, activities have been focused on the development of EPC (Electronic Product Code) for use in the field of RFID. Furthermore, GS1 has been supporting ECR (Efficient Consumer Response) and CPFR (Collaborative Planning, Forecasting and Replenishment) concepts.

Table 9.1 EDIFACT message types and their areas of use

Trade/ industry	Inquiry, offer, order change, order confirmation, order, just-in-time delivery, account statement, delivery schedule, despatch advice, partner reference data, price catalogue, invoice
Transport	Note of receipt, booking request, booking confirmation, fixed booking, forwarding and transport message frame, loading list, transport/forwarding order, transport data reporting/status report
Customs	Debit note, expanded credit note, expended payment order, credit note, payment order, payment advice
Banks	Customs duties, customs declaration, customs notice
Insurances	Open account, reinsurance

initiate the EDIFACT (Electronic Data Interchange for Administration, Commerce and Transport) standard in 1987. Widespread by now, this regulation is fit for use in all sectors of the economy and in public administration. Under EDIFACT, 40 sector- and user-specific subsets have developed by now.

All in all, there are around 200 different message types in use today. The most important subset derived from EDIFACT is EANCOM®-2002. An overview across the most important message types can be found in Table 9.1.

In the past, important characteristics of the different EDI standards were their range in the sense of their geographical applicability and the degree of their universality or their sector-specification. The internalization of business relationships, however, led to a loss of importance of standards that had been purely bilateral, national or sector-specific. (e.g. VDA or ODETTE for the German or European automobile industry).

Based on EDIFACT, the International Article Numbering Association (EAN) has conceptualized the so-called *EANCOM®* communication standard. It is designed for the global electronic information exchange for suppliers (manufacturers), investors (banks), service providers (forwarders), and customers (trade). Some important EANCOM® message types are collected in Fig. 9.2.

EDI has expanded rapidly in the last years. According to data from the European Economic Institute (EWI) there were more than 500,000 users by the year 2000.

At the same time, it can be assumed that EDI is still far from universal presence in day-to-day business and that it will still need some time for this. This is especially surprising with the data processing and transmission technology having reached a high functionality standard, with problems connected to practical EDI application solved or, respectively, solvable.[4]

[4] Cf. Weid (1995), p. 23.

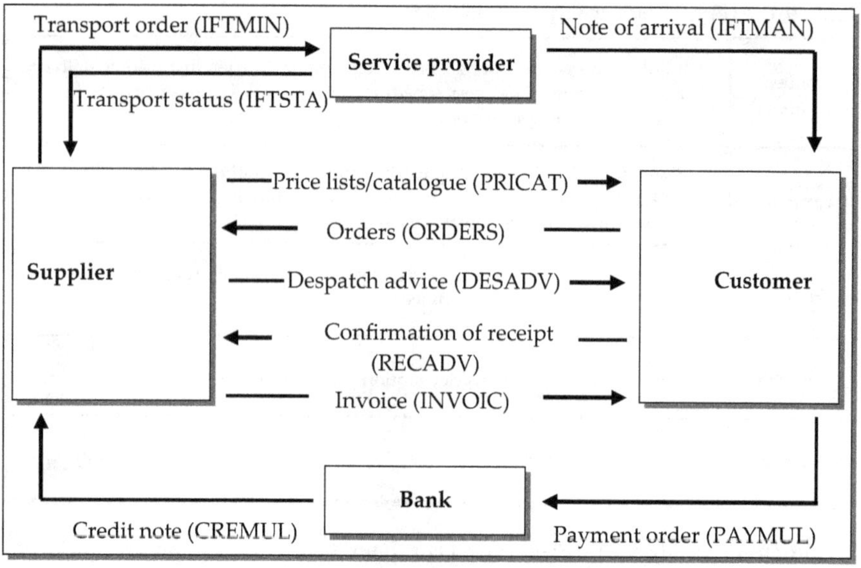

Fig. 9.2 Information flow in logistics with EANCOM® standards (Cf. GS1 Germany (2006a), p. 7)

The lack of wider expansion of EDI is caused partly by subjective assessment by the management regarding the advantages of its application. Often, costs and other disadvantages like concerns about unauthorized access are reasons against the introduction of EDI. Especially small and medium-sized companies are worried about the costs. For this target group, *web*-EDI solutions have been developed by now. They require only small investments in hard and software and only limited specialized EDI system know-how. With this, it is possible to profit from EDI advantages with a manageable additional manual effort for small data volumes.

Table 9.2 shows and contrasts positive and negative effects of EDI application.[5] Besides the costs for introduction and operation and the resulting saving potentials, there are indirect costs and saving potentials that are difficult to measure. Furthermore, the arguments against EDI application as well as the operative and strategic advantages in competition resulting from EDI application must be assessed as qualitative factors. Gaining a competitive advantage usually does not result from EDI application alone but from realizing a full strategy like ECR or CFPR that has EDI as an elementary prerequisite.

As a result of the spread of Internet and the available e-commerce solutions (Internet data exchange, e-procurement, Internet market places etc.), the web based XML standard has become very important.

One can assume that EDIFACT and EANCOM® will remain important for electronic data exchange because of their presence and standardization. By

[5] Cf. Weid (1995), p. 59 et seq.

Table 9.2 Advantages and disadvantages of EDI
(Cf. Weid (1995), p. 129 et seq. and Gleißner (2000), S. 133 et seq)

EDI employment	
Negative effects	**Positive effects**
Implementation costs	**Cost reductions** (Staff, equipment and capital commitment costs)
▪ Collection of information ▪ External EDI consulting costs ▪ Coordination with future communication partners ▪ Staff training ▪ Hard-and software for data processing (computer etc.) and communication (modem etc.) ▪ Software adaption/development for existing data processingsystems ▪ Costs for initial connection to VAN services and systems (ISDN etc.) ▪ Pilot or parallel operation in the introduction phase	▪ Handling of documents ▪ Recording incoming documents ▪ Print and copy costs for outgoing documents ▪ Transmission costs (postage, fax) ▪ Reduction of the stock with shorter information flow (shorter time for replenishment) ▪ Manual sorting, distributing, document storage ▪ Correction efforts in case of errors ▪ Content comparison of different data formats
Running costs	
▪ Software maintenance/development and error elimination ▪ License fees to central institutions ▪ Software maintenance ▪ Connection costs ▪ Staff training	
Other costs	**Saving potentials**
▪ Not exactly quantifiable	▪ Elimination of redundant jobs ▪ Simplification of process organization ▪ Reduction of the vulnerability to errors ▪ Flattening of hierarchies ▪ Moving decision-making competence to lower levels ▪ Integration and reintegration of processes ▪ Higher staff productivity (fewer routine jobs) ▪ Higher liquidity because of time advantages between invoicing and receipt of payment; possible lowering of costs of capital
Qualitative disadvantages	**Qualitative advantages or potential use**
▪ Data security hard to realize ▪ Documentation of EDI transactions ▪ Proof of complete and correct data receipt ▪ Unauthorized access (internal/external hackers) ▪ Insufficient reference data quality (products, addresses); missing management responsibility ▪ Relying of technical DP infrastructure ▪ Creation of barriers to entering/leaving a market with IT	▪ Using time potentials by shortening the times for delivery, order processing, and handling ▪ Increase in flexibility and efficiency in process organization ▪ Increase in employee satisfaction ▪ Higher transparency of business processes, improvement of information management ▪ Increase in the available reaction time and quality in the logistics chain by earlier and more exact information availability

(continued)

Table 9.2 (Advantages and disadvantages of EDI (Cf. Weid (1995), p. 129 et seq. and Gleißner (2000), S. 133 et seq) (continued)

▪ Resistance against the introduction of innovative technologies and processes ▪ Abandonment of person-specific know-how ▪ Insecurity about future transmission standards ▪ Break even point is not reached in smaller companies because of small data volume or number of transactions ▪ Pressure from larger trade partner forces to introduce EDI ▪ High introduction costs and a long time until they pay off ▪ Missing organizational requirements and willingness ▪ Creation of dependences on cooperation partners that are hard to correct ▪ Legal problems: –Formal requirements for closing of contract Inclusion of general business terms –Storing requirements for EDI documents	▪ Facilitation of global business activity ▪ Basis requirement for realizing ECR strategies ▪ General strategic advantages to competition without EDI

integrating XML, it will in the future be possible to significantly increase the number of companies connected to electronic data exchange.[6]

Table 9.3 shows a comparison of the two data exchange formats.

Table 9.3 Comparison data exchange formats XML and EDIFACT/EANCOM (GS1 Germany (2009a), p. 75)

XML	EDIFACT / EANCOM
(Meta-) markup language	EDI standard
W3C recommendation since 1998	ISO Norm 1998
Fixed syntax/flexible semantics	Fixed syntax/ fixed semantics
Machine-to-machine communication Human-to-machine communication	Machine-to-machine communication
Flexible data structuring by DTDs or schemes	200 message types at the moment
Flexible data layout with XSL	No specification for data layouts
Few experiences from EDI application	Sophisticated technical equipment necessary

[6] GS1 Germany (no year).

9.1.2 Identification Standards

Besides standardizing the communication between business partners, standardizing identification systems is also important. This standardization is necessary to connect the information flow to the physical flow of goods. This connection is seen as a labeling of the goods and transport units that can be read by all parties involved in the process chain. The most common identification standards will be explained in the following.

The *Global Location Number* (GLN) makes possible the identification of companies or company departments. This manufacturer number is a part of the EAN number and is assigned centrally by the national standardization organization.

The *Global Trade Item Number* (GTIN) enables the clear and global identification of products, sales and trade units as well as services. The number usually has 13 digits and describes the product and its specifications like color, size, weight, packaging unit etc. It consists of a base number identifying the manufacturer, an individual item number, and one check digit. The GTIN number is the key to access reference data (designation, weight, volume, class of goods etc.) in databases or the price of the article. The manufacturer assigns the item number. Case study 8.1 shows the fundamental data transfer upon scanning a GTIN number as a barcode at the till.

With the space for a 13-digit number not available on small products, there is also a shorter, 8-digit, GTIN number for such cases.

In North America, the 12-digit UPC (Universal Product Code) is used. It is fully compatible with the GTIN number.

Case Study 9.1: Information Flows in Distribution of Goods

At the till in the supermarket, the barcode of the wine bottle is moved across the reading area of the scanner; the GTIN number is read. This unique number is combined with the reference data (designation, weight, trading unit size, class of goods etc.) in the product-lookup (PLU) file. There it is combined with the stored price information. The item is shown in the till and the sales slip is printed out. At the same time, the number is passed on to the database of the supermarket or the trading company and stored there. It is then removed from the local inventory in the inventory control system. Then the data is further processed in the inventory control system of the trading company. After accumulating more sales of the same item, a delivery from the central warehouse of the company is triggered. These inventory changes in the central warehouse can also be aggregated. They are either transferred to orders from the suppliers or passed on as stock information to the manufacturer. The manufacturers or suppliers can then use the information for their own planning process; they may supply new items to the trading company or initiate new rounds of production (see Fig. 9.3).

The *Serial Shipping Container Code (SSCC)* is used to clearly identify shipping units on their way from the sender to the recipient. This makes possible

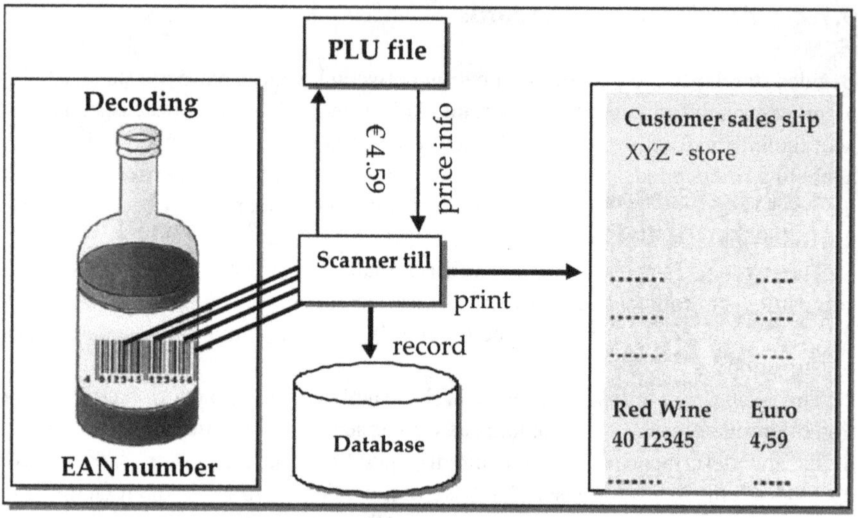

Fig. 9.3 Information flow in the distribution of goods with EANCOM® standards (GS1 Germany (2009b), p. 7)

the tracking of a logistical packaging unit or a carrier across the full transport process. Fast location tracking of the transport unit (Tracking & Tracing) is possible. The SSCC can also be used for callbacks and for quality management to check the punctuality of the delivery or for damages to the goods. However, the SSCC is also only a key which grants access to the necessary data of a shipping unit (sender, recipient, forwarder, weight, volume etc.) from connected databases.

The *Global Returnable Asset Identifier (GRAI)* is a special variant. It is used to identify reusable packaging and transport aids (pallets, barrels, boxes etc.). This facilitates the recognition, tracking, and inventory keeping of reusable transport means, the value of which has a considerable weight in logistics processes.

With the introduction of RFID technology, it has become necessary to ensure the international readability of transponder information. To achieve this, the *Electronic Product Code (EPC)* has been developed (see Sect. 9.2).

9.2 Identification Technology

To enable the quick and easy reading and transmission of the described standardized data formats, technically standardized identification systems are required. The goal is to transfer the tasks of recognition and identification from humans to automated technical systems that work with very high or full accuracy. It is necessary to acquire data quickly and accurately to pass on the gathered information, process it with centrally stored data and to continuously generate

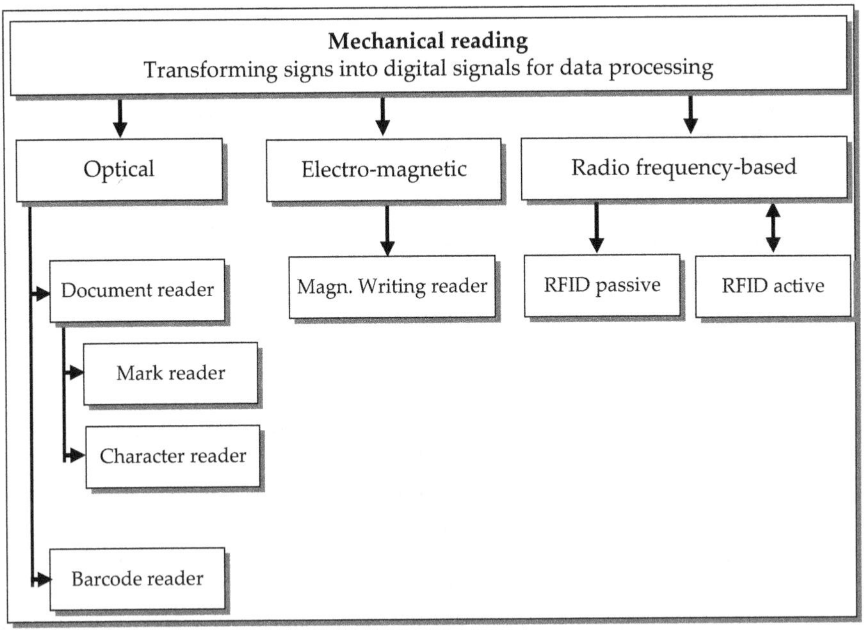

Fig. 9.4 Mechanical reading in automatic data acquisition

controlling information for the logistical process from there. In turn, the processed data should be available for all parties in the logistics chain in as little time as possible. Besides the quick availability of controlling information, automatic data acquisition has the advantage of rendering manual work like the recording of shipment and status information on paper lists unnecessary, thus making it possible to reduce staff costs.

There is a multitude of technical possibilities for the recording of goods today (see Fig. 9.4). The easiest way is to identify the item by measuring its weight or volume. Using information carriers is technically more challenging. Possibilities include the mechanic recording using contacts attached to the item, and the reading of magnetic stripes or opto-electronic information carriers like character, barcode or QR-code scanning. The most advanced recording method is the communication with electronic or electro-magnetic data carriers in the form of fixed or programmable electronic memory chips like RFID.

Besides mark readers, the easiest way for automatic identification in logistics is optical character reading (OCR) with the so-called OCR code (see Fig. 9.5). Reading devices are able to read labels attached to the items with a corresponding

Fig. 9.5 Characters for optical recognition

Fig. 9.6 EAN barcodes for individual items

combination of numbers. However, the data volume transferrable with this method is limited.

Using a *barcode,* the data volume and the reading rate can be increased. The barcode consists of a combination of vertical bars that can have different lengths, depending on the standardized data format. The barcode is read by infrared reading devices (scanners or cameras) to record the width of the bars and the distance between them. By using more scanners that are mounted in different positions, barcodes can be also read when they are uneven, attached to the sides or are only attached loosely (luggage on flights). To increase the reading rate, a blank area must surround the barcode. At the end of the combination of bars, there is a check number. To make the information readable for humans as well, they are often printed beneath the barcode in standard characters.

Today, the barcode in its different variants is the most common and the most standardized auto-identification technology in use.

For product identification, mostly the EAN-8 or the EAN-13 barcodes are used (see Fig. 9.6).

With its larger information volume, the EAN-128 barcode is mostly used in logistical areas such as material flow in production, in warehouse processes as well as for delivery control and shipment tracking in the distribution process. Furthermore, it is important for the tracing of batches or the expiry date in chemical, pharmaceutical, and food industries.

In contrast to the shorter product codes EAN-8 or −13, the EAN-128 contains the SSCC and the EAN as well as different additional characteristics that facilitate the reading and prevent misreading (see Fig. 9.7). A two- to three-digit application identifier before the actual number defines the information stored in the following character sequence. This prevents confusion and makes possible complex data processing processes on the downstream databases.

The EAN-128 code is often used with SSCC content for logistical transport processes (see Fig. 9.8).

Besides the one-dimensional barcode described above, more and more two-dimensional barcodes are used. For these so called QR-codes, information can additionally be generated from the horizontal arrangement of lines, dots and areas

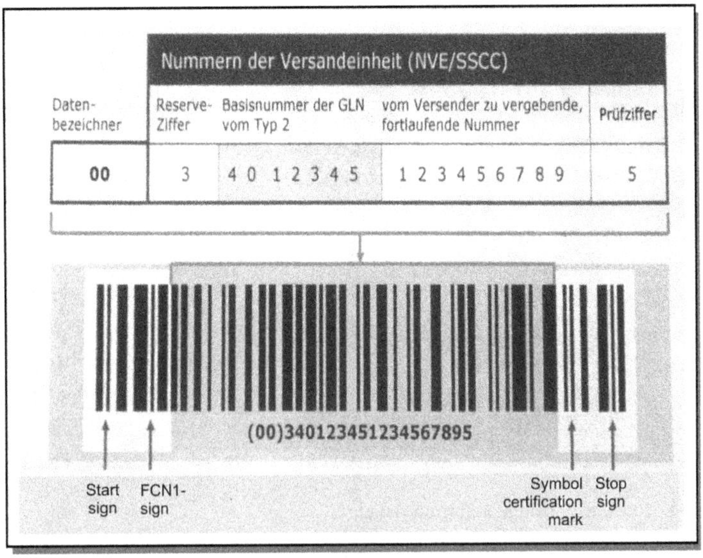

Fig. 9.7 EAN-128 as SSCC (Cf. GS1 Germany (2009a), p. 27)

Fig. 9.8 EAN-128 transport label – concrete example of SSCC (Cf. GS1 Germany (2009a), p. 25)

as well as from their distances to each other. They are more and more used in the fields of postage and address information, online tickets (see Fig. 9.9) and a wide utilization range with mobile devices like smart phones.

Fig. 9.9 2-D codes for postage and train tickets

Table 9.4 Advantages and disadvantages of barcodes

Advantages	Disadvantages
Small costs per unit/label	Sensitive against dirt, moisture, mechanical impact
Standardized reading units in use	Reading only possible visually
Standard widely used in logistics	Limited data volume
	No data addition possible in the process

The barcode is still the most common standard solution for identifying an object in the logistics chain. Its advantages and disadvantages are shown in Table 9.4.

RFID technology, which has been technically available for a long time already, used for military friend-foe identification and later for anti-theft devices in shops, will become more widespread in logistics as well. The most important argument for this is its significant performance increase compared to the barcode.

Its principal functioning is shown in Fig. 9.10.

So-called transponders[7] are attached to the goods, load carries, or containers. They consist of a microchip that stores the information, an antenna for sending and receiving, and housing. The housing may be a box or just consist of film, which is a special advantage for product identification for consumer goods. They are so flat

[7] Transponder is a combination of '*transmit*' and 'res*pond*'.

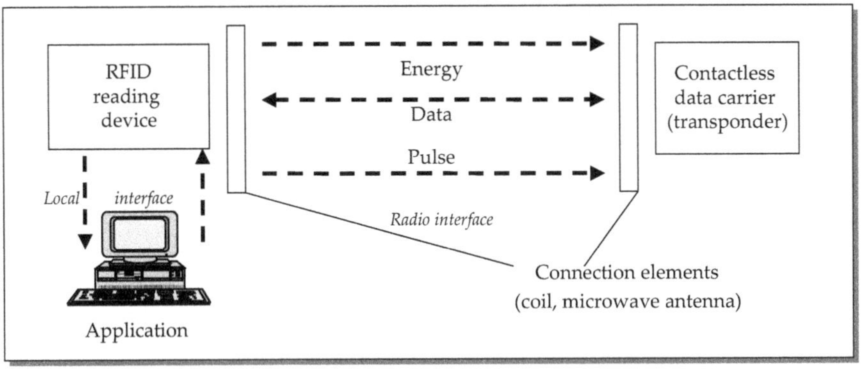

Fig. 9.10 Logic of the RFID data flow

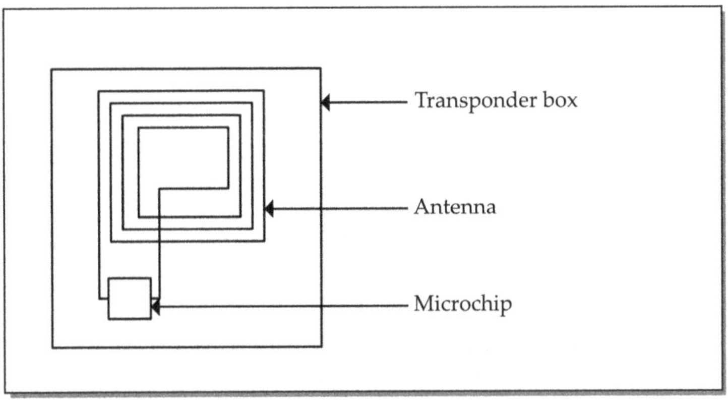

Fig. 9.11 Transponder (smart label) (Cf. GS1 Germany (2003), p. 13)

that they can be attached to the items as stickers (so-called smart labels; see Fig. 9.11).

With respect to the energy supply, transponders are divided into active and passive transponders. The active writing/reading device initiates the transmission or reading of the information on the transponder. Active systems use the energy from an integrated battery for controlling processes of the microchip and for independently sending information. Passive systems, on the other hand, receive their energy by induction – they drain energy from the (electro-) magnetic frequency field. The capacity of an RFID transponder can reach up to 64 KB. There are also models where the data can be complemented during the use or deleted afterwards to make room for new information.

The advantages and disadvantages of RFID technology are shown in Table 9.5.

Basically, there are two methods to store data on an RFID transponder. The first method is to directly store all relevant data of one article or a logistical unit on the transponder. This is a decentralized approach to data storing and, through this

Table 9.5 Advantages and disadvantages of RFID technology

Advantages	Disadvantages
Reusable	Sometimes high prices for transponders
Readable without direct visual contact and across large distances to the transponder (label or tag)	Potential disturbance of receipt or sending by metallic environment
High data volume storable	Higher data volume means more difficult control because of data protection
Writable with data during the logistics process	Uncontrolled operation criticized by data protection specialists
Relatively immune to dirt	Danger of data loss if exclusively stored externally on transponder
Wide range of data (e.g. temperature in the cool chain)	Processing of high data volumes necessary in downstream applications
Automation of processes	Complex technology
Localization of objects also within the warehouse	

independence from downstream databases, shortens the time for transmission and decision-making, creating an advantage. However, it requires high storing capacity of the transponder and bears a high risk for data loss.

The second method is to use the EAN numbering system, the same as for barcodes. To use the higher performance of the transponder compared to the barcode, the EAN codes can be integrated in the so-called *Electronic Product Code (EPC)*. In this area, EPCglobal – an initiative of the national standardization organizations – is working on defining an international EPD standard. Furthermore, a special EPCglobal™ server network is planned to organize the global information exchange of EPC numbers and corresponding reference and transaction data via the Internet.

9.3 Stock Management and Warehouse Control

9.3.1 Classification of IT in the Fields of Inventory and Warehousing

Due to their complexity, stock management and warehouse control are not imaginable without IT systems anymore. In the different areas, various operative application systems work together. They are generally divided into the four system levels that control the logistical process. The first and highest level is the *ERP system*. The stock management, the economic assessment of the inventory situation, the purchase controlling, and the initiating of reorders as well as the recording and management of customer orders take place here.

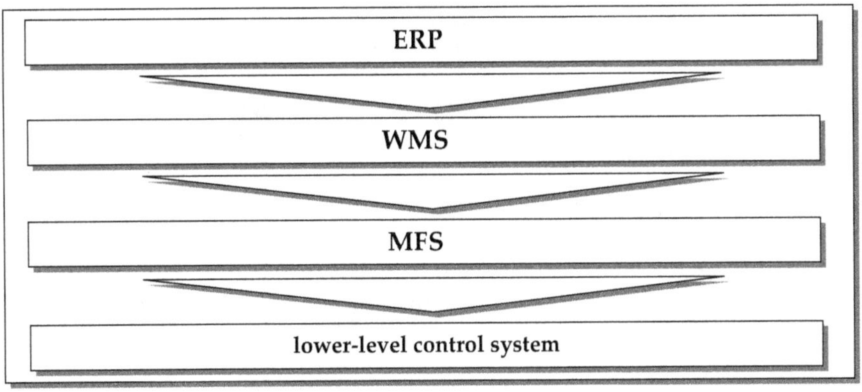

Fig. 9.12 Network of IT systems in the warehouse

The second level is the *warehouse management system (WMS)*. Its tasks are local stock management, the storage area management and assignment as well as the stocktaking and replenishment planning.

The *material flow system (MFS)* is the third level. It controls the internal transports, orders of goods coming into and going out of the warehouse as well as the individual procedures for replenishment, picking, and stocktaking.

The fourth, lowest and purely operative level is the *lower-level control* system by programmable logic control unit (PLC). Their elements control the handling technology and data transmission systems as well as peripheral devices like printers for receipts and barcodes.[8] The network of IT systems used in or bordering the warehouse is shown in Fig. 9.12.

9.3.2 ERP Systems

From the point of sale, the controlling and planning tasks of logistics processes are handled by so-called *Enterprise Resource Planning (ERP)* systems. Mainly in the trade sector, ERP systems considerably improve the information base for prognosis procedures with an item-based sales data recording and stock management in downstream, operative units. The sales forecast of a stock-managing point (warehouse) is not exclusively based on the own sales or delivery data anymore but can work with data from the demanding points (branches).

In detail, those are:

- Sold volume with respect to time and space
- Recording of inventory data
- Recording of goods leaving and entering
- Calculation of stock range and
- Assessment of delivery potential and presence of items

[8] Cf. Arnold et al. (2008), p. 4 ff.

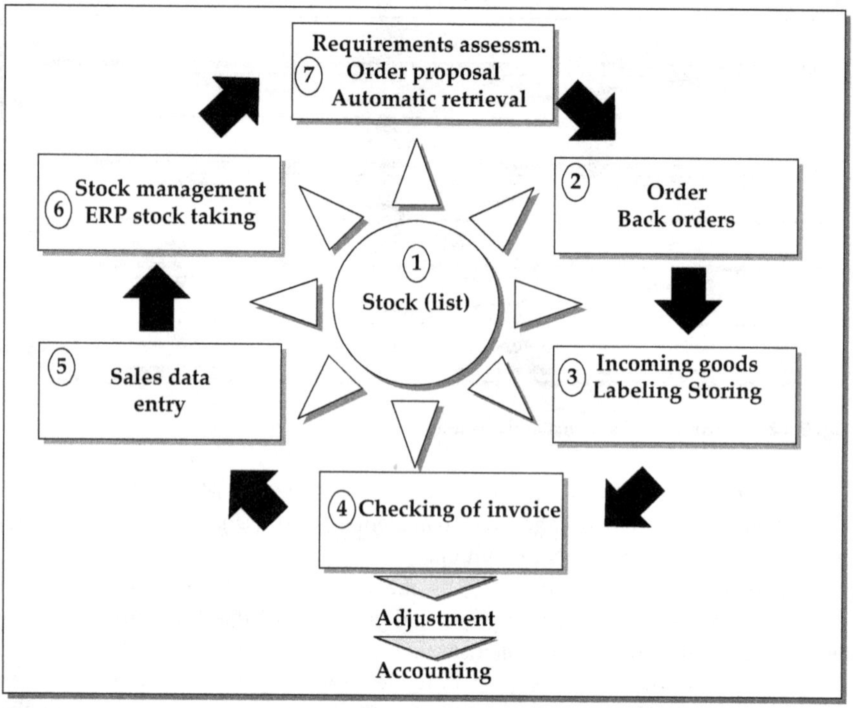

Fig. 9.13 Cycle of a closed ERP system (Cf. Hertel et al. (2005), p. 225)

The ERP system is based on the process logic shown in Fig. 9.13.

More exact information about stock, movement of goods, and current sales forecasts can reduce the level of uncertainty. Traditionally, this is achieved by storing goods, thus investing capital. The forecasts are the more exact the more they are based on the sales data of the end customers in the value chain. Prognosis methods can include many different influences, e.g. trends, seasonal factors, calendar-related factors, or marketing activities.

The utilization and exchange of information in ERP systems depend on the depth, i.e. the level of detail, of information that enters the system. The level of detail is divided into time-related (data from days, weeks, months), item-related (items and the item groups), and location-related (branch or site level of the locations). The information process is furthermore determined by the degree of automation of the data exchange and of the information exchange. From this, in turn, come the timeliness and quality of the information as well as the costs of the communication infrastructure.

With respect to the forms of ERP systems, four basic variants can be described, as shown in the following Fig. 9.14.

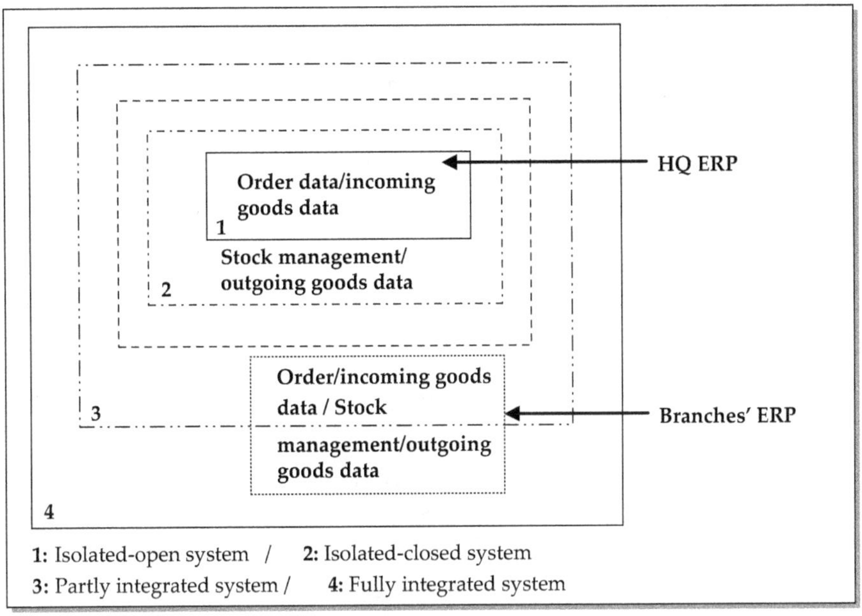

Fig. 9.14 Forms of ERP systems in trading companies (Cf. Hertel et al. (2005), p. 227)

As seen in the figure, in *isolated-open ERP systems*, only the order data or the data sets of the centralized incoming goods are recorded. Conclusions about realized sales and revenues can therefore only be made indirectly and after a delay. In *isolated-closed ERP systems*, not only incoming goods and order data is recorded but also the data from the centralized incoming goods of the branches. However, this is not connected to the order and stock data of the branches. *Partly integrated ERP systems* link central ERP systems to the order and incoming goods data of the decentralized branches system. *Fully integrated ERP systems* take this further. They also record information about the incoming goods and issue of goods that are connected via the individual systems. With this approach, information about the real stock and sales situation in the branches can be transmitted within the centralized system.[9]

9.3.3 Warehouse Management Systems

Systems for comprehensive warehouse controlling and stock management are called *Warehouse Management Systems (WMS)*. Their main tasks are recording goods received, assignment and management of storing areas, control of the relocating and picking processes, reorganization of the storing area occupancy,

[9] Cf. Hertel et al. (2005), p. 227.

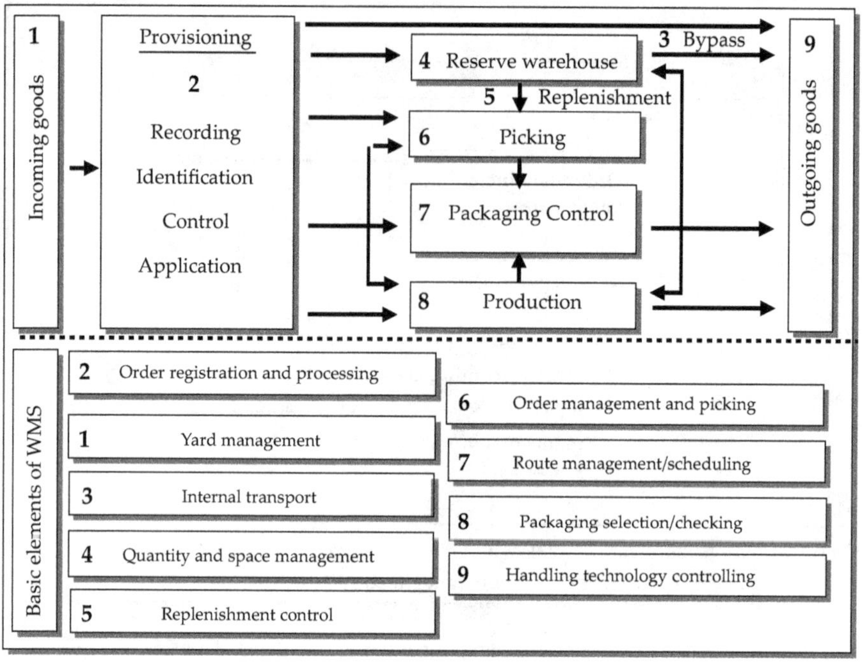

Fig. 9.15 Process-oriented view on warehouse management systems (Cf. Bode and Preuß (2004), p. 325)

replenishment control, recording outgoing goods, and stocktaking. The warehouse processes and the corresponding functionalities of a WMS are shown in Fig. 9.15. The allocation is made with the help of the corresponding numbers in the picture.

9.3.4 Material Flow Systems and Lower-Level Control Systems

Systems for material flow control provide for the smooth material flow in the warehouse. This includes the controlling of the storing, assigning of storing space, picking and replenishment controlling with respect to priority and/or efficiency. In this context, efficiency means short ways for storing and moving out, high storage space utilization, and short access times.

The *controls* or *lower-level control systems* (PLC – Programmable Logic Controller) are the *intelligent* units directly responsible for controlling the individual conveying systems in the warehouse. The lower-level control systems trigger the starting or stopping of the continuous handling systems, automated stacker cranes etc. and also initiate e.g. the dropping of goods from sorters on their destination. They are programmed based on a predefined set of rules (syntax) and are triggered by the material flow system. The connection between the WMS, the material flow control and the PLC is shown in Fig. 9.16.

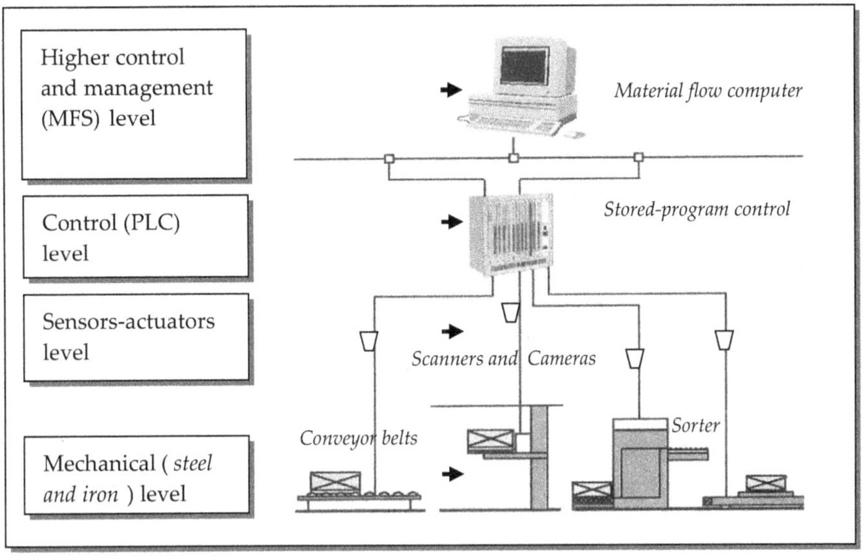

Fig. 9.16 IT structure in the warehouse (Cf. Caninenberg (2004), p. 34)

9.4 Transport Planning and Controlling

Transport systems can be divided into internal (within the company) and external transport systems (company to company or to consumer).

Internal transport systems are usually parts of production planning and control systems or of warehouse management systems, a detailed description of which shall not be given here.

Planning systems for external transports calculate the most favorable combination of means of transport and their utilization, using the best route and sequence of loading and unloading locations. One main task is to consolidate part loads on longer routes to achieve maximum utilization of the means of transport. Further tasks include the controlling of complete loads to optimize the routes and to avoid empty runs or include back loads in the planning (see Chap. 7). In road transport of goods, these planning systems are mostly used for complete, part load or groupage service.

Route planning or scheduling systems are IT applications that support transport fleet managers to efficiently plan routes for delivering or picking up goods. A number of customers, the demands and locations of which are known, are to receive a defined good from one deposit with a number of vehicles with certain capacities. The system gives suggestions about how the routes should be made in order to minimize total transport costs while adhering to certain secondary conditions (e.g. capacity and time restrictions).

The determinants of route planning are collected in Table 9.6.

Table 9.6 Determinants of route planning and scheduling

Distance determination	▪ Digital network method
Travel time calculation	▪ Digital road network ▪ Average speed ▪ Speed depending on the distance ▪ Speed depending on the streets ▪ Weather factors ▪ Traffic/road works
Loading/unloading/ waiting times	▪ Minimum standing time per customer (paper processing, maneuvering time) ▪ Time depending on the quantity ▪ Time depending on the customer
Other restrictions	▪ Driving and resting period regulations ▪ Regulations about working times for drivers and loading/unloading staff ▪ Different types of vehicles with different maximum loads; equipment necessary for delivery, e.g. hydraulic ramp, forklift) ▪ Empties/back loads ▪ Uncalculated waiting times at customer sites ▪ Driving ban on holidays, due to weather and in inner cities

Tactical route planning is concerned with standard or framework tours that take place at certain times (daily, weekly) and have a more or less defined order or deliver to a defined area.

Operative route planning means to create a daily planned route based on the current order situation or on framework routes that have already been defined. Such systems are linked to commercial systems of order management or ERP systems or are part of them.

The advantages from using computer-based route planning systems are collected in Table 9.7.

Transport fleet management systems are used for transport fleet controlling. Basically, they consist of an on-board (computer) unit (OBU) installed in the vehicle and central analysis software. The on-board units record the process data accumulating during employment and subsequently transmit it to the analysis system. The latter contains basic data about the vehicles (purchase costs, amortization, insurance premiums, service life etc.). The on-board unit record consumption and situation data (fuel consumption, mileage, tire condition, motor strain, GPS location data, street navigation etc.). This data serves to compile statistics about the

Table 9.7 Potential advantages of route planning

Potential advantages of trip planning	
For executives/management	▪ Rationalizing potential between three and five per cent of the transport costs ▪ Transparency across the full transport process ▪ Competent information for tactical and strategic decisions
For transport fleet managers	▪ Competent scheduling in the available time ▪ Transparency across the full transport process ▪ Reduction of administrative efforts ▪ Facilitation of the process ▪ Systematization of the planning process ▪ Reduction of risks during holidays or absence due to illness
For drivers	▪ Same workloads for all drivers ▪ Better adherence to delivery times ▪ More information about orders and trips
For customers	▪ Better delivery service by increasing the order time ▪ Better compliance with delivery restrictions

vehicle utilization and strain, indication for vehicle use (e.g. fuel-saving driving, temperature of cooling or freezing systems), repair intervals, the ideal time of use, and source-based cost information for individual transport activities, including the deduction of road tolls. Transport fleet management systems are offered by vehicle manufacturers, technical testing companies, and independent software vendors.

So-called *Fleet Monitoring Systems* support transport fleet management. These systems establish a connection between the driver and the fleet management (headquarters) via mobile communication. This connection can be used to send the current position as well as changes to the status (driving, waiting for unloading, unloading completed) from the vehicle. The fleet manager can assign new orders to the most favorable vehicle based on this information; the job can be directly assigned via mobile communication. The on-board units for road tolls in use today generally feature the components necessary for fleet monitoring: GPS locating with digital net as well as mobile communication with the GSM net.

Tracking & tracing systems make possible the more or less complete tracking of transported goods and vehicles and, at the same time, ensure later traceability. Information about the current location of a shipment are important for interventions in the transporting process or for initiating counter-measures like a substitute delivery in case of delay, loss, or changes to the location of receipt. Transparency of the status during the transport process is also an important quality factor for

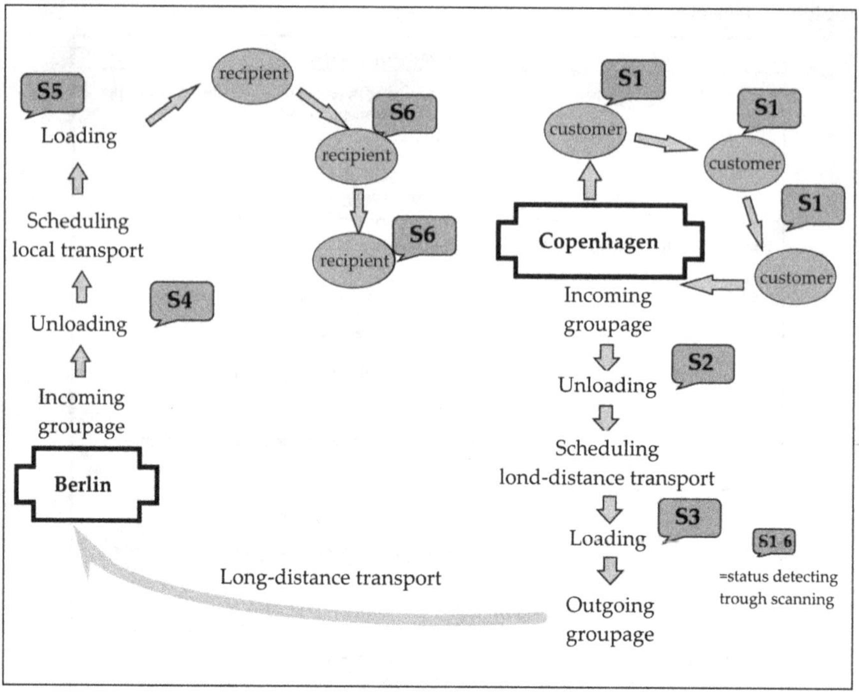

Fig. 9.17 The principle of tracking & tracing systems

logistical services. Furthermore, legal regulations about the traceability of foods or the cold chain can be adhered to. Data inputs in tracking & tracing systems are made by acquisition on certain key points, e.g. when leaving the warehouse, when handling the goods at the cross-docking point, when entering the receiving warehouse, and on delivery to the customer. Tracking & tracing systems are often used for tracking valuable or dangerous goods and in the field of express and parcel delivery. Furthermore, continuous satellite-based locating and tracking is possible; it is used e.g. for trucks, trains, or containers.

Figure 9.17 shows the changes to the status of a shipment during the transport process. Usually every status is documented by scanning and can be stored in the tracking system with the corresponding date and time.

9.5 Strategic and Operative Planning Systems

9.5.1 Site and Network Planning

Software systems are available for almost all planning processes introduced in Chap. 8. *Systems for site optimization* or *distribution network planning* give results based on mathematical or heuristic methods that can be transferred to draft decisions after a qualitative assessment. The quality of the results from IT applications largely depends on the possibilities to include parameters and influencing factors. The bases of such systems are always mass data in the form of individual business events (number of shipments, locations of dispatch and receipt, size, volume etc.) from which *ideal* warehouse locations or distribution networks can be calculated after different iteration processes under the same conditions. These optimization systems support management decisions; but with *soft* and non-quantifiable decision factors always having to be taken into account as well; they cannot replace them.

9.5.2 E-procurement

On the operative level, the Internet has been responsible for considerable changes to the purchasing process; this is described by the term e-procurement. The data exchange between suppliers and buyers had been realized via EDI. The Internet, however, has made possible unlimited and easy contact. All business processes take place electronically, ideally without documents. Decentralized ordering of goods by the buyer who also overviews the process is possible. All of this takes place on a virtual market place which is used as a transaction platform that ensures the information supply between sellers and buyers and supports the initiation and, where possible, the realization of transactions. Market places can be divided into horizontal – i.e. for trans-sector products and services, e.g. for energy and office material – and vertical, sector specific – e.g. for the automobile or healthcare sector – trading places. Solution concepts are mostly simple B2B approaches (market place or catalog solutions). Catalog solutions are favorable for standardized products with little complexity. The Internet ensures a high market transparency. Low prices or at least low price volatility and a regular demand are assumed. Sometimes, configuration via catalog systems is possible.[10] There is a certain price level where price comparison or explicit tendering is better. Such tendering processes assume a buyer market situation. The goal is to get the best acquisition price. The specific requirements are tendered. After the pre-selection of suppliers, the process is reduced to pure price negotiations. One problem is the taking into account of unequal suppliers or quality. During the tendering, the price development is downward. The best price at a certain point is shown to all suppliers who can see potential for improvements from this.

The admission to these B2B platforms is defined by selection criteria like sector specialization, market share, neutrality, transparency, and information services.

A further step towards integration is setting up a B2B desktop-purchasing system which makes possible the automatic execution of many internal processes with numerous market places and, if possible, all suppliers. The goal is to shape all workflows within the company as well as to include the solution in existing

[10] Cf. Stieglitz (2003), p. 266 ff.

ERP/SCM software. Such systems enable companies to unbundle the processes. The preparation of offers and the completing of order forms are no longer necessary. If the payment is made with a purchasing card or a credit note, the same is true for invoice checking, manual receipt of payment, or receipt archiving.

Beyond the purchasing sector, the term *e-logistics* has been established for the strategic planning and realization of logistics systems for Internet-based business. The operative and administrative controlling of logistics processes for supply to end customers in Internet trade, cargo pools for setting up and bundling transport space, but also pure information logistics services can be subsumed under the term e-logistics. In this context, information logistics means the information supply and the management of information flows to realize business processes.[11]

9.5.3 Enterprise Resource Planning

Based on Material Requirements Planning (MRP I) systems, systems for sequential material requirements and capacity planning, called Manufacturing Resource Planning (MRP II), developed mainly in production areas. A further development was Production Planning Systems (PPS). Following an integrated approach, *Enterprise Resource Planning (ERP)* systems go beyond production and make possible the supervision of all company resources. The functionalities of ERP systems primarily have an operative character. However, they also supply data for strategic planning decisions. Using these transaction-oriented ERP systems, company resources like raw materials, finished goods, staff, and financial means are connected to achieve the company goal; and efficient suggestions are made for the process and dimensioning of transactions. Following guidelines by the management and initiated by the order habits of the customers, ERP software solutions manage and control all business processes of a company in the commercial sense.

An ERP system can include the function areas sales, order processing, purchasing, materials management and logistics, finance and controlling, human resources, quality management and, finally, production. It is crucial that the data storage and the process handling in an ERP solution takes place cross-functionally. ERP systems specifically adapted for logistics connect, manage, control, and monitor the logistical sub-processes from the order handling to the delivery or even to the invoicing and monitoring of the receipt of money. WMS and ERP systems can be independent software packages, e.g. in grown IT structures in a company. However, they should be closely connected to each other. In case of purchasing new equipment in this area, it is sensible to opt for an integrated WSM/ERP solution.

[11] Cf. Straube (2004), p. 69 ff.

9.5.4 Supply Chain Planning

Leaving the organizational parts of an individual company that maps the different business processes – maybe across several sites – with ERP, Supply Chain Management (SCM) is the bordering field. SCM systems work on three different levels:

- Supply Chain Design (SCD)
- Supply Chain Planning (SCP)
- Supply Chain Execution (SCE)

The aim of SCM systems is to design, plan, and control the flow of materials and the cross-company processes during the complete logistical chain, from raw materials and production, to trade and, eventually, to the customer. Doing so, a multitude of connections between interfaces with different characters takes place (see Fig. 9.16). Like ERP systems, SCM systems tend to have an operative character.

The main task of *Supply Chain Design* is the designing of logistics networks and the valuing of the necessary investments.

In the field of *Supply Chain Planning,* IT systems realize consumption or requirements planning, stock planning, transport planning etc. in the logistics chain and calculate them for coming periods. Such calculations and plans for logistical capacities derived from them are realized by specialist *Advanced Planning and Scheduling Systems (APS).*

Supply Chain Execution (SCE) systems use IT functionalities to support the operative realization in the fields of order processing, production, warehousing, and transport.

The connections between Supply Chain Design, Supply Chain Planning and Supply Chain Execution are shown in Fig. 9.18.

Case Study 9.2: Enterprise Resource Planning for Supply Chain Processes

This case study describes the implementation and the employment of an ERP system at a warehousing services provider. Because of the detailed description, topics from the other chapters come up as well. This can be seen as proof of the wide-reaching character of IT systems for logistical tasks.

1. The Company

Aarau storehouses plc, in Switzerland, offers logistics services as well as integrated logistics solutions for nationally and internationally (cross-borders) active clients. Besides classic logistics services like transporting, warehousing, and handling, services like stock management, quality control, packing, labeling, and customizing are offered. The company realizes crucial parts of their clients' logistics processes.

These offers are connected with certain requirements:

- Universal business processes, across company borders between industry, logistics and trade

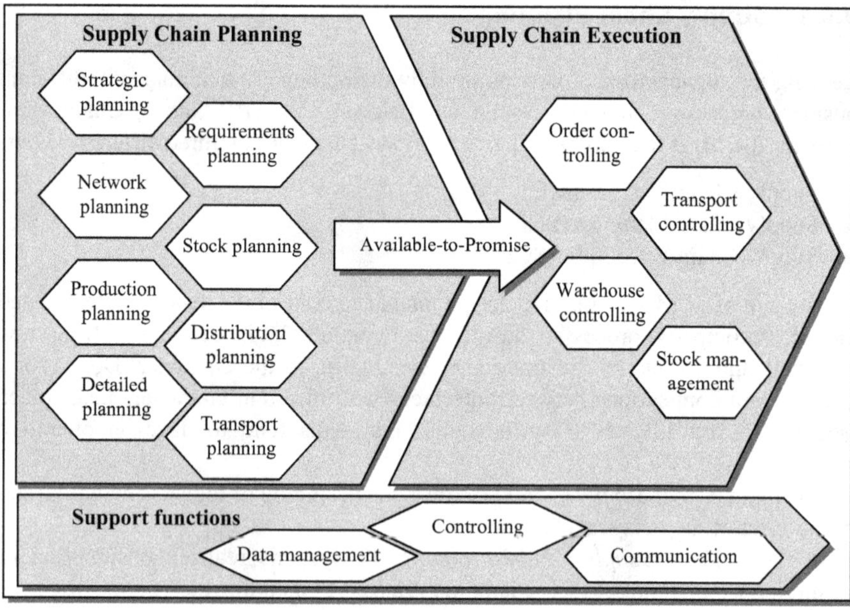

Fig. 9.18 Task model SCM systems (Cf. Kuhn and Hellingrath (2002), p. 13)

- Quick, batch-oriented handling of the goods, e.g. according to expiry date or quality criteria
- Flexible and reliable realization of customer requirements like picking and packing

In these areas, Aarau storehouses specializes in logistical challenges in the field of food, beverages, and tobacco. Customers in this field include Masterfoods, Nestlé, Royal Canin, Gustav Gerig and Ricola. For Masterfoods, Aarau storehouses supply goods to retailers (partly using distribution centers) and bulk buyers everywhere in Switzerland. This includes delivering full pallets as well as displays according to customer demands. Furthermore, customers in the non-food sector include Elcotherm, Swatch, Vespa, Piaggio, FORS Liebherr, Sibir, Titan, and Electrolux. Based on decades of experience in the fields of warehousing, customizing, and distribution of foods, the company vision is to offer individually tailored logistics services that fit into the existing logistics processes to customers active nationally and internationally. The company presentation says:

Aarau storehouses is a modern logistics service provider focusing on food logistics with complete services and wide know-how. Our goal is to build long-term relationships with our partners as their logistics service provider.

2. The Sector: Significance of Information Technology and E-Business

Information technology (IT) and the electronic exchange of data between business partners play a central role in logistics. IT is used for different purposes. Besides normal business processes, these are especially the increase of efficiency for individual processes, the provision of prerequisites for new function and the process integration with clients.

As a contract-based service provider, Aarau storehouses is especially interested in offering their clients high additional value. This includes transparency, e.g. by the access to current stock data and *permanent stock-taking,* a very high level of flexibility and planning security as well as a smooth integration of the IT system with that of the clients.

3. The Integration Project

3.1 Starting Situation and Reason for the Project

The reason for starting the project was the wish of a client to be able to do a complete tracing of goods based on batches. The background for this was a recently introduced regulation to ensure the traceability of batch-based quality information (EU regulation 178/2002, General Food Law). This means that a continuous flow on information must be ensured for the supplier of the primary products to the producer and the trading. Logistics must therefore be able to record the batch information from the producer and deliver them, across all following procedures to the places where they are eventually for sale.

3.2 Implementation Partner (Supplier of Business Software)

The *GUS group* offers business applications (ERP solutions) focusing on the so-called life science industries (pharmaceutical, chemical, food) as well as complete solutions for distance trade and logistics systems. It develops, distributes, and implements complete solutions for ERP, quality management, e-commerce, and logistics.

3.3 Business Partners

Masterfoods is active worldwide as a producer of sweets, snacks, and food brands as well as for products for pets. Retailers (directly or via distribution centers) and large customers are supplied to in Switzerland.

4. Specification of the Integration

Aarau storehouses deliver food producer's full product ranges to retailers and large customers in Switzerland. This case study takes the customer Masterfoods as an example for these services. The logistics outsourcing includes the tasks order management, stock management, batch tracing, picking and customizing as well as transport (see Fig. 9.19).

For these processes, Aarau storehouses is fully integrated into the communication processes between the producer and the client. Integral communication elements like stock information, orders, activity data (necessary because of the production of displays for special offers) as well as despatch advices are distributed to the partners using an information hub, made possible by the integrated logistics solution at Aarau storehouses.

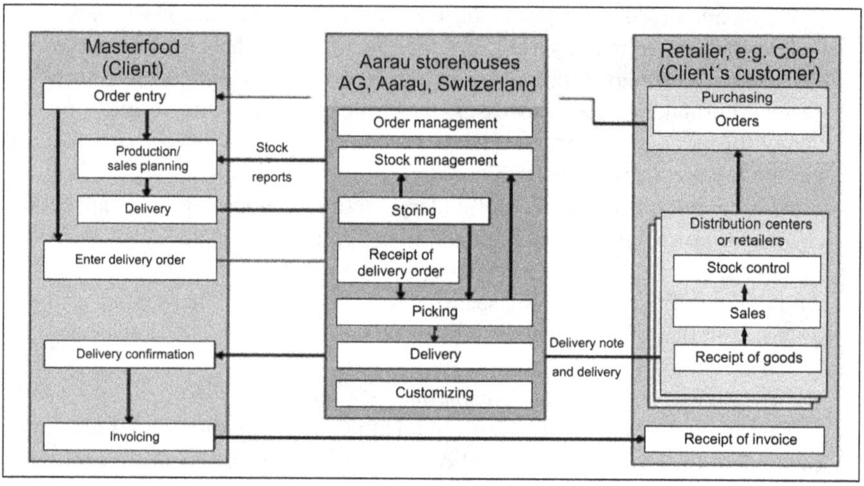

Fig. 9.19 Integration of Masterfoods and retailing

- Based on the sales planning between Masterfoods and the retailer, Aarau storehouses daily receives shipping order information from the Masterfoods headquarters.
- From this, Aarau storehouses creates the picking and delivery jobs for internal logistics and distribution in Switzerland.
- Based on the delivery data, the retailers receive dispatch advices.
- Masterfoods receives regularly updated stock information (permanent stock-taking).
- Uncalculated waiting times at customer locations:

Based on the stock data and the sales planning, Masterfoods generates production orders in their own sites and delivers to Aarau storehouses.

Aarau storehouses stores a stock from few days to 3 months, depending on the product. Usually, the products are delivered within 24–48 h after an order requiring high delivery precision and quality.

The quality orientation made possible by the integrated logistics and IT concept is aimed most of all at the batch traceability required in the food, beverage and tobacco sector. As a result, Masterfoods can see which product was delivered to which customer with the batch number of a product. This also makes possible the quick and targeted realization of possible product recalls.

4.1 View Process

Aarau storehouses divides the logistical process into receipt of goods/storing, order management, picking, and customizing. Customizing includes services like creating product display with a defined mixture of goods. Additionally, quality management accompanying the process and batch tracing takes place. The task, jointly completed by Masterfoods and Aarau storehouses, secures a high level of delivery precision. It has to be noted that the stock management works according to FeFo: first expire, first out. The respective product with the nearest expiry date is delivered first; the date of receipt is not taken into account. This strict monitoring of batches and expiry dates ensures that no product that is due too early will get into the client's sales channel. The basic processes can be individually adapted or expanded for each customer to fit into their own logistics processes. The following describes the processes order management and picking, as they have been adapted for Masterfoods.

4.2 Order Management and Picking

- Three times every day (5:00 a.m., 10:30 a.m., 4:30 p.m.), shipment order data is sent from Masterfoods to the ERP system of Aarau storehouses, using a leased line.
- If the goods are in stock, the order is passed on to picking and transport pre-planning. Otherwise, an out-of-stock message is generated and sent to Masterfoods.
- In pre-planning, the orders are assigned to a delivery zone. Detailed route planning takes place after this in the transport department.
- The dispatch message is automatically sent by EDI or fax to the retailer.
- The processes shipping orders and reported stock data is reported to Masterfoods with each daily closing.
- Picking and handling of reorders.
- From the order data from Masterfoods, picking orders are generated and realized in a two-shift system.
- Sorted to optimize routes, the picking orders are transmitted via radio data transmission to the picking machines. The ERP system, in dialogue with the radio terminal, automatically confirms the orders and gives the remaining quantity to update the stock information.
- If a minimum stock quantity is not met in the picking warehouse, a relocating order from the main warehouse (according to FeFo) is automatically sent to the displays of the forklift drivers.
- Every physical movement (storing, replenishment, outgoing) is monitored within the warehouse area by the ERP system. This guarantees complete control over the flow of goods and the expiry dates.
- Every picking order is completed with the automatic printing of a pallet label (EAN 128) at a printer in the area.

4.3 The Application

The described business processes and the corresponding cross-company communication largely rely on the adapted, batch-oriented ERP system *GUS ERP CHARISMA* as the central logistics system. The system performs the following tasks:

- Keeping of customer data, order data, stock lists as well as stock information, batch information and performance data
- Automatically transforming incoming orders to picking or customizing orders and controlling the picking

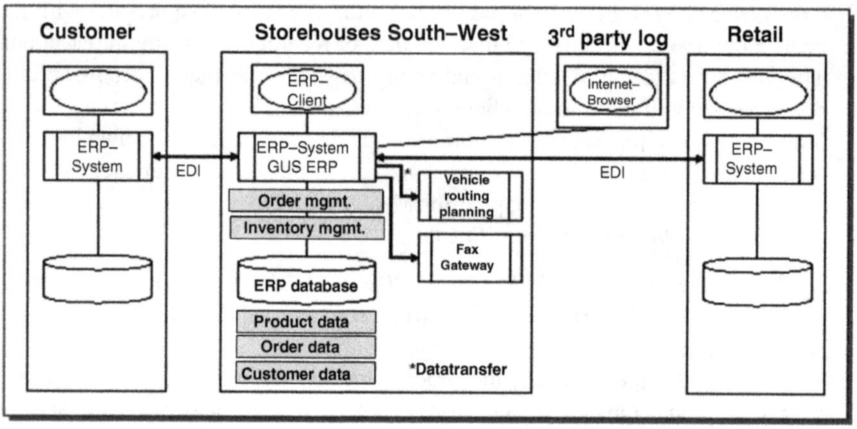

Fig. 9.20 Application overview

- Updating the stock numbers after every moving in or out of items (continuous stock taking)
- Regularly reporting stock information to the client
- Sending dispatch messages to the retailers via EDI or fax.

Additionally, Aarau storehouses gives the delivering forwarders the possibility to reserve a time window for their delivery on an IT platform accessible via the Internet. This considerably facilitates the planning for Aarau storehouses and for the forwarders.

4.4 Software

GUS ERP CHARISMA is configured standard software for logistics management, adapted to the need of the customers. Lagerhaeuser Aarau uses the version developed for the iSeries platform. This version requires additional programming for the adjustment of processes necessary for every customer. However, CHARISMA can be used besides GUS-OS ERP for Life Sciences, the GUS Group's newly developed open business software. This allows for a workflow-based process adaption or expansion without changes to the program code (see Fig. 9.20).

4.5 Pallet Identification

The clear identification of pallets and the management of additional information about the pallets are necessary for the batch tracing and a precise stock and quality management. This information is exchanged electronically between Masterfoods, Aarau storehouses and the retailers. Additionally, it is attached as EAN 128 labels to the pallets for shipment identification. EAN 128 labels encode the EAN 13 codes of the products on the pallet, additional information like batch number and expiry date as well as the Serial Shipping Container Code (SSCC) for identifying the pallet itself. Designing the codes and EAN 128 labels, it had to be taken into account that especially the large customers in retailing

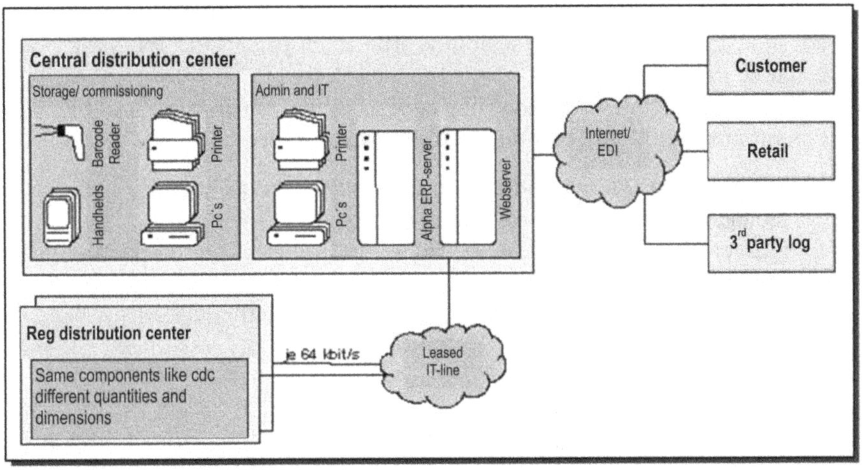

Fig. 9.21 System overview

have, respectively, own requirements for the code structure within the relatively open standard.

4.6 Reference Data

It is essential for the realization of batch tracing and a precise stock management that even parts of pallets can be clearly identified. This requires information about the layout of pallets (reference data) to know e.g. how many layers a pallet has and how many units (boxes) are stacked per layer. This data is generally provided to Aarau storehouses. However, the incoming goods do not always correspond to the expected formats. In such cases, the goods are standardized during the checking of incoming goods. This makes possible the realization of processes according to the requirements in the following logistics chain.

4.7 The Technology

The ERP system at Aarau storehouses is operated at the Schafisheim site on an IBM iSeries AS/400 system. Due to the starting situation, a centralized solution with access via Internet or VPN was chosen. GUS ERP Charisma can be fully operated via Internet access. Additionally, various password-protected areas are available (see Fig. 9.21).

The security of EDP applications and the availability of data and information is a top priority for Aarau storehouses. To achieve this, all relevant data is mirrored on the central system in the Aarau storehouses company headquarters in Lausanne, in the central distribution center in Spreitenbach as well in the regional warehouse in Schafisheim. Communication with customers and retailers takes place by Internet and EDI, in some cases still via fax.

5. Evaluation

Upon introduction of the system, some adaptions had to be made to the processes, mostly in picking. A higher discipline is required for recording remaining quantities and handling sub-quantities. Despite this, the process adaption could be realized smoothly. The requirements set by Masterfoods could be met.

The realized integrated logistics concept means an increase in efficiency, more transparency, and cost reduction for Masterfoods:

- Storing space Masterfoods needs does only have to hold the preliminary products and the connected interim spaces
- Delivery to retailers is realized from a central location, despite the goods coming from different sites
- Processing time in the warehouse can be significantly reduced
- Direct transfer of order data prevents mistakes
- Error rate was reduced to few per mils
- Delivery reliability was increased
- Batch traceability was completely realized; EU regulation 178/2002 is constantly met
- Average weight per order was significantly increased, reducing the transport costs
- Permanent stock taking renders the expensive yearly process partly unnecessary
- Logistics costs can be analyzed based on transactions down to the level of the delivery note. With this, the client has a tool for detailed cost analysis of their sales channels

6. Success Factors

The most important success factor is for the logistics service provider to record and map the customer requirements as precisely as possible. This means that the business model of the service provider must be designed in a way that it can flexibly take over sub-processes from different clients' logistics and create synergy effects despite the differences. Aarau storehouses has designed their business processes and ERP system in a way that it can be connected to or even integrated in the process and system architecture of their clients.

The key to this is an information system that automates, records, and controls the individual processes. With every sector, every client and even every client's customer introducing requirements to the complete system, there is not just one correct process for the food industry but there are many. And these variants create the differentiation potentials of the clients.

The competence of Aarau storehouses is, on the one hand, to include the diversity but, on the other hand, to make it into standardized and, in the sense of a defined quality, unified sub-processes. The business software is a central component for this; the diversity can only be mapped if the data model is correspondingly expandable and if the processes can be mapped flexibly.

Review Questions

1. What is information logistics?
2. Explain the term 'communication standard'.
3. Describe the EDI principle.
4. Outline the informational connections in ERP systems.
5. What are the differences between open and closed ERP systems?
6. What are the central functions of a warehouse management system?
7. State the difference between the fleet monitoring- and tracing & tracking systems and route planning.
8. What is the difference between enterprise resource planning and supply chain planning?
9. What is the number of the shipping unit?
10. What are the advantages of transponders?

Additional Literature

Arnold, D., Kuhn, A., Isermann, H., & Tempelmeier, H. (2008). *Handbuch Logistik*. Heidelberg: Springer-Verlag.

Franke, W., & Dangelmaier, W. (Ed.). (2006). *RFID – Leitfaden für die Logistik*. Wiesbaden: Gabler.

Kopfer, H., & Bierwirth, C. (Ed.). (1999). *Logistik Management: Intelligente I + K Technologien*. Berlin/Heidelberg: Springer.

Scheer, A. W. (Ed.). (2008). *Rechnungswesen und EDV*. Heidelberg: Springer.

Schütte, R., Vering, O., & Wiese, J. (2000). *Erfolgreiche Geschäftsprozesse durch standardisierte Warenwirtschaftssysteme*. Heidelberg/Berlin: Springer.

Stadtler, H., & Kilger, C. (2007). *Supply chain management and advanced planning*. Berlin: Springer.

Straube, F. (2004). *E-Logistics – Ganzheitliches Logistikmanagement*. Berlin: Springer.

ten Hompel, M., & Schmidt, T. (2009). *Warehouse management*. Berlin: Springer.

ten Hompel, M., et al. (2004). *Software in der Logistik 2004*. München: Huss-Vlg.

Wildemann, H. (2002). *Distributionslogistik*. München: TCW.

Investment and Financing in Logistics 10

Learning Objectives

Establishing and maintaining the logistics systems is capital-intensive. This is true for systems in both macro-logistics and micro-logistics. Therefore, issues of investment and financing play a central role in logistics. A cross-company view on supply chain management, on additional financial and investment figures, and – as a consequence – on the capital flows and payment processes in the financial supply chain offers additional approaches to optimize fragmentation and collaboration within value chains.

This chapter aims to familiarize the reader with the management of capital requirement, capital utilization, and net working capital. Different forms of investment financing will highlight the strategic significance of financing. Established and novel concepts relating to this will be presented and service providers for financing stock inventory, logistics real estate, and logistics movables will be introduced. By considering the various agents in logistics chains, the reader will be provided with guidelines as to the optimization of financing and of financing processes, and insights into the strategic and operational relationship between financing and logistics will be given.

Keywords

- Fixed assets
- Current assets
- Working capital
- Supply-chain finance
- Cash-to-cash-cycle
- Capital tie-up
- Leasing
- Pay on production

H. Gleissner and J.C. Femerling, *Logistics*, Springer Texts in Business and Economics,
DOI 10.1007/978-3-319-01769-3_10,

10.1 The Basics

10.1.1 Managing Capital Requirement, Origin of Capital and Capital Utilization

Investment and financing are areas of financial management within a company. Essentially, finance is the management of *capital requirement*, *origin of capital* (financing), and *capital utilization (investment)*.[1] Financing is defined as the procurement of required capital to make necessary operating investments. Depending on the kind of capital procurement, we can distinguish between external and internal financing. In cases of *external financing,* a company raises capital through individuals, banks, or through the capital market. *Internal financing* distinguishes between self-financing, provisions-based financing, and restructuring of assets through released capital. It is vital that internal financing be closely aligned with the operational performance processes and sales processes. As a consequence and in contrast to the management of external financing, internal financing must not solely be managed by the financial department of a company. Instead, the management of working capital, of capital requirement, and of capital utilization needs to be integrated into the management and leadership process of all other operational functions along the value chain in a goal-oriented manner. This also includes logistical functions and supply chain management, as illustrated in Fig. 10.1.[2]

Forms of financing and financial instruments can be differentiated according to a variety of criteria. In cases of external financing, these instruments are commercial credits and bank loans, credit substitutes, and special forms of financing of current assets. A company may also rely on short-term financing in the form of e.g. bank overdrafts, trade credits, or loans to customers (commercial credits) (see Sect. 10.2.1). Regardless of the forms of financing, all *credit and capital relationships* of a company are usually recorded on the balance sheet on a specific balance sheet date. The balance sheet itemizes a company's assets and liabilities and shows its equity and capital as the sum of all financial obligations towards owners and creditors. The asset side of a balance sheet discloses how the assets have been utilized while the liabilities side reveals the source of funds, i.e. it lists all claims of equity providers and debt capital providers.

Investments are made to acquire or provide capacities in the areas of procurement, production, distribution, and disposal, which are all closely intertwined with logistic functions. We can distinguish between real investments (manufacturing plants, semi-finished and finished products), immaterial investments (investments in research and development) and financial investments (equity rights, legal claims). Further distinctions can be drawn between original or initial investments for the (initial) generation of capacities and investments in expansion, replacement,

---

[1] Cf. Peridon et al. (2009), p. 10.

[2] Cf. Pfohl et al. (2003), p. 21.

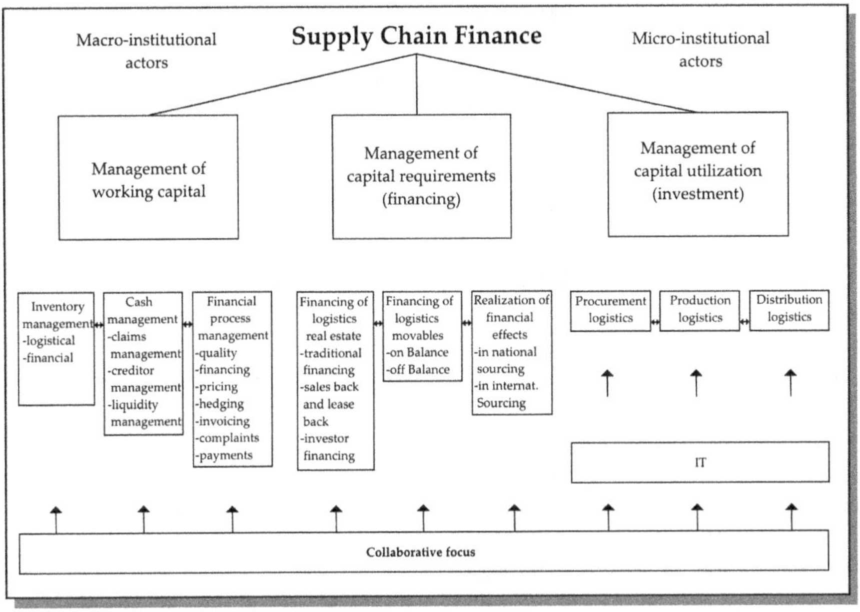

Fig. 10.1 Financial processes in logistics and supply chain management (Cf. Jehle and von Haaren (2008), p. 4)

and maintenance. The latter types may also be categorized as rationalization investments, conversion investments, or diversification investments, depending on the desired changes in the provision of services and the service portfolio.[3]

The most important financing requirements of logistical investments are:

- Inventory and transport
- Purchase, construction, and furnishing of logistics real estate
- Utilization of logistics real estate

If logistical capital spending decisions are being made, investment calculating procedures may help assess the profitability of an investment in, for example, fixed assets. These procedures can either involve static investment calculations (pay-off method, cost comparison method) or dynamic investment calculations (annuity method, internal rate of return method, net present value method). If investments are made in current assets – e.g. to enhance the service level of delivery through increased stock – the effects on the company value need to be calculated. Instruments of the cash flow statement or the discounted cash flow calculation (DCF), for instance, may be used to this end.[4] Yield analyses also needs to be applied to supply chains. A key figure in assessing the effectiveness and analysing

[3] Cf. Thomen and Achleitner (2006), p. 602 et seq.

[4] Cf. Pfohl et al. (2003), p. 12.

the profitability of supply chains is the *return on investment* (ROI) (see case study 11.2). Especially assets in the form of current assets decrease profits as they entail interest expense from capital tie-up. In addition, they also lock up capital which then cannot be used for other purposes. Logistics costs have a bearing on profit margins, too. These are costs such as order processing costs, transport costs, warehousing costs, and inventory costs. The speed of capital turnover is determined by the financing alternatives for the assets needed.

10.1.2 Managing Fixed Assets and Current Assets

Fixed assets comprise all durable assets (goodwill, company value), long-term financial assets (interests or bonds), and tangible assets (properties, buildings, plants and equipment, machinery). Their purpose is to support the on-going business. From a logistical viewpoint, fixed assets mainly include logistics real estate, machinery (high-bay warehouses, packing machines), equipment (conveyors, picking systems) and means of transport (vehicles, ships, planes) (see Sect. 5.4.2).[5] The management of fixed assets aims at the least possible capital tie-up in these tangible assets. This can be achieved, for example, through:

- Efficient utilization of existing fixed assets
- Amortization of assets which are not necessary for business (anymore) or which do not yield the capital costs
- Avoiding an outflow of funds as a result of a purchase

Current assets comprise all items that remain in the company for a short period of time, such as the inventory of finished and semi-finished products, bank balances, or cash balances. The optimization of current assets has its primary aim in reducing stock and (customer) claims so that tied-up capital is freed up. Reducing current assets allows a company to at least partly finance their capital requirements from their own resources. Approaches in reducing current assets include[6]:

- Efficiently structured and improved operational financing processes and money processes
- Integrated business planning in regard to customers, suppliers, financial institutes, and investors
- Goal-oriented collaboration across all departments involved in the value chain

The *working capital* is calculated as current assets minus short-term liabilities from deliveries and services and other (possible) current liabilities. Figure 10.2 shows the working capital as displayed on a company's balance sheet.[7]

[5] Cf. Gomm (2008), S. 106.

[6] Cf. Hofmann (2005), p. 206 et seq.

[7] Cf. Klepzig (2010), p. 18.

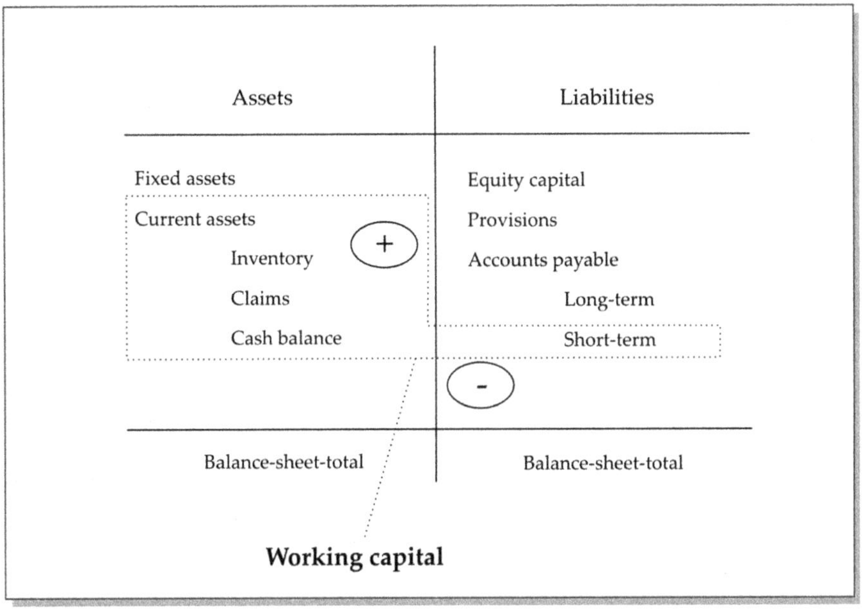

Fig. 10.2 Working capital as shown on the balance sheet (Klepzig (2010), p. 19)

The working capital is the share of current assets which needs to be financed by the interest return on capital. Managing the working capital is of great importance since a company's profitability is greatly influenced by it. *Working capital management* seeks to keep the working capital's portion of locked-up capital as small as possible. The following core processes – which can largely be classified as pertinent to logistical functions – and the measures associated with them may provide considerable help in optimizing working capital[8]:

- *Order-to-cash:* shortening of the timespan between placement of an order and receipt of money. Measures that are applied to this end mainly include optimized financing and payment conditions, i.e. improved and expedited order processing, claims management, and complaints management
- *Total supply chain management:* integration of the different value-added steps into a single integrated and optimized value-added chain by means of shortened cycle times, integration of sales planning, and optimized production and procurement processes

[8] Cf. Klepzig (2010), p. 11 and p. 38 et seq.

- *Purchase-to-Pay:* optimization of the timespan between purchase and payment by means of changes made to financing and payment conditions, through a favourable choice of suppliers, through optimized order processing and audit, through optimizations in the incoming goods department, and by means of liabilities management.

10.1.3 Financial Supply Chain Concept

While supply chain management has so far only been concerned with the optimization of the flows of material, goods, and information (see Sect. 3.2), the concept of financial supply chains also includes cash flows during production and along supply chains. The management activities dealing with cash flows are termed *supply chain finance* or *financial supply chain.*[9]

Examining the flow of financial means allows us to outline the non-monetary service processes of a supply chain from a financial viewpoint. Trade relationships with suppliers and customers usually entail cash flows that are in reverse to the flow of goods. Financial supply chain management aims to[10]

- Increase the transparency of financial flows
- Improve the analysis and planning of cash flows
- Reduce tied-up current assets
- Optimize financial process costs

Furthermore, the number of parties and persons involved increases when examining *financial supply chains*. Traditional supply chains are made up of stakeholders such as suppliers, producers, logistics service providers, customers, and consumers. Financial supply chains, however, may also include internal players such as the financial department and external players such as banks (large banks, direct banks, investment banks), financial service providers (leasing or factoring companies) and investors (credit institutions, insurance companies, enterprises, private persons, institutional investors).[11] It is important that all of these players be considered in a supply chain since payment defaults may occur at one stage if delivery has failed at the previous stage. This would have an effect on the financial flows of all stakeholders involved in the supply chain.

Along with the examination and assessment of investments and finance functions, an analysis of the fixed and current assets and of the capital costs should be carried out and cash flows and returns should be tracked. A suitable instrument for this is the *cash-to-cash-cycle* (cash-conversion-cycle, cash-flow cycle, cash-to-

[9] Cf. Skiera and Pfaff (2004), p. 1399; Gomm (2008), p. 58 et seq.

[10] Cf. Brandt (2004), p. 118; also Sarbach (2006), p. 12.

[11] Cf. Hofmann (2005), p. 206 et seq.; Gomm (2008), p. 155 et seq.

cash-cycle-time),[12] which essentially measures the time of capital tie-up. The cash-to-cash-cycle also serves to determine the following figures[13]:

- *Days Payables Outstanding:* the time between the receipt of the supplier's bill until its payment
- *Days in Inventory:* the time during which purchased resources are involved in the production process until they are sold as semi-finished or finished products
- *Days Sales Outstanding:* the time between the product's sale and the receipt of payment from the customer

Figure 10.3 shows an outline of a cash-to-cash-cycle. In this case, raw and process materials as well as semi-finished products are being purchased, followed by the production of commodities and their sale. The timeline displays the corresponding stages of order placement, procurement, supply, and delivery. The elements of the cash-to-cash-cycle take effect after the goods have been supplied and upon receipt of the bill. Payment takes place at a later point. In between are *days in payables*, which indicate the outstanding liabilities. Invoicing is effected

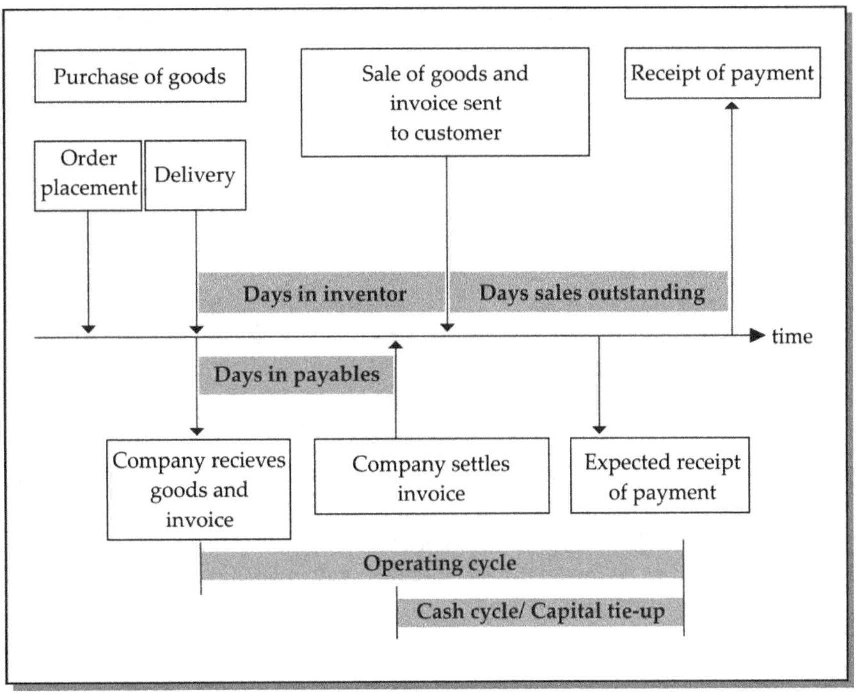

Fig. 10.3 Outline of the cash-conversion-cycle (Brandt (2004), p. 117)

[12] Cf. Pfohl et al. (2003), p. 11.
[13] Cf. Saarbach (2006), p. 29.

upon delivery and the payment due is determined. In between invoicing and the effected payment are the *days in receivables*, indicating the outstanding debts.[14]

The cash-to-cash-cycle is calculated by adding *days in inventory* and *days sales outstanding* minus *days payables outstanding*. Thus, the timespan can be determined within which a company has invested money in a specific business case, i.e. the time during which the current assets are tied-up.[15] For companies it is important to know how long a specific amount of capital cannot be used otherwise so that no additional tie-up of current assets unnecessarily compromises the company's liquidity. This tie-up keeps companies from obtaining financial income which could have been realized using the tied-up capital.

In order to reduce the working capital, *cash-out* activities should be carried out as late as possible while *cash-in* activities should take place as early as possible. Both of these activities are dependent on the way in which the processes that trigger payment are stipulated.[16] On the procurement side this becomes possible if the supplier is to be paid at the latest possible point, which can, for example, be achieved through:

- Supply of high-quality materials only shortly before demand arises
- Passing of ownership of materials from consignment warehouses only shortly before demand arises
- Payment is not effected immediately but only as part of the monthly payment run
- Payment is only effected after assembly of the delivered part or after sale of the finished product

Conversely, cash-in is most beneficial if the customer pays at the earliest possible point, which can, for example, be achieved through:

- Adherence to delivery dates
- Delivery of faultless products that do not give cause for complaint
- More frequent payment runs
- Cash before delivery or delivery against deposit
- Early dunning

The entire supply chain can thus be optimized within the financial supply chain through the optimization of a specific company's internal financing in the sense of a cash-to-cash-cycle. However, this does not mean that each company would strive to shorten their cash-to-cash-cycle. Instead, collaborative solutions need to be found

[14] Cf. Klepzig (2010), S. 46 et seq.

[15] Cf. Gomm (2008), p. 125

[16] Cf. Klepzig (2010), p. 47 et seq.

which shorten the cash-to-cash-cycle of the supply chain as a whole, even though individual companies may not be fully optimized this way. The potential for increased efficiency of the capital tied-up in the supply chain may thus be realized through[17]:

- Optimized cash management by cooperatively maxing out credit lines
- Introducing credit note procedures throughout the supply chain to expedite financial flows
- Improved capital requirement forecasts for all cash flows

10.2 Financing Deliveries of Goods and Inventory

10.2.1 Overdraft Facilities and Supplier's Credit

The financing of raw and process materials as well as of semi-finished and finished goods necessary for production and the transport thereof is called operating resources funding or inventory financing. Inventory financing is usually carried out by granting short-term credit lines in the form of overdraft facilities, supplier's credits, or consumer credits.[18]

By means of a *current account,* banks or financial institutions provide a borrower with funds. The borrower is granted a credit line which constitutes the credit limit. This credit can be used on demand by the borrower. On-going payment transactions repay the amount that has been used of this credit to the bank. Interest is only payable on funds that have actually been used. During its period, the loan can be used up to its limit at any time again.

With regards to inventory financing, we can further distinguish between a committed *inventory credit facility* – where the inventory is assigned to the bank as security – and an uncommitted short-term operating credit, which can be used flexibly. Inventory financing is always shown on the liabilities side of the balance sheet, regardless of its form.

A *supplier's credit* (also called accounts receivable loan) allows for the short-term use of financial means. Upon purchase the supplier grants the buyer a certain time for payment free of interest. Usually, this is a period of grace of 30–90 days or in individual cases even considerably more. A supplier's credit is usually granted informally and without collateral. This type of credit, however, generally constitutes an expensive financing alternative.

Consumer credits in the form of deposits are a type of credit that is used apart from supplier's credits. These credits are especially popular in the capital goods and construction industries. The customer pays pre-agreed instalments of the purchasing price (advance payment, part payment) upon order placement and/or after production.

[17] Cf. Pfohl et al. (2003), p. 47.

[18] Cf. Thommen and Achleitner (2006), p. 556 et seq.

As opposed to supplier's credits, the instalments incurred are paid by the customer. Apart from their financing function, customer credits are advantageous for the supplier in that they increase the likelihood of the good's being accepted by the customer.[19]

10.2.2 Factoring, Forfaiting, Cash Forwarding

Factoring is a so-called credit substitution and constitutes a financial instrument that is to support suppliers in realizing claims towards their customers more quickly. Factoring involves a bank‘s or a factoring company's (factor) purchase (purchase of receivables) of mostly short-term claims incurred through goods deliveries or services rendered. The bank or factoring company takes on all debts resulting from these deliveries or services. Up to 90 % of the sum receivable may be immediately financed in advance. After collection of the accounts receivable by the factor, the seller of the receivables obtains the remaining 10 % minus a factoring fee charged by the factor. Often times factoring companies are also in charge of the accounting, invoicing, and specifically of dunning. Figure 10.4 illustrates the factoring processes.

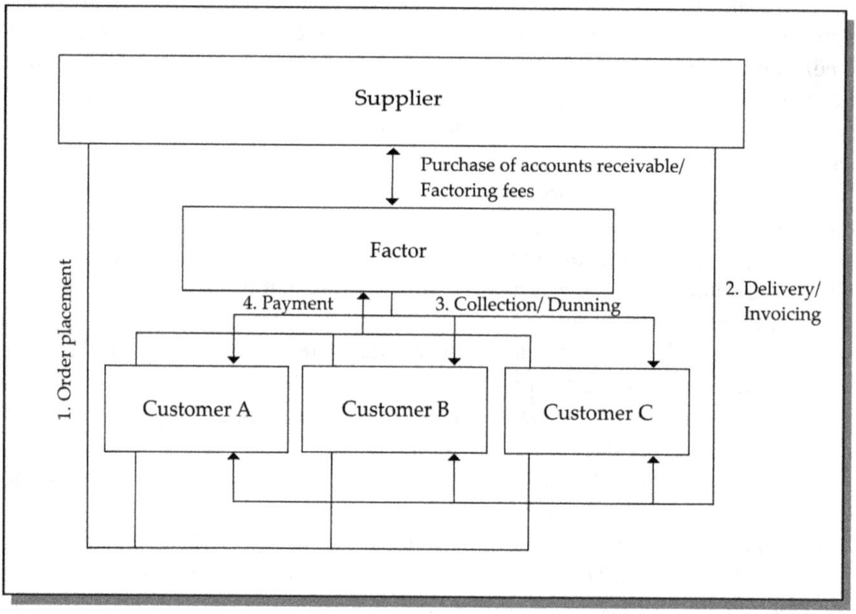

Fig. 10.4 Outline of the processes and functions of factoring (Cf. Thommen and Achleitner (2006), p. 561)

[19] Cf. Becker (2008), p. 171.

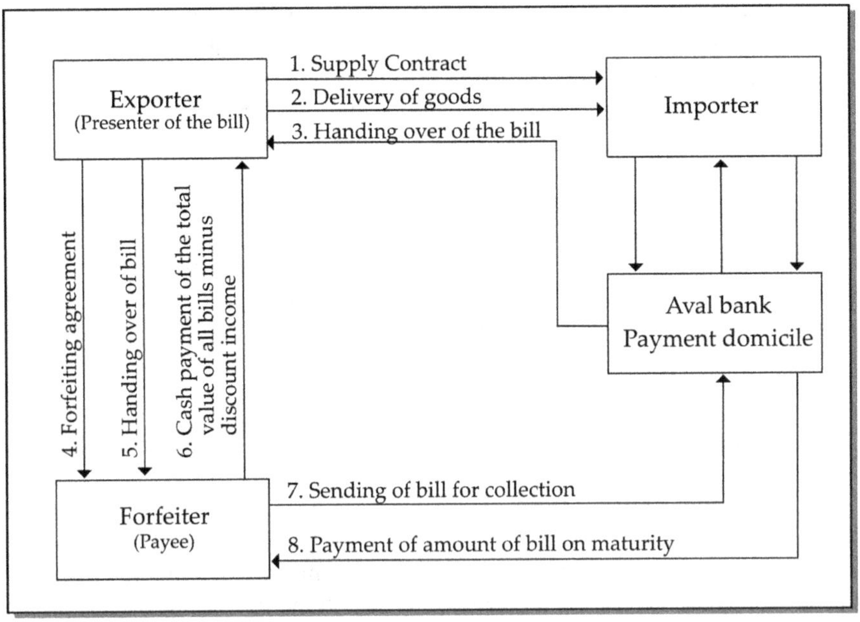

Fig. 10.5 Outline of the processes and functions of forfaiting (Cf. Thommen and Achleitner (2006), p. 564)

Another form of financing tailored to inventory financing is *forfaiting*. This form of financing is mainly employed in medium-term and long-term export financing. It involves the sale of export receivables to a special institution (forfaiter) without recourse. The default risk and foreign exchange risk are borne by the forfaiter. The level of costs is accordingly high. Figure 10.5 shows the processes of forfaiting.

Central to forfaiting are *aval credits* (bills of exchange) which act as suretyship. These are used nationally and internationally if delivery obligations, warranty obligations, and payment obligations need to be backed by collateral security. Mostly, companies obtain these credits from a bank since none of the companies involved (exporter, importer) wants to take the risk associated with advance payments. The forfeiter acts as payee and collector of the receivables from the exporter's delivery. The exporter obtains a bill of exchange from the importer, which is made to the exporter (promissory note).[20] In adition to aval credits, there are *customs guarantees* and *freight guarantees* which do not require collateral securities in cash (warranty guarantee).

[20] Cf. Thommen and Achleitner (2006), p. 563.

Cash forwarding is a variation of factoring. As opposed to conventional factoring, cash forwarding involves an undertaking of 90 % of the invoice amount upon acceptance of the goods from the supplier.[21]

Apart from factoring companies and banks, logistics service providers also offer additional claims management services in collaboration with banks. These may include accounting services, which are offered by debtor companies especially set up for these purposes. Outsourcing these functions to logistics service providers offers more transparency for the acceptance of goods, collection, delivery and invoicing. Service providers such as Deutsche Post World Net (DPWN) offer such services in collaboration with Postbank under the label *Financial Logistic Solutions.* [22]

10.2.3 Inventory Management and Off-Balance Sheet Inventory Financing

It is the primary aim of inventory management to reduce working capital by reducing inventories (see Chap. 7). Inventories essentially comprise raw materials, process materials, and finished goods, which are stored in warehouses as part of a procurement, production, and distribution system of a company (see Sect. 8.3). Reduction of inventory should bring about a reduction of capital costs, which need to be used for the (pre-) financing of inventories. If inventories are to be reduced to lower the capital costs, however, one must consider the risk of decreased delivery capability. Approaches in reducing inventory are[23]:

- Negotiations with supplier to enhance delivery service
- Differentiation of delivery service according to products
- Differentiation of delivery service according to customers and customer profitability

At the same time, inventory cannot be reduced within a supply chain simply by transferring the storage functions to upstream or downstream steps in the value chain, i.e. to suppliers and customers. Instead, it should be the aim to intensify the relationships between suppliers and consumers. This may be effected, for instance, through cross-company approaches in inventory financing, as opposed to companies relying on their individual financing approaches within a value chain. This can potentially result in lower interest costs.

Well-directed *inventory allocation* can result in financial advantages at certain points of the supply chain.[24] This process involves goods being received by the supplier at the latest possible time or delivered to the consumer as early as possible.

---

[21] Cf. von Eisenhardt-Rothe and Jütte (2003), p. 163.

[22] Cf. von Eisenhardt-Rothe and Jütte (2003), p. 163.

[23] Cf. Pfohl et al. (2003), p. 38 et seq.

[24] Cf. Pfohl et al. (2003), p. 39; Klepzig (2010), p. 129.

A suitable means to reduce inventory through inventory allocation on the supplier's side is to establish a consignment warehouse or to implement vendor-led inventory management (e.g. VMI) (see Sect. 7.5).

Inventory financing can take place both on the supplier's and on the consumer's side. On the respective company's balance sheet, the inventory is shown as part of the working capital and capital tie-up. Inventory financing may also be taken on by logistics service providers. This is particularly sensible if procurement and distribution logistical services are also carried out by service providers. Depending on the level of risk-taking, the logistics service provider may assume ownership of the inventory (authority to disposition) or carry out interim financing (*vendor hub financing*).[25] A vendor hub is a warehouse managed by a logistics service provider. This includes the management of transfer of ownership from supplier to consumer (see Sect. 7.4.2). Vendor hub financing can utilize factoring.[26] Such concepts, however, require partnership-like concepts within which customers, suppliers, logistics service providers, and financial service providers cooperate to reach the most favourable solutions possible. Usually, this results in the set-up of a new company which takes charge of the balancing of accounts (*off-balance inventory financing*), along with the entire goods management (purchase, logistics, debtor management).

10.3 Financing of Logistics Real Estate

10.3.1 Self Financing and External Financing

Logistics real estate may be provisioned through purchase, rent, or leasing of an existing object or through the building of a new one. If the property is owned by the company, they have the sole right of disposition over it for an unlimited period of time. *Property financing* of logistics real estate can be self-financed or externally financed. Self-financing enables a company to set up the property or sell it at any time without having to make its size, location, and facilities dependent on a bank. This, however, requires the company to be able to finance the property exclusively from their cash flow or from an increase in capital. Moreover, the capital tied up in the property will not be available for other investments. Property financing can be carried out using self-financed equity capital (e.g. accumulated profits), externally financed equity capital (e.g. stocks), or borrowed capital.

A common way of externally financing logistics real estate is *mortgage financing* at a fixed interest rate with annuity amortization.[27] This involves repayment rates which cover the interest payments and the external capital that needs to be paid off. These repayment rates may be linked to business success and thus, for

[25] Cf. Stenzel (2003), p. 145 et seq.

[26] Cf. von Eisenhardt-Rothe and Jütte (2003), p. 168.

[27] Cf. Stenzel (2003), p. 142.

example, to the property's capacity utilization. This way they can follow the cash flow development (*pay as you earn*).[28] This makes it possible to adapt the financing to the peculiarities of a logistical project, as is the case in contract logistics, for example. Banks and credit institutions are lenders for mortgage credits. The financing is collateralized by an encumbrance (mortgage, land charge/annuity land charge) on the property.

For industrial companies and trade companies, *investment* in their own logistics real estate used to be the most common form of provisioning real estate. The actual utilization of the real estate used to be the main reason for this kind of investment, which means that logistics real estate was only rarely purchased for capital investment purposes. Companies still use approximately 50 % of all logistics real estate in Germany for their own logistics.[29] The remaining 50 % of existing real estate are run by logistics service providers. Alternative ways of real estate provisioning are available for logistics service providers. If an existing property with sufficient capacity is owned by the client, the service provider will rent and make use of the real estate for the duration of the contract. Should the contract stipulate that the logistics property needs to be provided by the service provider, the service provider may draw on their own property resources. If these resources do not suffice, the service provider may also provision the real estate within the framework of a (new) project. The basic decision here is whether to buy, rent, or lease the property. Alternatively, a joint venture between the client and the logistics service provider is conceivable.[30]

10.3.2 Rent and Leasing

Rent is to be understood as a landlord's permission for a tenant to use a property (rental object) for the duration of the rental period in exchange for money.[31] Commercial rental contracts – as used for logistics real estate – are usually made for a period of several years. The contractual design criteria include, above all, periods of notice, renewal clauses and rent adjustment clauses. In practice, a variety of rent options have established themselves, as for example *sale and rent back, build and rent,* or *buy and rent*.[32]

Logistics real estate may also be provisioned based on a lease contract (plant leasing). Leasing is a form of tenancy which is 100 % externally financed. Owners, tenants or lessees of the property may be industrial enterprises, trading companies, or logistics service providers. The lessor is a leasing company acting as intermediary. The lessor purchases or builds the property on behalf of the lessee by means of

[28] Cf. Bjelicic and Kostka (2000), p. 167.

[29] Cf. Nehm et al. (2009), p. 46.

[30] Cf. Nehm and Schryver (2007), p. 237 et seq.

[31] Cf. §535 (1) and (2) German commercial code.

[32] Cf. Ockens (2003), p. 206.

a bank loan. It is assumed that the lessor will at least partly earn back the capital expenditure plus interests and profit. The lessee merely has to pay lease instalments. This excludes the lessee's reliance on credit lines and providing of collateral through the company's assets, which may significantly improve a company's rating for other financing transactions.

From a fiscal point of view, it is important in what way *lease contracts* affect the balance sheet and taxation. Depending on whether the investment risk is borne by the lessor or lessee and to whom beneficial ownership is assigned, we can distinguish between operative leasing and finance leasing.[33] In cases of *finance leasing*, the brunt of the risk is borne by the lessee and the leasing item will show on the lessee's balance sheet. All other cases are referred to as *operate leasing*, which means that the lessor will include the item on their balance sheet. Therefore, only operate leasing offers the possibility of off-balance financing and thus of balance-sheet contraction. Regardless of the specifications in the contract, an item cannot be accounted by the lessor if the property is tailored to the lessee's needs (*specialty leasing*) and only the lessee could use it in an economically reasonable way upon expiration of the lease contract. However, most (standard) logistics properties can be used in a variety of ways which is why specialty leasing is not usually carried out.

Depending on the initial situation and the goals of the lessee, we can make a distinction between new-building leasing, buy-and-lease, and *sale-and-lease-back*[34] (see Sect. 7.3.2). Especially the sale-and-lease-back of logistics real estate is an up-to-date option for logistics service providers. It entails the lessor's purchase of an existing property from the lessee in order to subsequently lease it back to the lessee. Since the outsourcing of logistics services to logistics service providers is associated with increasingly shorter contract periods, service providers need to align their capacities more and more with the duration of service agreements. With regard to the provision of logistics real estate, this requires them to rent or lease real estate according to demand and to ideally sell their own real estate in order to subsequently rent or lease it back.[35]

10.4 Financing of Logistics Movables

10.4.1 Internal and External Financing

Logistics movables comprise equipment and chattels of storage and transshipment facilities. This includes conveying technology, racks, packing machines, and other (intralogistical) objects, any kind of vehicle in logistics facilities (industrial trucks),

[33] Cf. Becker (2008), p. 190.

[34] Cf. Feinen (2003), p. 191.

[35] Cf. Stenzel (2003), p. 141; Mahler (2008), p. 40 et seq. and p. 199 et seq.

and every means of transport of the different transport modes in road freight traffic, sea freight traffic, inland waterway traffic, and air freight traffic (see Sect. 4.2). Specifically equipment such as high racks, conveying technology, and picking installations are frequently tailored to the real estate in which they are installed and operated. Despite their building-like character, high racks are therefore not considered real estate but are classified as machine.[36]

The *forms of financing* for logistics movables are largely the same as for logistics real estate (purchase, rent, leasing). However, flexible financing of provisioning and usage are even more important for logistics movables. On the one hand, the employment of the different capacities is dependent on capacity utilization. On the other hand, the service provision for logistic processes and the resulting cost structures greatly depend on the technological standard of the equipment and vehicles used.

Thus, changes to e.g. the IT infrastructure (release changes, cloud computing) or changes of engine (soot filter, hush kits) for vehicles and planes have a significant impact on the economic operating life (investment security). For economic reasons, it is therefore not advisable to finance, for example, vehicles and IT systems from one's own resources.

As with logistics real estate, logistics service providers need to consider the operating life of logistics movables with a view to contract periods. Here, high flexibility when renting or leasing items has to be weighed against increased costs for usage.[37]

10.4.2 Leasing

The *leasing* of mobile capital equipment (equipment leasing) involves the lessor's ordering of goods which had been requested by the lessee. Make, special design options, and supplier are determined by the lessee. Lessee and supplier also agree on a price. Procurement costs and the financing of the leasing item are borne by the lessee.

The distinction between *credit purchase* and leasing is much more difficult to draw for logistics movables than for logistics real estate. This is partly due to the fact that there are numerous mixed forms (hire purchase, externally financed purchase) and partly because the lessee has the option of purchasing the leasing item at a later point. The lease contract may already stipulate this (leasing with purchase option). Depending on the leasing item, such option rights are frequently made use of for logistics movables. This is especially true if the market value of the leasing item is higher upon expiration of the contract than the purchasing price was at the beginning of the contract. Thus, utility vehicles are acquired by the user and used for different purposes, depending on their mileage and distance travelled.

[36] Cf. Mahler (2008), 104.

[37] Cf. Steinmüller (2007), p. 553.

As opposed to hire purchase, leasing involves that – for the term of the contract – the investment good remains the property of the lessor, who also bears the residual value risk. The lessee only pays for the usage of the leasing item and need not include it on the balance sheet (off-balance-sheet treatment). The advantages resulting from this include improved liquidity since no capital is tied up, transparent cost structure (fixed payments), and fiscal advantages (tax-deductible operating costs). In addition, leasing makes it possible to keep abreast of the latest technological developments since switches to new products and technical solutions (e.g. hardware upgrades) may also be implemented during the term of the leasing agreement for a fee (migration fee) and at changed conditions.[38]

(Full) service leasing is an extended form of leasing, where the lessor offers additional services apart from financing. Examples of this are vehicle leasing and fleet leasing, where different services related to vehicles and fleets can be individually combined and offered to the customer. Among these services are[39]:

- Maintenance and repair
- Provision of replacement vehicles
- Payment of vehicle tax, toll charges and other charges
- Accounting of fuel cards and gas station services
- Analysis of gas consumption and other vehicle data
- Offering insurance services
- Handling of claims

Another form of service leasing is *fleet leasing*. With this type of leasing, an entire fleet consisting of several vehicles is the subject-matter of the leasing agreement. Fleet leasing can include services such as the provision of replacement vehicles as well as services related to fleet management. This is due to a trend followed by more and more transport companies – especially from the road freight transport sector – resulting in an outsourcing of their fleet management. Even aircraft fleets and ships can be the object of leasing agreements. This is referred to as *large-scale-project leasing,* for which a leasing company is often especially set up. Such a company only leases the specific object and deals with all procurement and financing processes.

10.4.3 Build-Operate-Transfer Models

Another way to finance fixed and current assets in logistics and supply chains is *build operate transfer*. This is a financing model based on public private partnership that has become very common among private-sector companies for infrastructure projects in the energy sector, waste disposal industry, telecommunications sector, and for airports or road construction.[40] The manufacturer acts as the client of the

[38] Cf. von Eisenhardt-Rothe and Jütte (2003), S. 155 et seq.

[39] Cf. Mangold (2001), S. 223 et seq.

[40] Cf. Donier (2006).

operating company. Investment in production facilities and technological infrastructure is made by the operating company (SPC – Special Purpose Company) which is the owner of the production plant. The operating company is also in charge of designing, building, and operating the plant, including maintenance, replacement investment, and spare parts logistics. Since the investment is non-capitalized on the client's balance sheet (according to the German Code of Commercial Law HGB), the costs associated with the provision of the services vary for the client.

Other stakeholders involved in an SPC are investors, plant manufacturers, project managers, financial service providers, and works managers. The set-up of an operating company constitutes a case of *project financing*, which means that payment is usually effected on availability or upon production (pay on production, pay to production).[41] Minimum order quantities are frequently associated with build operate transfer. This offers the client the advantages of low capital tie-up for stocks and fixed assets as well as prefinancing by the operating company.

Build operate transfer has not only been applied to technical equipment and production plants but has also increasingly become applicable to supply chain management. The aim is to allocate logistic tasks to specialized service providers within the logistics chain. These tasks may include storage, transport, and goods handling as well as the entire inventory management.

Case Study 10.1: Build-Operate-Transfer Models in the Industry

Automobile Industry and Utility Vehicle Industry

Since 2002, parts of the production facilities and conveying technology at the Ford plant in Köln-Niehl have been operated through a build-operate-transfer model by *EISENMANN AG*, Böblingen. These include, amongst other things, the final-assembly line for the models *Fiesta* and *Fusion*. The build-operate-transfer model comprises the operation and maintenance of all conveying systems. Invoicing takes place on the basis of readily assembled cars.

Production at *MMC Smart* in Hambach (see case study 6.3) is based on a build-operate-transfer model which involves numerous suppliers. Due to the low vertical range of manufacture of only 10 % at MMC Smart, system suppliers do not only undertake 90 % of production but they have also borne roughly 50 % of the production plant's investment costs amounting to 400 million Euros.

The painting plant at the utility vehicle production site of *MAN* Latin America is also operated through a build-operate-transfer model by EISENMANN. EISENMANN is in charge of varnishing the driver's cab, which includes full responsibility for the plant and personnel.

Airline industry

A build-operate-transfer model has been realized between *Flughafen München GmbH* and *Lufthansa AG* for Terminal II at Munich airport. Lufthansa AG took on 40 % of the investment costs. In return, the terminal was designed according to Lufthansa's specifications.

[41] Cf. Gomm (2008), p. 246 et seq.

Review Questions

1. Which concepts suitable to finance logistical fixed and current assets do you know?
2. Distinguish between capital costs and capital tie-up and explain how they can be reduced.
3. Explain the concept of Supply Chain Finance.
4. Which stakeholders can provide financing in logistics?
5. Outline the significance of the cash-to-cash-cycle.
6. Assess equity financing and external financing from a financial perspective and with a view to financial risks.
7. Name the differences between factoring and forfaiting.
8. What are the advantages of pay on production?
9. Explain the importance of capital budgeting, using the example of an intralogistical investment.
10. How are build-operate-transfer models different from purchase or leasing?

Additional Literature

Bowersox, J. D., Closs, D. J., & Cooper, M. B. (2012). *Supply chain logistics management.* New York/London: McGraw-Hill.

Buxton, C. (1999). The financial-logistics connection: Supply chain financing. In *Finlistics* (no. 23, 1999).

Feinen, K. (2002). *Das Leasinggeschäft, 4. neubearbeitete Edition.* Frankfurt: Knapp.

Hartley-Urquhart, R. (2006). Managing the financial supply chain. *Supply Chain Management Review, 10*(6), 18–25.

Hofmann, E. (2011). *Supply chain finance solutions: Relevance – propositions – market value.* St. Gallen/London: Springer.

Jacob, D., & Schröter, N. (2009). *Bedeutung von PPP für ein modernes Infrastrukturmanagement im öffentlichen Sektor.* Wiesbaden: Springer.

Palmia, S. (2009). *Financial supply chain.* New York: McGraw-Hill.

Skiera, B., Pfaff, D., & Weiss, J. (2004). *Financial supply chain management.* Bonn: Galileo Press.

Stock, J. R., & Lambert, D. M. (2001). *Strategic logistics management.* Boston: McGraw-Hill.

Timme, S. G., & Williams-Timme, C. (2000). The financial-SCM connection. *Supply Chain Management Review, 4*, 32–43.

Wildemann, H. (2010). *Betreibermodelle: Leitfaden zur Berechnung, Konzeption und Einführung von Betreibermodellen und Pay-on-Production-Konzepten, 9 Edition.* München: TCW.

Logistics Controlling

11

Learning Objectives

Theoretical and practical insights in modern accounting suggest that the rating of logistics services and costs as well as comprehensive controlling in logistics and supply chain management are a central prerequisite for Logistics Controlling and Performance. The following chapter gives an overview about the current situation and the development of cost and performance accounting in logistics. Additionally, key figures and key figure systems will be introduced with respect to logistical tasks. Apart from these basics, more elaborate approaches to comprehensive supply chain controlling and performance management in the field of logistics will introduce the reader to current challenges in company practice. The author introduce the prevalent German appellation *Controlling* as a synonym for *Managerial Accounting* which is rather used in Anglo-American literature

Keywords

- Cost and performance accounting
- Functions and the process of Controlling
- Key figure systems
- Key Performance Indicators (KPI)
- Supply Chain Controlling
- Cost drivers and cost effects
- Process costs and process performance
- Balanced scorecard

H. Gleissner and J.C. Femerling, *Logistics*, Springer Texts in Business and Economics,
DOI 10.1007/978-3-319-01769-3_11,

11.1 Conceptual Basics

In today's complex and widely connected world and its multitude of quantitative and qualitative data and information, successful company management requires tools to rate, plane, and control processes. All measures to acquire, develop, assess, and rate such control information and transfer them to concrete courses of action can be summed up as the instrument of *Controlling.* In this context, Controlling is not only to be understood as an instrument of control but as one of information and management for executive levels.[1] Planning, control, and analysis as well as information derived from them are elementary functions of Controlling for management (see Fig. 11.1). These Controlling functions can be used in all areas and on all levels of a company. They are as useful for the whole company as they are for the logistics department.

In academics as well as in business practice, the term Controlling has a very wide area of application and various contents. Additionally, the form of Controlling depends largely on the size of the company. The acquisition, preparation, and control of the data and information happen in the company process according to an hierarchical order: from the individual operative and administrative business processes, and the accounting for cost and performance accounting to, eventually, the Controlling, resulting in detailed assessment and proposals for action.

Missing data to evaluate processes and an overrating of the significance of data about business process results or an unreflecting focus on insights from Controlling instruments have repeatedly led to wrong management decisions in the past.

This shows the challenges Controlling faces in business practice. Its quantitative results clearly have a supporting function in business process rating and controlling. However, it should not limit management decisions with respect to their flexible

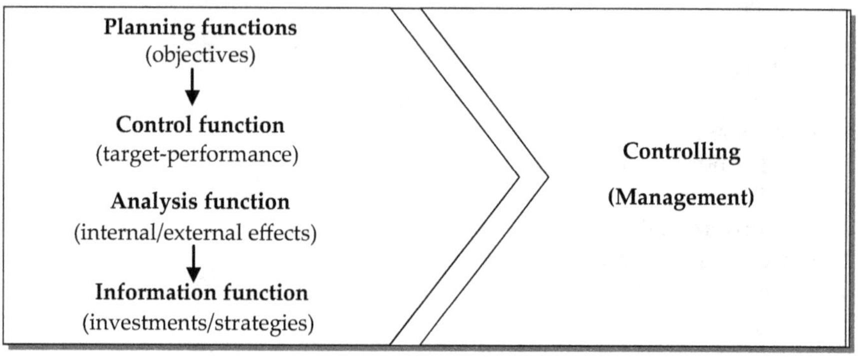

Fig. 11.1 Basic functions of controlling

[1] Cf. Weber (2002b), p. 5.

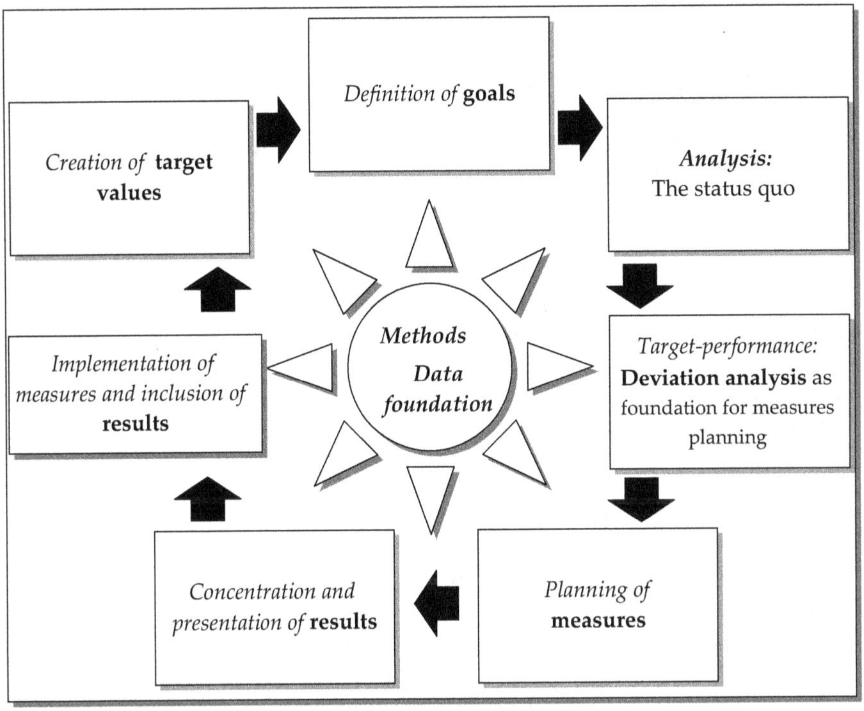

Fig. 11.2 The controlling process (Cf. Kiesel (1997), p. 346)

reaction to market realities and an additional qualitative assessment of market and customer requirements.

Controlling is a continuous process that works with mathematical and statistical methods and can be supported by numerous IT systems. Usually, these systems are connected with or integrated in other IT systems in the company. Controlling processes are often designed as a control circuit as shown in Fig. 11.2.

Looking at the field of employment for Controlling in a logistical context, a distinction has to be made between logistics controlling as the task of one department and the Controlling for logistics service providers. The latter is the same as general Controlling for companies and will not be described in detail here. The following refers to aspects from Controlling as a concrete function.

In addition to the realization of services, the rationalization of transport, handling, warehousing, and additional services is a continuous task of logistics.[2] This means that the core task of logistics controlling is to support the management in controlling and managing logistical processes. Contributing to cost reduction and efficiency increase is as important as the aspect of control.

[2] Cf. Klaus (2002), p. 30.

The use of Controlling instruments for logistical processes in business practices is not yet adequate. The problem that occurs most often is the unsatisfying level of precision in assigning logistics costs to the logistics services that caused them.[3] One reason for this is the high effort and the difficulty that is necessary to acquire data for logistical processes. A second reason is the increased difficulty to establish stable systems for data acquisition because of tasks changing quickly with varying customer requirements. Due to the strong influence of shareholders and financial investors, individual parts of the logistics process may be affected by events or measures such as mergers or outsourcing.

Owing to the diversity of logistical processes, it is not possible to define Controlling concepts that apply to logistics as a whole. Such concepts must always be tailored to the special requirements and the individual situation in which logistics processes take place.[4]

The goals of logistics controlling can be defined as:

- Formulating and specifying logistical goals and their integration in the context of the whole company
- Budgeting and creating targets for logistics
- Making available instruments with operational value to create targets and, subsequently, measure success
- Supporting and coordinating strategic and operative logistics planning like e.g. investment decisions
- Regulations and information transparency for cost and performance accounting for logistics and its processes
- Foundation for cost control and efficiency increase (performance) in logistics
- Formulating further management information, e.g. concerning the organizational structure or staff leadership

Analogue to the process- or network-oriented logistics in supply chains, additional logistics controlling must be active across steps in the value chain. To assess the economic efficiency of a full supply chain or a complex logistics network, Controlling instruments must be universally used. Such cases are called Supply Chain Controlling; it is a controlling instrument of Supply Chain Management.[5] The fact that the Controlling process must be viewed across numerous steps in the value-adding process which means across companies or even countries is another challenge for the unified source-based assignation of logistics services and costs.

---

[3] Cf. Straube et al. (2005), p. 26.

[4] Cf. Weber (2002b), p. 13.

[5] Cf. Stölzle and Otto (2003), p. 29 et seq.

11.2 Logistical Costs and Performance Accounting

With the processing of cost and performance information, first data aggregation takes place via the cost types, cost center, and cost unit accounting. Connecting the performance (output) to the resulting costs (factor input) and their source-based definition takes place in accounting (see Fig. 11.3).

To improve the quality and significance of Logistical Costs and Performance Accounting, it makes sense to differentiate in detail the logistical cost types as early as when they are acquired. In the field of transport, for example, it is possible to differentiate between the cost types parcel service, forwarder, or own transport fleet. As cost centers, transport costs can, for example, be differentiated between purchasing transports, transports between sites of the own company, and delivery transports to customers.

Cost Center Accounting connects individual costs (e.g. from production) to the full costs; this also means calculating the share of logistical costs in the costs for a certain product. More differentiation means a higher quality of the source-based detection of the logistics costs for a certain product. The term product, in this context, can have a wide range of meaning. Usually, it means products as the result of industrial production. However, it can also mean *service products* (transport, storing orders) that are the subject of Cost Center Accounting.

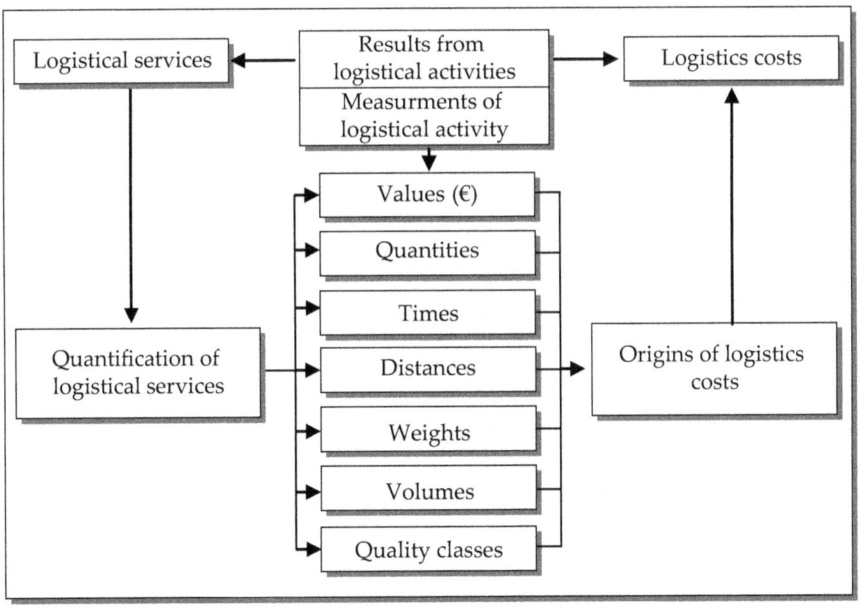

Fig. 11.3 Characteristics of logistical activities (Cf. Reichmann (2006), p. 420)

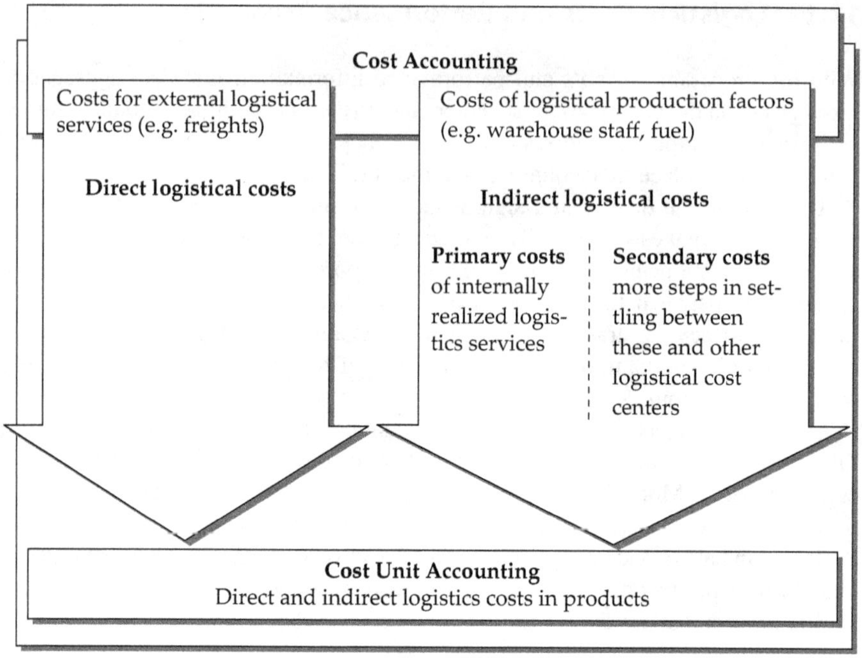

Fig. 11.4 Traditional integration of logistics in cost accounting (Cf. Weber (2002b), p. 107)

These results are directly included into calculating the contribution margin or into price calculation. They may also be the foundation of activity-based costing (see Sect. 11.4.1) (see Fig. 11.4).

Basic elements of logistics costs are:

- Order processing costs
- Transport costs for covering distances in the logistical process
- Warehouse costs, for covering time in the logistical process
- Stock costs
- Control and system costs, for logistical control, administration and IT used in logistics

The following Table 11.1 outlines the logistical cost blocks and their characteristics regarding their variability for changes in the logistics volume.

Logistics costs are mostly accounted for in external accounting from a perspective of taxes or financial accounting. However, logistics performance must be dealt with by internal accounting. The key to a high-quality result is the precise recording of performance in logistical processes. Prior to such a performance recording, the goal of this measure should always be clarified.

Table 11.1 Basic scheme of logistics costs

Category	Description	Price elasticity	Type of costs
Order processing	Receipt of customer orders, realization of orders (picking, packaging, transport), customer contact during the process (tracking & tracing), order completion	Fixed/variable	Staff, material costs
Transport (internal)	Own transport fleet	Fixed/variable	Staff (e.g. driver), material costs (e.g. write-down for truck, fuel)
Transport (external)	Forwarder/service provider	Variable	Material costs (e.g. invoices from forwarders)
Warehouse capacity	Warehouse/equipment	Step-fixed/variable	Staff (e.g. technicians), material costs (e.g. rent, loan, write-down)
Warehouse handling	Staff	Fixed/variable	Staff, material costs (e.g. write-off forklift)
Stock costs	Interests on fixed capital	Variable	Material costs (e.g. interest on credits)
Control, management, system costs	Logistics management, warehouse accountant, warehouse management system	Fixed	Staff (e.g. warehouse manager), material costs (e.g. writedown management system)

The following questions have to be clarified, and the following regulations have to be made – also in cases where performance recording happens automatically or with IT support:[6]

- Definition of the performance to be recorded
- Process-oriented description of the performance and the extent of recording
- Frequency of the recording (every hour, daily, randomly etc.)
- Location and sources of performance recording

Table 11.2 Portion of logistics costs, compared by sectors (Cf. Straube et al. (2005), p. 301)

Sector/market	Portion of logistics costs (% of turnover)
Food	8.0 %
Chemistry	6.9 %
General engineering	6.0 %
High Tech/ Electronics	5.2 %
Automotive	5.1 %

[6] Cf. Weber (2001), p. 67.

A study by the *Bundesvereinigung Logistik* (German Logistics Association) gives ideas about the size of logistics costs in different sectors. On average, the share of logistical costs in the complete turnover of a company is between 5 and 8 % (see Table 11.2).

11.3 Key Figure Systems

The data acquired from Logistical Cost and Performance Accounting often have a large volume. To be able to use this information for control or analysis as continuing control information or as a foundation for management decisions, it has to be compressed. Usually, the results from compression are key figure systems. All information and key figures about the overall system company and its parts are collected in a reporting system. In this context, the term *Performance Management* is often used. It shows the performance of the logistics system and the costs connected to this.

In particular, ratio systems describe the result of logistics from a compilation of actual data. However, they can also be prepared as planning or target values for a certain business period. In order to gain significance, key figures must never be created and assessed individually but should always have the form of a system consisting of more key figures. They should be comparable with respect to their content, the calculation method, and the time period during which they were acquired.

In a company in general, but also in specific for the logistics department, a key figure system performs the following tasks[7]:

- Quantification of company and logistics goals as guidelines to comply with the budget, increase efficiency, or lower costs in logistical processes and as a foundation for target agreements with the management (operationalization function)
- Continuous comparison of the logistical processes (actual values) with the given target values as a part of internal control (control function)
- Systematic analysis of the deviation between actual and target values as well as of the periodic or aperiodic internal vulnerability and efficiency analysis (analyzing function)
- Analysis of external developments
- Analysis of the company situation compared to that of other companies or comparable logistical processes
- Support for decisions about countermeasures in case of deviations between target and actual values; assessment of possibilities to expand the logistical capacity (regulation function)

[7] Cf. Grochla (1983), p. 51.

To ensure valid comparability, the following principles should be kept in mind for working with key figures[8]:

- Clear quantification of key figures and the corresponding basic data
- Significance through connection or comparison with other key figures
- Appropriate relation between information value benefit and effort/costs of determination i.e. selection of a limited number of significant key figures
- No standardized key figures but key figures tailored to the company's requirements
- No individual key figures but sensible key figure systems for internal, inter-company, or periodic comparisons
- Clear (graphic) presentation of the key figures
- Explanation and interpretation of the key figures, e.g. with respect to internal and external factors of influence

Figure 11.5 shows the requirements for key figures regarding calculation and administration.

These requirements for the acquisition of key figures show again that high accuracy must be ensured for these activities and that reasonable comparisons

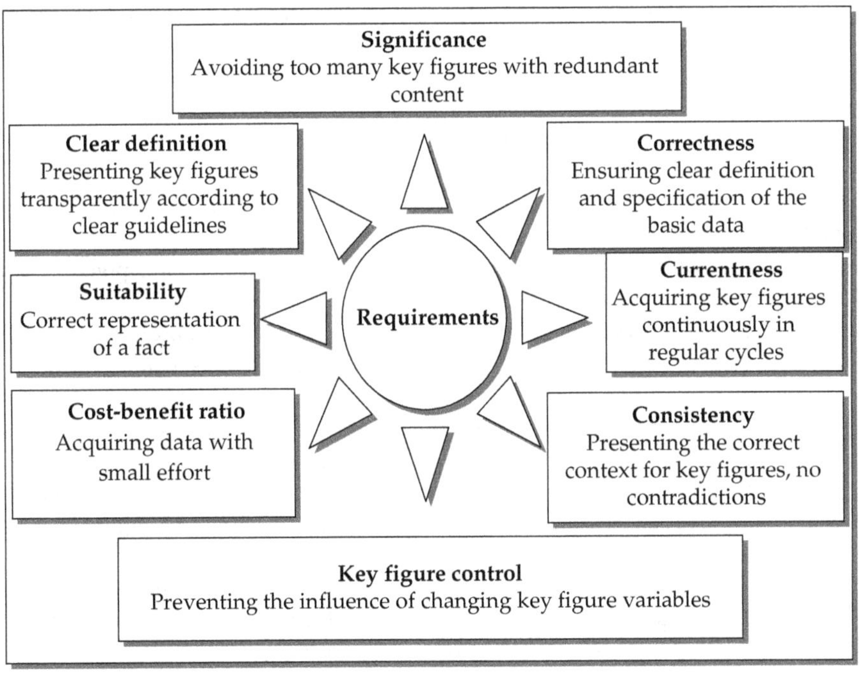

Fig. 11.5 Requirements for key figures (Cf. Friemuth et al. (1997), p. 98)

[8] Cf. Grochla (1983), p. 61 f.

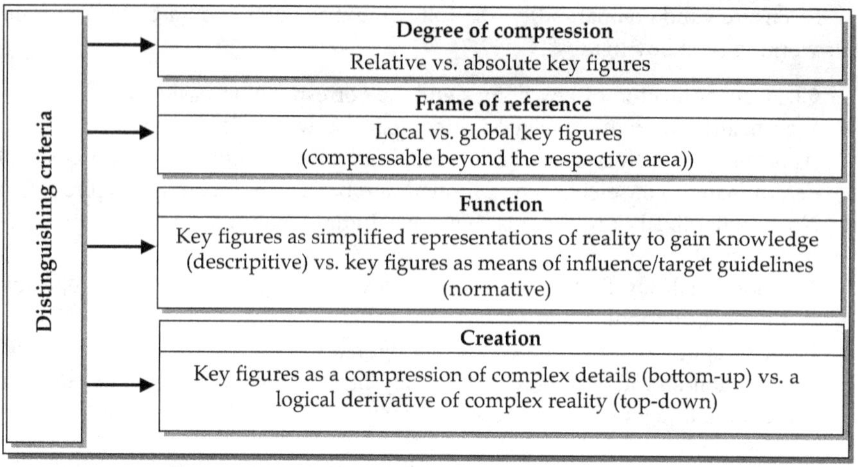

Fig. 11.6 Distinguishing criteria for key figures (Cf. Weber (2001), p. 229)

and results are only possible with key figure systems, i.e. in a combination of several key figures. On the other hand, the number of key figures should be limited to ensure clarity and to win clear insights for success-critical processes or process elements in logistics. In this context, the term *Key Performance Indicator* (KPI) has been established.

The criteria presented in Fig. 11.6 can be distinguished for the differentiation of key figures. Relative key figures are clearer but to concretely describe facts, absolute key figures are essential. The frame of reference can also be different and is often connected to the hierarchical level for which the corresponding key figures are to be a foundation for decisions.

However, to assess the logistical efficiency, linking relative and absolute key figures can be useful. Furthermore, a distinction between key figures that are used for continuous efficiency control and key figures that are to be principle targets for logistics planning can be made. Largely connected to this is the question how the key figures were created. Mostly, there is a compression of information from the operative level (bottom-up) upwards to the company as a whole. If the key figures have a guideline function, this happens the other way round (top-down). Logistical key figure systems can be divided into four categories, with respect to their content (see Fig. 11.7).

Structural key figures describe a logistics system by its size, form, and performance potential. They are absolute key figures without further input or output reference.

Productivity key figures show the performance capability of logistics.

Economic efficiency key figures rate efficient factor input. They assess the logistical performance with the necessary costs. These key figures are connected to monetary values in currency units that mostly refer to defined periods.

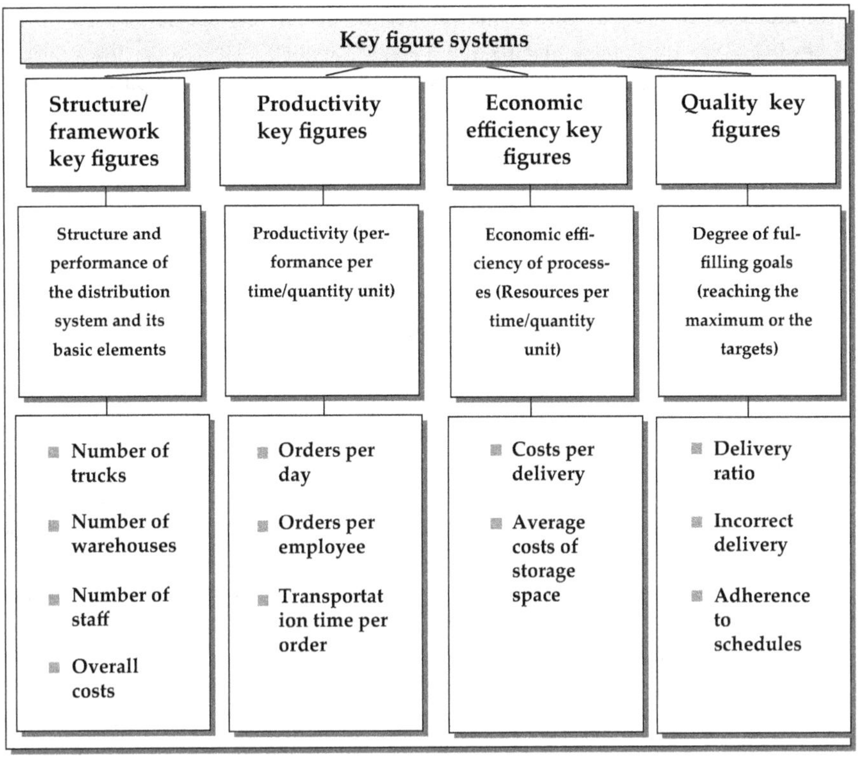

Fig. 11.7 Logistical key figure system (Cf. Gleißner (2000), p. 96)

Quality key figures show the degree to which goals in logistics performance were reached. Usually, a comparison with the target values from the planning phase takes place.[9]

Case Study 11.1: Key Performance Indicators

Lafarge Roofing Components GmbH in Oberursel, Germany, produces parts for rooftops – e.g. roof gutters, ventilation elements, roof windows, etc. – for the globally active sales companies of the Lafarge Roofing Group. It has production plants in Germany, Estonia, South Africa, and Malaysia. The global purchasing of commercial goods like solar modules for rooftops is an integral part of the operation. The two central warehouse locations are in Germany.

The central logistical task is bringing the production output or the commercial goods to the central warehouses and, subsequently, delivering to the warehouses

[9] Cf. Schulte (2009), p. 646 et seq.

of the sales companies worldwide. The terms of delivery are usually *ex works*, i.e. the costs must be borne by the sales companies. To control the logistics system, a key figure system is employed. It is divided into three parts:

- Costs
- Delivery Capacity Indicators
- Working Capital

The goal of this representation is the creation of cost transparency for the logistical processes in the organization.

The most important *framework key figures* are the absolute costs per time unit, i.e. the sum of all incoming and outgoing freight (shuttles from the plant to the warehouse, delivery to the sales companies); the capacity costs (rent for the external warehouse, write-downs for the internal warehouse); the handling costs (costs per incoming/outgoing goods for the external warehouse, staff costs for the internal warehouse, packaging); and the costs for central management functions (strategic purchasing, planning and scheduling, warehouse management). The costs are compared to the corresponding planning values and those from the previous year. In case of significant differences, additional ad-hoc analyses are conducted to clarify the reasons for this. Doing so, values, e.g. quantity, productivity, and price effects are shown separately (*productivity and economic efficiency key figures)*.

The most important *quality key figure* is called *WOTIF* (Work on Time in Full). It shows to which percentage the goods were ready to be picked up in a warehouse location at the agreed time. It is calculated for every production site or commercial good and can be broken down to the level of individual items. Also in this case, significant deviation from the target values is analyzed. Subsequently, controlling measures are drawn up with the production planners of the sites or the suppliers.

Another key figure to assess delivery quality is the *planning accuracy* of the individual sales companies (deviation ordered quantity to actually picked-up quantity in per cent per time unit). If it deviates from the standard tolerance, it can also influence the WOTIF key figure. The key figure planning accuracy is the subject of controlling meetings with the sales companies.

The actual stock quantity is compared to the target values and those from the previous year to assess working capital (see Chap. 10). For this, raw materials, semi-finished products, and finished products are assessed individually per production site. The separate assessment of the actual values corresponds to the different persons responsible for different stocks within the organization. For example, the raw material management department or the production-planning department in the plants is responsible for the raw material and semi-finished goods stock; the logistics-planning department is responsible for the so-called make-to-stock items.

Days of Stock (DOS) is another key figure for stock controlling. It is calculated as follows:

$$\text{DOS (days)} = \text{size of stock}(€)/\text{turnover}(€) \times 360(\text{days})$$

Compared to planning values and values from the previous year, this key figure provides information about the warehouse handling frequency or the range of the stock.

When the results from the Logistical Cost and Performance Accounting are prepared and available for Controlling, the structure described above is sufficient to create a multitude of key figures about the logistical process or the logistical environment. That is why a systematic approach to working with key figures is necessary; their use should be limited to a small number of key figures that are directly aimed at the target KPI.

While working with KPI's certain limits to the use of key figures should be considered:[10]

- Key figure inflation, i.e. when too many key figures with largely the same content are created; often, the creation effort and the significance do not have a favorable ratio
- Mistakes because of imprecise definition and specification of the basic data
- A lack of consistence of key figures; contradicting statements or key figures that clearly do not have a connection to each other
- Problems with the direct or indirect key figure control, i.e. it is possible to manipulate key figures by changing activity variables, e.g. during the acquisition, in a way that critical insights can be covered

To avoid key figure inflation, to make the results from Controlling easy to use for the management, and to present interpretation approaches to the management, the Balanced Score Card (BSC) system was developed.[11] The BSC is a structured and balanced key figure system that usually looks at a company from four different perspectives with a cause-effect relationship:[12]

- Financial perspective (turnover, ROI, logistics costs, transport fleet costs etc.)
- Customer perspective (degree of delivery service, incorrect delivery ratio, loss ratio etc.)
- Business process perspective (order processing time, average warehouse handling time, average stock etc.)
- Learning and development perspective (fluctuation ratio, employee satisfaction, proposals for improvement etc.)

[10] Cf. Schulte (2009), p. 667 et seq.

[11] Cf. Kaplan and Norton (1997), p. 7 et seq.

[12] Cf. Vahrenkamp (2007), p. 432.

Initially designed for top-management as a basis for assessment and decision-making for the full company process and its context, the BSC is nowadays used for logistical tasks as well.

With traditional key figure systems usually comparing internal or external contents and/or different performance periods, the BSC is the first system to connect KPIs from different areas of the company process to its environment. The goal of such an alternative way of compiling KPIs – that show certain or potential factors of influence – is to be able to recognize connections or effects in other areas of company activity.[13] If, for example, there is a cost increase per

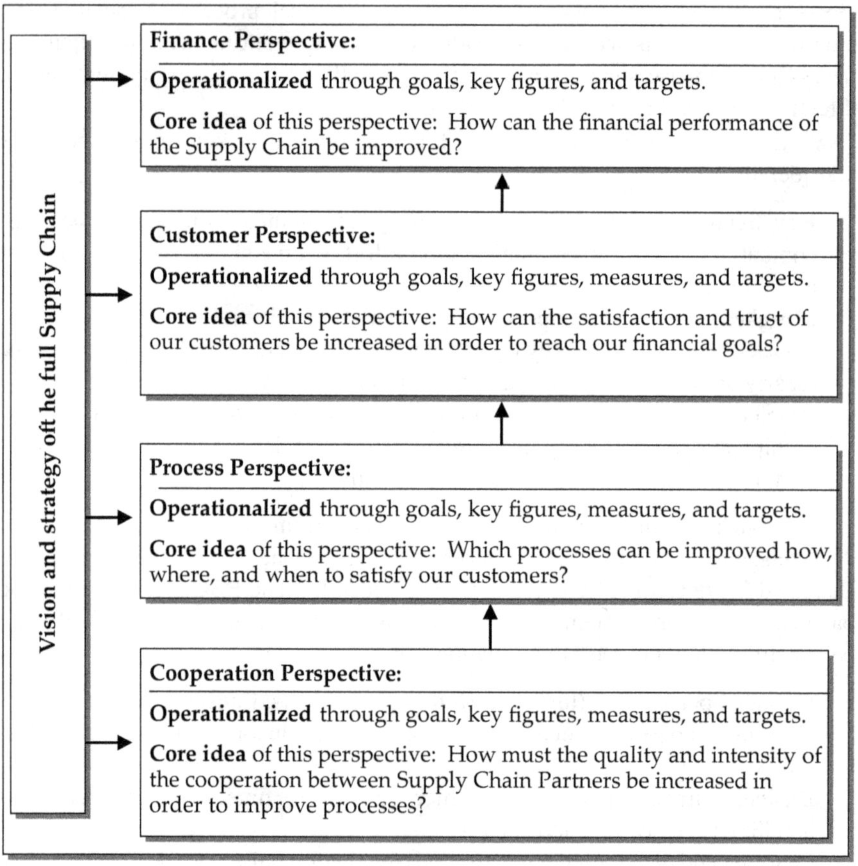

Fig. 11.8 Balanced score card of a supply chain (Cf. Rödler et al. (2003), p. 42)

[13] Cf. Weber and Schäffer (2011), p. 193 et seq.

handled shipment for a defined period without a decrease of shipment volume or an increase in external costs in logistics, this can be explained by a growth rate of the KPI in the HR department *Costs per Staff Hour* due to a higher number of employees on sick leave.

A characteristic of the BSC is its compact and clear layout (which may also be graphical). This is achieved through the aggregation of the individual perspectives to one minimum that is relevant for decision-making (KPI).[14] The BSC of a Supply Chain shown in Fig. 11.8 shows the quality and intensity of the Supply Chain parties' cooperation and that customer satisfaction is directly connected to this.

Case Study 11.2: ROI Scenario Calculation

The *DuPont System of Financial Control* is one of the most well known key figure systems. It is aimed at the company goal profit maximization and works with the Return on Invest (ROI), i.e. the amount of money gained or lost on invested capital, as its top key figure. The ROI is divided into its individual elements and mathematically fully linked.[15] The individual elements are the adjusting screws of management activity. This visualizes the effects of measures on the ROI – also in the field of logistics – and the proportions thereof.

Figure 11.9 shows an initial scenario.

The effects of individual measures in logistics on the ROI will be calculated in the following.

(a) *Decrease in operative logistics costs*
From the initial scenario, it is now assumed that the management is successful in reducing logistic costs by 10 %. The share of logistics costs in other costs is 20 %.

(b) *Reduction of stock*
It is assumed that the management is successful in reducing the stock and the necessary operating capital by 20 %.

(c) *Increase in sales*
The company is able to increase its sales by 10 %. This was only made possible by investing 10 % more into other costs and material costs.

Calculate the changes to the ROI for all three scenarios! Use the procedure from Fig. 11.10.

[14] Cf. Karrer and Petzold (2004), p. 91 et seq.

[15] Cf. Weber and Schäffer (2011), p. 190. und Meyer (2011), p. 141 et seq.

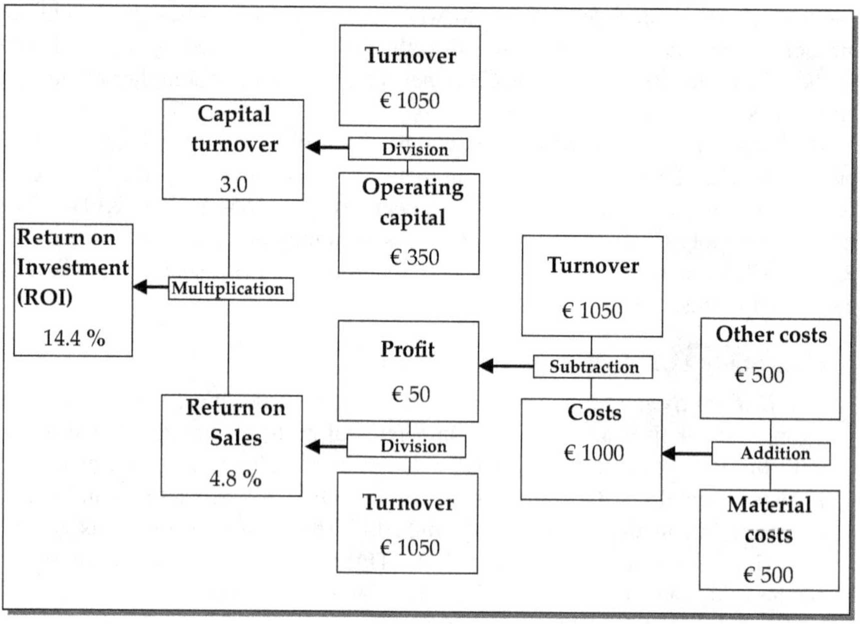

Fig. 11.9 DuPont system of financial control – initial scenario (Cf. Meyer (2011), p. 142)

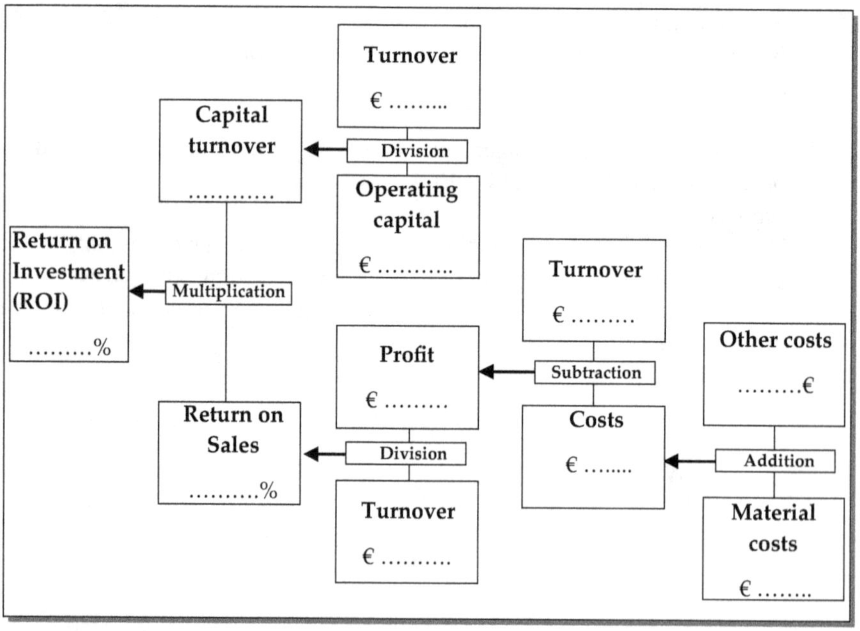

Fig. 11.10 ROI calculation in the DuPont system

11.4 Further Controlling Concepts

11.4.1 Process Cost Accounting

From traditional cost accounting, cost center and cost unit accounting, definition problems arise, especially for the source-based allocation of operating cost. Logistics costs more often appear with a character of overhead costs which usually means flat rate settlement. This leads to an imprecise cost settlement that is not tolerable.

In the fields of production and operating costs for production, this inefficiency in cost and performance accounting has already been analyzed. For improvements in cost allocation, the concept of *Activity-Based Costing* was invented in the USA. Similar to this but different in the process and the range of use, the concept of *Prozesskostenrechnung (process cost accounting)* was developed in Germany. It assesses and analyzes trans-department processes and services in indirect fields.

An orientation towards the process is now more and more common in logistics or for logistical tasks. This makes necessary performance and cost accounting that is divided into the process steps. Thus, process cost accounting becomes an instrument especially suitable for Controlling.

Prior to this, the *process analysis* takes place. Existing logistics processes are divided into main processes, sub processes and, eventually, activities (see Fig. 11.11).

Resources are necessary for activities to take place; these resources cause costs. The amount of the costs can be seen from the number of elements to be moved. In the example in Fig. 9.11, these are the elements to be counted (parcels, packages,

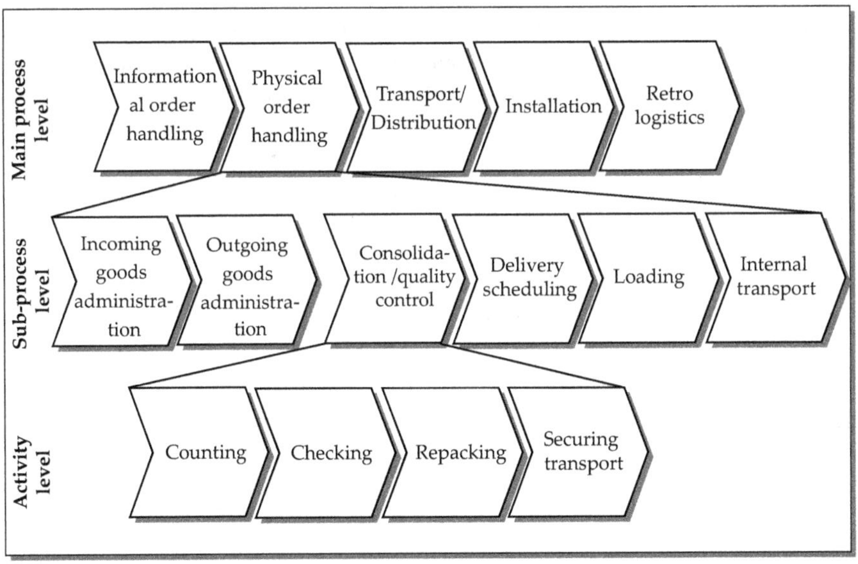

Fig. 11.11 Analysis levels of process cost accounting (Cf. Delfmann and Reihlen (2003), p. 9)

pallets etc.). These elements are called cost drivers. For example, one logistics employee needs 20 s to count the items (elements) for one order. The costs for the employee can be calculated by Euros per second of his or her monthly wage and to Euros per counted item. If such an analysis takes place for every process step that contains different activities, the exact process costs can be calculated.

It is obvious that for a customer order that contains only one package or packing unit of an item, different costs will accumulate for picking, provisioning, and delivery than for an order with a full pallet of the same item. Then again, an order with half a pallet of the same item and five packing units of different items will create different costs. It is widely common to charge the customer a flat rate for every order. This is the point where process cost accounting comes in: it enables detecting the individual sets of costs for calculating the logistics costs for the customer order.

11.4.2 Supply Chain Controlling

As already mentioned, logistics nowadays is not anymore limited to isolated, functional elements such as transport or warehousing but sees itself as a designer of logistics networks in Supply Chain Management. Controlling tasks must be adjusted to this. Tasks like planning, efficiency control, and controlling for processes across value creation steps and company limits are parts of *Supply Chain Controlling*. It focuses on identification, assessment, and optimization of the overall benefit of supply chain processes; subsequently, the profits from these benefits can be allocated to the supply chain actors based on the source or the performance.[16] The instruments for this are the same as the usual Controlling instruments. Realizing Supply Chain Controlling across value creation steps and company borders, however, is not without problems. In everyday work, the provision of the necessary data material by the individual supply chain stakeholders is often difficult. Often, there is a lack of openness and trust between partners; exchanging internal data about performance or, most of all, costs then does barely take place within reasonable time. At this point, relationship management with instruments for increasing trust must be used (see Sect. 3.4.2)

11.5 Costs and Cost Effects; Performance and Performance Effects

The costs and cost effects and the performance and performance effects in warehouse and transport nets show how complex the mapping of logistical cost and performance facts can be (see Chap. 8). In planning warehouse and transport networks, great emphasis is placed on the *tradeoffs* between the network structures

[16] Cf. Neher (2005), p. 29.

and the logistical cost and performance figures because they make possible the assessment of the performance capability and the economic efficiency of the structures.

Cost figures of warehouse networks can be:

- Warehouse costs (staff, rent costs)
- Handling costs (handling, picking, packing)
- Stock costs (interests, obsolescence, storage space costs)
- Transport costs (between warehouse steps, transports to the branches)
- Process costs (controlling, coordination)

The *warehouse cost structure* mostly consists of fixed costs. Variable costs include *handling costs* like incoming goods administration, moving goods in and out of stock, internal transport, picking, and packaging tasks. *Warehouse costs effects* mostly occur when changes to the warehouse structure are made. If the number of warehouses is reduced, there will be higher capacity requirements for the remaining warehouses (space, design, technical equipment). On the other hand, cost reductions can occur through balancing effects in employment of staff as well as through synergies in the management.

Stock costs are a considerable cost driver because of the high capital employed for stock. With stock management largely taking place based on prognosis about potential demand, and replenishment times subject to fluctuations, the risk for incorrect or excess stock increases; it is a considerable *stock cost effect*. The danger of failures (Out of Stock) necessitates maintaining safety stocks, which, in turn, leads to higher stock costs.

Transport costs occur in the form of freight costs for deliveries to production plants, warehouses, or branches. The *transport effects* are mainly influenced by the transport volume structure and the transport cost structure. The transport volume structure depends on how widely vehicle utilization for warehouse deliveries can be realized by bundling. All components of transport cost rates based on distance and volume influence the transport cost structure. If the number of warehouses is increased, there is a tendency for transport costs to warehouses to increase. This effect grows with the elasticity of the transport rate regarding the quantity because lower delivery quantities per warehouse have a stronger effect on the transport costs. On the other hand, delivery costs to branches decrease with a growing number of warehouses for transport costs with decreasing volume because the number of expensive transport distances decreases.

As shown in the paragraphs on stock, warehouse, and warehouse supply costs, *central warehouse network structures* have numerous cost advantages. The necessary capacity for the same handling volume is higher in a full system with a decentralized structure than with a centralized structure. A growing number of warehouses, in turn, leads to higher fixed costs. However, it must be noted that the overall costs are considerably influenced by individual cost categories. Furthermore, individual values are difficult to quantify because of the general difficulties in collection, definition, and allocation of logistics costs. Figure 11.12 summarizes the tendencies of cost effects of warehouse and network structures.

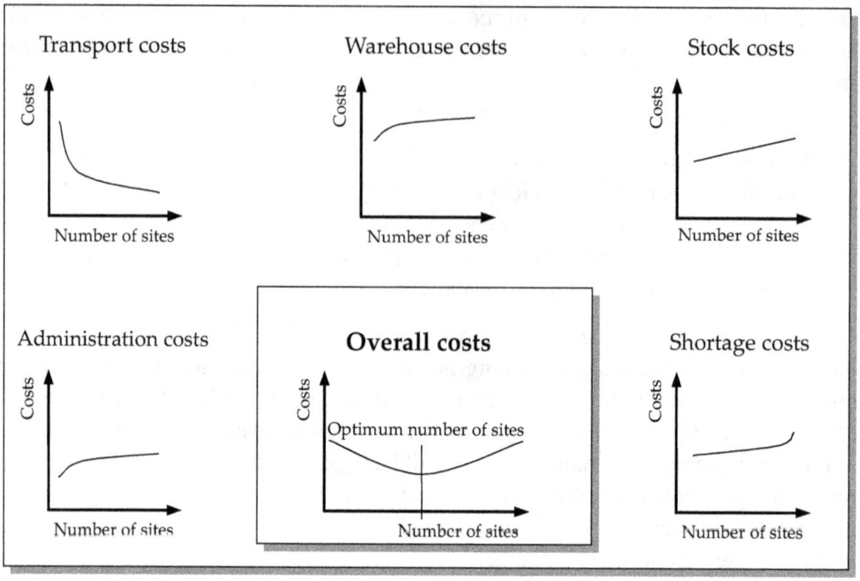

Fig. 11.12 Cost effects of warehouse and network structures (Cf. Pfohl (2010), p. 61)

Mapping *performance effects* is equally difficult. Bundling effects through centralization are contrasted with longer delivery times resulting from growing transport distances to the branches. Additionally, the fluctuation in transport reliability grows with larger distances, which makes necessary higher safety stocks.

Logistics controlling must support these – and more – interconnections of logistical decisions.

Review Questions

1. Which functions must Logistics Controlling include?
2. What does a Balanced Score Card show? Outline a BSC for a logistics service provider.
3. Why is process cost accounting a suitable instrument for logistics?
4. What are cost drivers? Name cost drivers in logistics.
5. What should be components of Performance Management for distribution logistics?
6. Name three central KPIs for purchasing logistics.
7. Connect Supply Chain Controlling to Supply Chain Management. What is the benefit for Supply Chain Management?
8. Outline productivity and quality key figures in logistics.
9. Name key figures for stock management.
10. Which logistics approaches can a company use to improve its ROI?

Additional Literature

Davis, C. E. et al. (2011). *Managerial accounting*. Danvers: Wiley

Garrison, R. et al. (2011). *Managerial accounting*. New York: McGraw-Hill.

Gladen, W. (2008). *Performance Measurement – Controlling mit Kennzahlen, 4. Aufl.* Wiesbaden: Gabler.

Hansen, D. R. et al. (2011). *Managerial accounting*. Mason: South-Western College Pub.

Horvath & Partner (Ed.). (2000). *Balanced scorecard umsetzen*. Stuttgart: Schäffer-Poeschel.

Horvath and Partners (2006). *Der Weg zu einem wirkungsvollen Controllingsystem*, 6. Aufl. München: Deutscher Taschenbuchverlag.

Jahns, C. (2003). *Balanced scorecard – Arbeitsbuch*. St. Gallen: SMG Publ.

Kaplan, R. S., & Norton, D. P. (Ed.) (1997). *Balanced scorecard*. Stuttgart: Schäffer-Poeschel.

Keebler, J. S., Durtsche, D. A., Manrodt, K. B., & Ledyard Michael, D. (1999). *Keeping score: Measuring the business value of logistics in the supply chain*. Oak Brook: Council of Logistics Management.

Küpper, H.-U. (2008). *Controlling – Konzeption, Aufgaben und Instrumente, 5. Aufl.* Stuttgart: Schäffer-Poeschel Verlag.

Lambert, D. M., & Burduroglu, R. (2000). Measuring and selling the value of logistics. *International Journal of Logistics Management, 11*(1), 1–17.

Lynch, D. F., Keller, S. B., & Ozment, J. (2000). The effects of logistics capabilities and strategy on firm performance. *Journal of Business Logistics, 21*(2), 47–68.

Meyer, C. (2007). *Betriebswirtschaftliche Kennzahlen und Kennzahlensysteme*, 4. Edition. München: Wissenschaft & Praxis.

Müller, A. (2000). *Strategisches Management mit der balanced scorecard*. Stuttgart u.a.: Kohlhammer.

Oliver, S. M., et al. (2009). *Managerial accounting*. Upper Saddle: Pearson Education.

Preißner, A. (2007). *Balanced Scorecard anwenden: Kennzahlengestützte Unternehmenssteuerung*, 2. Edition. München: Carl Hanser Verlag.

Reichmann, T. (2006). *Controlling mit Kennzahlen und Management-Tools – Die systemgestützte Controlling-Konzeption* (7th ed.). München: Vahlen.

Schneider, C. (Ed.). (2004). *Controlling für Logistikdienstleister, Konzepte, Instrumente, Anwendungsbeispiele, Trends*. Hamburg: Dt. Verkehrs-Verl.

Stölzle, W., & Otto, A. (Eds.). (2003). *Supply Chain Controlling in Theorie und Praxis*. Wiesbaden: Gabler.

Weber, J. (2002a). *Logistik-und supply chain controlling* (5th ed.). Stuttgart: Schäffer-Poeschel.

Weber, J. (2002b). *Logistikkostenrechnung: Kosten-, Leistungs- und Erlösinformationen zur erfolgsorientierten Steuerung der Logistik*. Heidelberg: Springer.

Weber, J., & Schäffer, U. (2000). *Balance scorecard & Controlling, Implementierung – Nutzen für Manager und Controller – Erfahrungen in deutschen Unternehmen*. Wiesbaden: Gabler.

Weber, J., & Schäffer, U. (2008). *Einführung in das Controlling* (13th ed.). Stuttgart: Schäffer-Poeschel.

Business Models and Industry Solutions 12

Learning Objectives

Logistics plays a huge part in improving a company's competitive position. In many cases it is prerequisite for the creation of new business fields and for opening up additional markets. This chapter deals with various existing business models in the industry, trade, and service sectors; with the main focus on the logistical requirements and structures, which will be further explained through some exemplary industry solutions. In particular, the chapter illustrates the effects on logistics resulting from internationalization and globalization, from increasing division of labor and the forms of collaboration caused by this, and from new procurement concepts and sales channels brought about by e-commerce. The reader will gain insights into the consequences business decisions can have on logistics – and why the functions of logistics should be taken into account at an early planning stage.

Keywords

- Industry models, industry solutions
- Stationary trade, mail order business
- E-commerce
- Internationalization, globalization
- Automobile, textile, and electronics industries
- Procurement, production, and distribution strategies
- Contract logistics

H. Gleissner and J.C. Femerling, *Logistics*, Springer Texts in Business and Economics, DOI 10.1007/978-3-319-01769-3_12,

12.1 Business Models

A model is a simplified representation of reality. Hence, a business model is a simplified representation of a commercial activity where several business partners come together to exchange material or immaterial goods commercially. Business transactions in the form of business processes – such as product offering, product delivery, order and payment processing – can form a business model and are necessary for this exchange.[1]

A business model not only can be the description of an individual company but also of an entire industry or trading sector. This is also called industry solution. In this sense, the term business model is applied to mature industries, in which a dominant business model has become prevalent. Nevertheless, these models are subject to ongoing innovation and change, which is fuelled by the need for increased economic efficiency and cost reduction. Selected business models and industry solutions with their respective logistics systems will be presented in the following.

12.2 Logistical Industry Solutions

12.2.1 Industry Solutions in Manufacturing

Logistical industry concepts are as manifold in the manufacturing sector as the individual industries themselves. Thus, only a few industries and their logistics systems can be presented here. A key industry is the *automotive industry* including the *supplying industry*, which has taken on a leading role in the development and application of logistics concepts since the beginning of logistics. Many logistics concepts have been adopted from other industries, such as just-in-time delivery or vendor managed inventory, area freight forwarding, external procurement warehouses, and supplier parks (see Chap. 7).

The main logistical determinants of the automotive industry are:

- Strong fragmentation of value chains
- Decentralized manufacturing in networked production systems
- Increasingly globalized supply chains

Fragmented value chains generate widely ramified procurement-supply chains between car producers and their direct suppliers. These suppliers are also termed tier-1 suppliers. The integration of pre-suppliers (tier-2, tier-3 etc.) in these chains is still the exception rather than the rule, which is why industry-specific and optimized value chains in this field have hardly developed yet. In the future, the automotive

[1] Cf. Berning (2002), p. 16.

industry will direct their efforts towards this issue,[2] thereby making this sector ever more reliant on logistics.

The automotive industry greatly relies on purchased components, materials, and primary products, which entails highly complex supplier–buyer relations. For this reason, automobile manufacturers consistently seek to reduce their supplying resources in terms of components, product lines, and locations.[3] While in the past e.g. European manufacturers collaborated with 500–1,000 suppliers, some producers have already reduced this number to 350. The medium-term plan is to only collaborate with 30–50 suppliers. This strategy of *Single Sourcing* is considerably driven by purchasing pre-finished modules and systems.[4]

A module is defined as a self-contained functional unit that may consist of several or many (individual) parts and components. A system is a functional and technical-developmental unit oriented towards one main function. A system supplier's responsibilities, however, comprise more than merely procurement, assembly, and test operations. They are also integrated into a company's R&D, product designing, and logistics. These suppliers are also called *value-adding partners*.

The procurement strategies of *modular and system sourcing* lead to a regrouping of the supply chain (so-called *tiering*). The supply chain is enlarged by an additional tier made up of module and system suppliers and now comprises components, module, and system suppliers. Module suppliers autonomously coordinate the flow of materials and components of the upstream suppliers in the supply chain and assemble the pre-finished modules from the components.

One example is the cockpit of a car, which consists of an instrument carrier, heating and air-conditioning, all control elements, a steering column and steering wheel, safety installations (airbag), radio and navigation systems, and so on. Another example is the chassis frame, where either front or rear axle module, including undercarriage technology (springs, dampers) are available, as well as wheel systems including rims, tires, pressure control, and so forth (see Fig. 12.1).

The procurement of modules and systems is not confined to the automotive industry. Other sectors have also pursued this procurement-supply concept for some time.[5] Thus, computer manufacturers purchase complete hard drives and screens, builders procure entire bathrooms and door, window, and facade elements. Watch producers are supplied with clockworks while in the food industry spice blends and ready-made fruit supplements for yogurt production are purchased as complete modules.

With increased scope of delivery, modules correspond to the criteria of just-in-time procurement (i.e. quality, diversity, and volume), which is why the supply of modules takes the greatest share of just-in-time deliveries.[6] In this case, the

---

[2] Cf. Straube et al. (2005), p. 104.

[3] Cf. Wildemann (2010). p. 86 et seq.

[4] Cf. Ihme (2006), p. 276.

[5] Cf. v. Eicke and Femerling (1991), p. 59.

[6] Cf. Ihme (2006), p. 296.

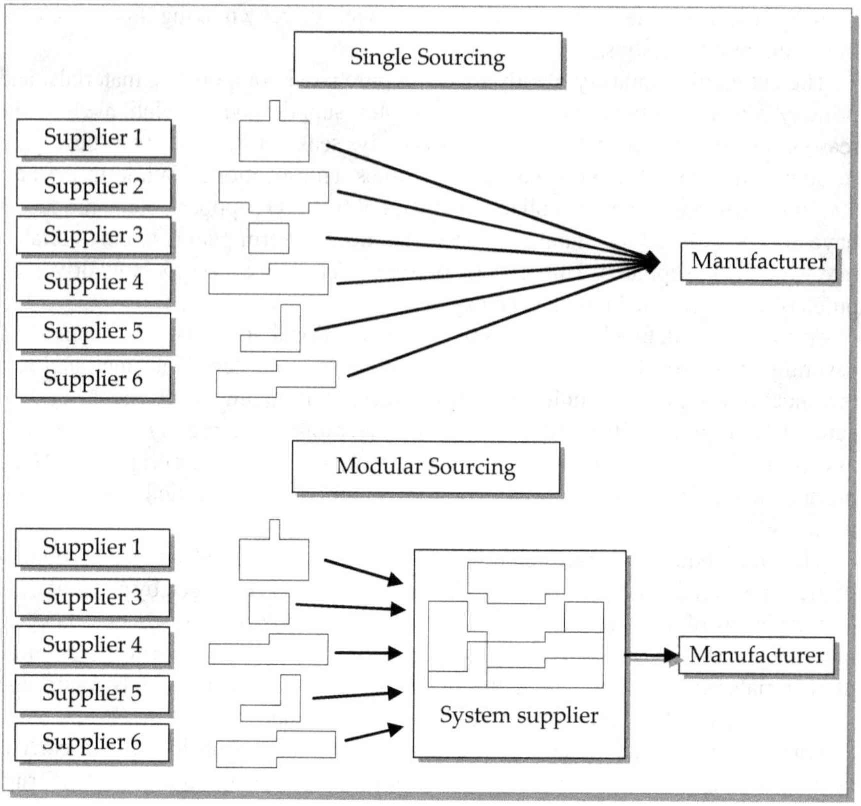

Fig. 12.1 Procurement structures in the automotive industry (Cf. v. Eicke and Femerling (1991), p. 33 et seq)

modules are directly delivered to the assembly line of the automobile plant. The loading aids needed to carry out these procurement processes have to meet special requirements. They are a special type of load carrier with appliances that allow carriage and fixation of the modules to avoid damage during transport and to facilitate handling during the assembly process.

Standards in the materials flow, such as the use of standardized loading aids and packaging, have long been established in the automotive industry, irrespective of the purchased goods. On the other hand, these cross-company efforts for standardization are complicated by OEM standards, which in many cases apply to containers for just-in-time deliveries. Likewise, electronic data exchange is largely standardized in the automotive industry, of which the VDA (German Association of the Automotive Industry)/ODETTE or EDIFACT standards are some examples.[7] However, more recent data exchange protocols, such as XML, are increasingly superseding these

[7] Cf. Ihme (2006), p. 206 et seq.

standards (see Chap. 9). If the information and communication technologies used between producers and their direct suppliers can be regarded as good, upstream suppliers and the involved logistics service providers are also in considerable demand of such technologies.

Apart from the acquisition of external suppliers, the European automobile industry is also characterized by a network of production and assembly sites that is strongly interwoven. The *Volkswagen Group*, for example, operates over 50 production sites world-wide. Likewise, premium car manufacturers have numerous production sites at their disposal, often located in the most important markets. On the one hand, foreign exchange fluctuations can thus successfully be offset. On the other hand, foreign markets are in many cases much more important than the car producer's own domestic market. *BMW*, for instance, now sells more cars in the USA than in Germany. This has led to the set-up of a new production site in Spartanburg, North Carolina, where the X5 and Z4 series are built and the X3 series will be built in the future. These cars are entirely produced there. With a car-specific vertical range of manufacture of about 30 %, most parts and components are procured through the supply network of the BMW Group. Numerous suppliers have relocated to the plant's proximity since it is a common strategy in the supplying sector to adopt the internationalization strategies of their producers. The engines are procured through BMW's production network and mostly manufactured at the engine works in Graz/Austria. Other car manufacturers employ similar procurement and production strategies.

Along with procurement and production, the importance of distribution is growing for the automotive industry, too. Distribution of cars is mostly effected through multi-level sales systems, which are brand-exclusive to many manufacturers. Via sales companies the producers deliver the cars to their own subsidiaries, to contractors, and to importers in the individual markets.

In the past, markets were supplied with cars according to a push strategy which corresponded to existing production facilities, i.e. a certain stock of cars was provided to the car dealers.[8] Due to the great variety of types and designs, this strategy of car distribution is now only suitable for volume manufacturers (*Opel, Ford, VW*). Premium manufacturers (*Audi, BMW Group, Daimler, Porsche*), have for some time pursued higher production flexibility, which enables them to adopt a pull strategy where the entire car production is aligned with customer orders.[9] To ensure high adherence to deadlines despite a great variety of types and designs and thus fluctuating production complexity, the production sequence follows a pearl chain principle, which does not allow for a change of design or delivery date after a certain point.[10]

Logistics service providers assume an important role for the automotive industry as well. With regard to procurement, they manage the transport-logistical and

[8] Cf. Sommer (2003), p. 240.

[9] Cf. Sommer (2003), p. 242.

[10] Cf. Feldkamp (2001), p. 175.

warehouse-logistical tasks. Furthermore, they are responsible for supply and disposal of parts and components for production and assembly sites, as well as for supply of spare parts and the distribution of the finished cars to car dealers or customers.

Just like the automotive industry, the *electronics and computer industry* displays a strong division of labor. Providers of electronic products are, however, more and more focusing on product development and sales while outsourcing their production to specialized companies, so-called Electronic Contract Manufacturers.[11] Production predominantly takes places at sites in Asia.[12]

Hewlett Packard (HP), one of the leading manufacturers of computer systems, operates a world-wide production network; partly with own plants and partly with contract manufacturers. The individual production sites have their own geographical sales channels and are responsible for the regional distribution. Procurement of materials and primary products is done through global supply chains which are used by all sites. HP centralized their European logistics operations in Germany at their site in Böblingen near Stuttgart. A high-performance network of Europe-wide logistics service providers guarantees delivery of products within 24–48 h.[13]

In the consumer electronics sector, *Philips* has undergone similar developments of their logistics structure. Increased cost pressure, shortened product life cycles, increased product value, and higher service requirements of customers necessitated the set-up of a Europe-wide, international distribution system.[14] Due to this, national storage and distribution centers were given up and so-called platforms were created. These platforms are cross-docks, where pricing and transshipment of goods as well as consolidation of individual orders and possibly of part loads to complete loads takes place (see Fig. 12.2).

Another important sector is the *chemical industry* and its branches, such as petro chemistry. Compared to the production of packaged goods, chemical production processes exhibit some special features which have an immediate bearing on the design of the logistics systems.[15] There are procedural processes with special requirements for the flow of materials and goods. For one thing, the materials may be continuous or discrete, which invites questions regarding (intermediate) storage, provisioning of stocks and delivery services. Due to the chemical and physical nature of the materials, which may either occur in gaseous or liquid aggregate states (e.g. as acids) or in solid form as bulk goods, special facilities and vehicles are necessary for transport and storage.

These include tanks, silos, pipes, conveyors, tank/silo wagons and special freighters, all of which need to adhere to hazardous goods regulations.

[11] Cf. Vahrenkamp (2007), p. 172.

[12] Cf. Schorb et al. (2007), p. 625 et seq.

[13] Cf. Schmid (2001), p. 145.

[14] Cf. Lammers and Neubauer (2005), p. 52 et seq.

[15] Cf. Grunow (2001), p. 323 et seq.

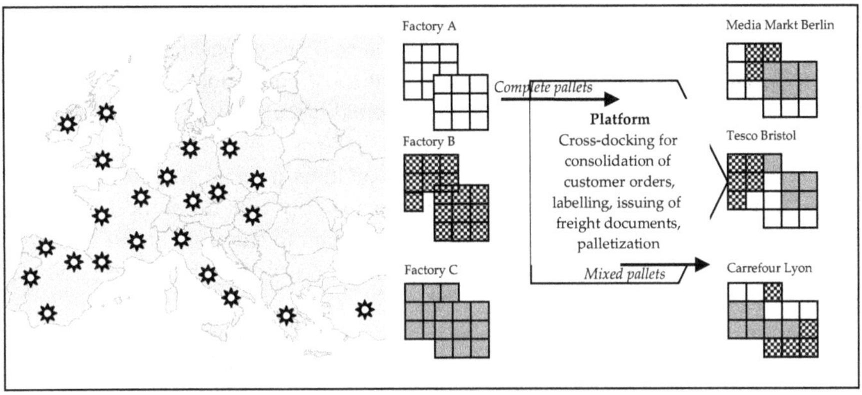

Fig. 12.2 Platform concept of Philips (Cf. Lammers and Neubauer (2005), p. 53)

The special requirements of this business leads to a low level of outsourcing on the one hand and causes a restrictive market offer on contract logistics services on the other hand. Growing market pressure, globalization and hence newly structured supply chains promoted the outsourcing process within chemical industry with contract service providers. A number of logistics service providers specialized in the chemical industry are available for storage and transport. Of major importance are the transport modes of water, road, pipeline and rail.[16]

12.2.2 Industry Solutions in Trade

Several business models can be distinguished in the trading sector, which falls into the categories of stationary trade, distance mail-order trade, and e-commerce. At a first glance, the following business models can be listed for stationary trade:

- Subsidiary and non-subsidiary (retail) stores
- Department stores
- Consumer markets
- Self-service department stores
- Specialist stores and discounters

The structure and number of supply points and the range of products influence the storage-logistical and transport-logistical structures of these business models to a great extent.[17] The physical nature of the goods (size, volume, weight, sensitivity)

[16] Cf. Buchholz et al. (1998), p. 87.

[17] Cf. Hardt and Kasch (2007), p. 641 et seq.

as well as the design and packaging have a bearing on the handling of the goods. In particular, the logistical characteristics of these types of businesses are[18]:

Specialist stores without subsidiaries do in most cases not operate a logistics system of their own. Generally, they are supplied directly by the producers or by wholesalers.

The logistics of *department stores* depends on their articles' properties and their range of products and is either effected as direct supply with decentralized storage or differentiated according to products in a decentralized manner. In this way, inventory risks can be minimized and costs may be saved by bundling goods for the supply of the stores.

Consumer markets and *self-service markets* are often directly supplied. However, more and more parts of their product ranges are supplied from central warehouses. Concepts of consolidation, such as cross-docking, are increasingly employed as well (see Sect. 8.3.3). Subsidiary chains, for example *Aldi*, *Lidl* or *Deichmann*, are normally supplied directly from central warehouses.

Large *discounters* operate more than 60 central warehouses in Germany to minimize the distances and thus the transport costs to the individual stores. This indicates that the logistics of discounters is rather cost-oriented and characterized by the simplicity of their inventory management and logistical processes.[19] All the more so as the product range of discounters is steadily growing and may comprise approximately 700–1,000 products, which are mainly fast moving consumer goods.

Depending on their sales strategy and store size, *full-range traders* offer up to 15,000 articles, a large share of which are slow moving consumer products. The logistics systems of full-range traders and discounters have to meet different requirements due to the contrasting movement of their goods (fast/slow moving consumer goods).[20] The picking of slow moving goods is usually carried out on picking vehicles, thereby largely separating the bundle units. Logistics systems for fast moving goods, however, are rather designed for large quantities which are mostly compiled to loading units on pallets. The storage systems vary accordingly. Full-range traders pick their goods from the easily accessible lower shelves while the buffer stock is placed on the upper shelves. The shipping of pallets with only one type of article renders the storage and warehouse design of discounters rather simple.

There are also a number of product-related business models, such as *textile trade*. These models strongly rely on global procurement and production structures, mainly to make use of considerable (labor) cost advantages abroad.[21] The logistics chains are in these cases often regulated by the trading companies. Thus, local forwarders are commissioned with the pick-up and composition of shipments in the country of origin. They then consolidate the shipments into container loads and

[18] Cf. Merkel and Heymann (2003), p. 170 et seq.

[19] Cf. Brandes (2003), p. 202.

[20] Cf. Auffermann (2007), p. 26.

[21] Cf. Christopher et al. (2009), p. 112 et seq.

carry out pre-carriage as well as on-carriage. The same applies to the takeover by logistics service providers in the target country, which are responsible for separation, processing and tagging of the textiles as well as their delivery to the stores.[22] These structures in the textile industry, however, are undergoing changes since successful business models in textile trade are subject to shortening fashion cycles with ever increasing demand for broad product ranges. This requires logistics systems to offer shorter pre-carriage, processing and delivery times. The concept of supplying stores directly from Asian production sites, for example, is supposed to meet these requirements.[23] Store-related picking, packing for export, and consolidation in containers for main carriage are all carried out at these production sites. After de-consolidation of the containers in the country of destination, the goods packages are delivered to the stores using the networks of the respective delivery services. In comparison to traditional logistics systems, transit times are reduced from 12 to 6 days for air freight and from 45 to 39 days for sea freight. The logistics costs can be cut by close to 15 %.[24]

In *mail-order* or *distance trade*, no delivery to business sites or stores takes place since the goods are delivered directly from the warehouse to the customer. Warehousing structures are normally organized centrally in the mail-order business.[25] Delivery services are critical for the success of a mail-order business. To achieve the highest possible delivery service level, different warehouse and shipping systems are used for different components of a product range. The handling takes place in shipping hubs which are often highly automated to keep operating costs low. For example, shelving racks, conveyors, cross-belt sorters, and parcel machines are thus utilized. Handling capacities of more than 160 million articles at one location per year can be achieved in high-rack warehouses.[26] Reliable and flexible delivery – which is paramount for mail-order systems – is greatly reliant on the technological equipment for storage and picking processes, on the organization of the *order processing* and on the means of transport used. The use of (partly) automated picking systems (Pick by Light, Pick by Voice) can positively influence the picking reliability and overall speed. On the other hand, manual picking strategies (pick lists) provide a high degree of flexibility against fluctuating demands as well as speedy processing of urgent orders.[27] Different concepts are offered for *delivery*. These include 24 h service, desired date delivery, defined delivery windows, and evening or Saturday delivery. Moreover, further delivery and collection concepts such as parcel shops integrate kiosks, gas stations, or lottery shops in the delivery process. Parcel delivery systems at gas stations are

[22] Cf. Buchholz et al. (1998), p. 90 et seq.; Nothardt et al. (2007), 688 et seq.

[23] Cf. Clausen et al. (2007), p. 26 et seq.

[24] Cf. Clausen et al. (2007), p. 28.

[25] Cf. Kloth (1999), p. 53.

[26] Cf. Witten and Karies (2003), p. 190.

[27] Cf. Femerling (2003), p. 217.

completely independent of opening hours. Track and tracing services make for a better co-ordination of the delivery between customer and sender, by which more accurate advance notifications can be issued and more reliability can be achieved. The *distribution of large goods* – i.e. of *white* and *brown* goods such as washing machines, refrigerators, TV sets, furniture and kitchens – is especially demanding for mail-order businesses.

In approaching these challenges it is important to bring parallel distribution channels together. In many cases distribution is effected within established structures. White and brown goods while high-volume goods are delivered ex works.[28] These delivery structures, however, lead to varying service agreements between sender and customer, which is met with less and less acceptance.

12.2.3 Industry Solutions in the Service Sector

Logistics for service providers is a business field for industry solutions that goes beyond physical logistics processes. These service providers offer, for instance, service for banks, insurance companies, public administration (back office), as well as the administrations of industrial and trading companies. Accordingly, logistical goals – such as providing the right persons, data, documents, information or even materials at the right time, in the right quality and quantity at the right place of demand – also play an important role in these areas. In particular, document and information logistics are becoming increasingly important for the service sector as they constitute an essential factor of production for immaterial products. Services of physical and electronic postal logistics as well as physical and digital archiving need to be rendered, including the necessary IT systems. Services like these are more and more offered by logistics service providers.[29]

Postal logistics includes:

- Mail processing (incoming mail, outgoing mail, distribution)
- Letter shops (printing, enveloping, dispatch, courier services)

  Archive logistics includes:
- Document reception, indexing, and management
- Document shredding according to legal regulations
- Archive management including scanning and integration into document management systems
- Process management using document management systems with workflows

Other service-sector industry solutions are available for health services. For these purposes, procurement and distribution logistics are most important in hospital wards where consumable articles (surgical dressings, syringes, medication etc.) and

[28] Cf. Gleißner (2003), p. 202 et seq.

[29] Cf. Peters (2003), p. 109 et seq.

reusable surgical instruments are needed. This includes purchasing processes, such as price negotiations, framework contracts, standardization of product range and scope of delivery and procurement processes, such as repeat orders, central warehouse storage and fine distribution on the wards.[30] Service providers specialized in this segment are responsible for the coordination of the supply chain between suppliers, producers, distributors and the hospital administration.

12.2.4 Industry Solutions in E-Business

Todays business models are significantly shaped by the Internet. *Electronic management of business processes* not only helps optimize traditional business processes but also forms the basis for entirely new business models. Prominent examples of this are online marketplaces where supply and demand are aligned in virtual space, where pricing processes are organized and where transactions are supported by information-technology.[31] *E-business* is above all carried out within business-to-business (B2B) models for transactions both between companies and their value-adding partners and between companies and their business clients. *E-commerce* is mainly employed within business-to-consumer (B2C) models, which describe transactions between companies and end consumers. Depending on the organization of the (business) processes, e-business is also used to support both procurement processes (Buy-Side or *E-Procurement*, *E-Purchasing*) and sales processes (Sell-Side or *E-Commerce, E-Sales*).

B2B and *B2C business models* describe markets or sales channels that are based on marketplaces on which providers and consumers meet. Therefore, as a first step it needs to be analyzed which markets or sales channels are to be targeted.[32] In doing so, it is crucial to determine whether business is being made with an end consumer or trading partner (wholesaler, retailer, subsidiary retailing business) and which of the basic models of B2B and B2C need to be employed. Providers of E-commerce can be:

- *Multi-channel providers;* established producers and trading companies which make use of the Internet as an additional sales channel
- Exclusive *Internet providers* which use the Internet as their only sales platform
- Traditional *mail-order businesses* which offer Internet solutions to their customers as a new communication and ordering medium, along with letter, fax, and telephone

One of the most fundamental consequences of e-business is in any event a further expansion of the markets. This is true both for the procurement markets, in which the globalization of purchase and procurement is further accelerated, and for the sales markets, where entirely new sales channels can be opened up. Since product ranges

---

[30] Cf. Pintsch (2004), p. 252.

[31] Cf. Femerling (2003), p. 211 et seq.

[32] Cf. Femerling (2003), p. 208.

are offered on the Internet, a substantially larger distribution area than with traditional mail-order trade is to be serviced. This directly affects delivery services, especially with regard to delivery times. The advantages of global sourcing may lead to increased costs in the procurement process due to longer transport distances, more frequent transshipment and more complex processing (customs clearance, import documents). On the other hand, cost advantages can be realized by utilizing the Internet for purchasing activities (*e-procurement*) in the form of Desktop Purchasing (DP) systems, especially for the purchase of standardized products.

E-business makes direct selling possible for producers but also for individual sales stages. If such (intermediary) sales stages are skipped, the order structure is inevitably affected. Few substantial orders, which had previously been delivered to businesses, are then superseded by many small orders being sent directly to end customers. This has direct effects on the logistics costs. The order processing costs rise since economies of scales do no longer apply for storage and picking facilities. Costs per transport item increase because deliveries involve many stops, few items are delivered at each stop, and several delivery attempts sometimes have to be undertaken, especially to private addresses (*last-mile problem*).[33] The logistical challenges are thus largely the same for distribution logistics as they are for the mail-order business,[34] where primarily small shipments need to be delivered to customers within a narrow timeframe. However, e-commerce systems additionally have to meet customer requirements with regard to speediness and reliability. This is subsumed under the term *fulfilment*, which comprises the entire physical, informational, and monetary order processing.[35] To this end, all existing systems of parcel services, mail organizations or forwarders and carriers are called into action. Equally important to order processing is the processing of return deliveries, subsequent deliveries in cases of shortfall quantities, wrong deliveries, or technical complaints.

All the statements made about the *structures of distribution networks* (see Chap. 8) basically hold true for the B2B and B2C concepts discussed here. Especially important are the substitutionary relations between the number of warehouses and storage stages and the delivery time. The more warehouses are operated in a specific area, the shorter become the possible delivery times, and the higher are the infrastructure costs of the network. Considering the costs, a centralization of the warehouse structure is suitable especially also for e-commerce solutions and can thus frequently be observed in B2C systems.[36] A decentralized network of warehouses, however, brings with it the advantage of extremely quick delivery times. Segmentation of the network structure specific to the product range is also favourable for the e-commerce sector. Depending on the product and customer requirements, parts of the product range can thus be stored centrally while other articles are stored in regional warehouses or are delivered via stock-less transshipment points. Moreover, more and more producers

[33] Cf. Witten and Karies (2003), p. 193.

[34] Cf. Bretzke (1999), p. 228.

[35] Cf. Schubert (2001).

[36] Cf. Lasch and Lemke (2002), p. 3.

and suppliers as well as wholesalers are integrated with their distribution logistics. Such differentiations may appear sensible if e-commerce offers from (centralized) European distribution networks are to be taken advantage of. From a cost and delivery-service perspective, however, part deliveries should be avoided.

12.3 International and Global Business Models

International and global business models may refer to procurement, production, and distribution and show effects on the respective logistics systems. The frameworks for international business models are continually changing while internationalization strategies and logistics strategies can be mutually dependent. Amongst others, the key factors for this development are[37]:

- Continued globalization through open borders
- Development of new growth markets in Eastern Europe and Asia
- Increasing offer range of logistics service providers
- High volatility and tendency towards falling transport prices
- World-wide differences in factor costs and, above all, labor costs

In designing *international distribution systems*, the contents about warehouse management and transport network planning explained in Chap. 8 can be extrapolated to the challenges of global logistics systems. Most important in doing so are the target values of delivery service and logistics costs.[38] Regarding the delivery service, it is of major importance to reach destinations within a reasonable time while distributing internationally. The requirements for this vary in different geographical regions. For example, within the EU, in large cities or agglomeration areas the same requirements apply, regardless of national borders. Compared to national structures, the number of stakeholders and institutions in global logistics systems is on the rise.

Thus, several transport modes and a multitude of forwarders and logistics service providers are integrated into world-wide logistics chains.

Existing national logistics structures often distribute goods internationally, as is the case of *Eurologistics systems* or when opening up new markets. This, however, entails a significant (re)structuring of the existing institutions and structures.[39] Several phases can be distinguished. In the first phase, finished products are exported from the country of origin to the country of importation. Distribution in the individual regions is carried out by importers, wholesalers or wire a own central warehouse of the manufactors (OEM). Depending on the state of development, this results in a multi-tier distribution system with one central warehouse for one sales region and regional warehouses for the respective countries. In another phase, the production or

[37] Cf. Neher (2005), p. 34 et seq.

[38] Cf. Freichel (2002), p. 267.

[39] Cf. Freichel (2002), p. 264.

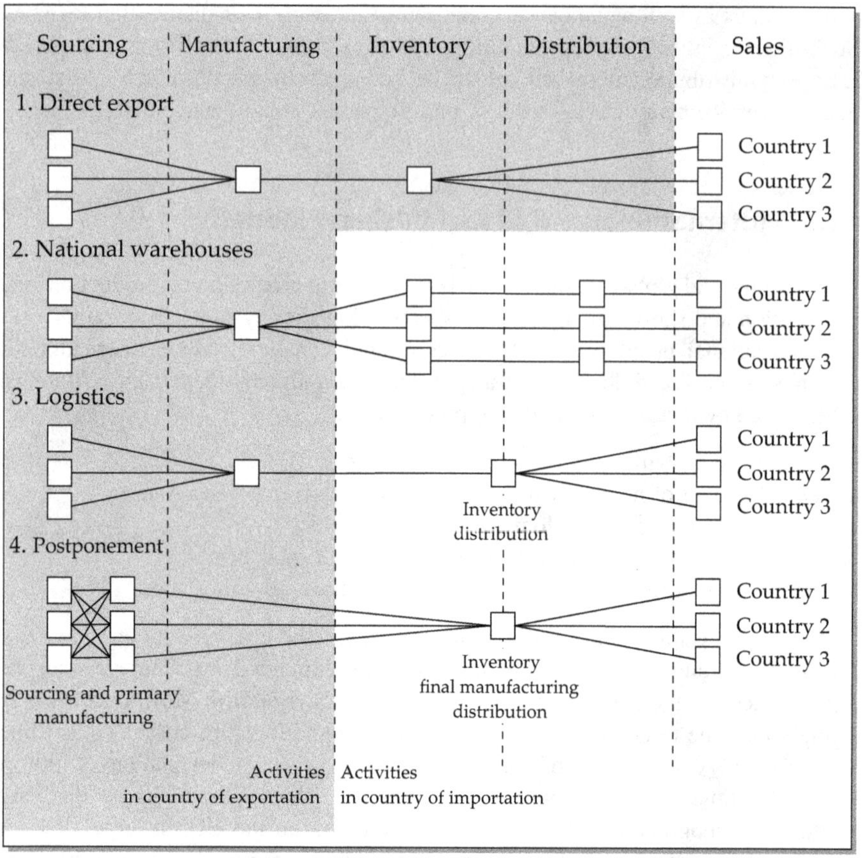

Fig. 12.3 Models of international logistics systems (Cf. Neher (2005), p. 41)

at least the assembly of (country-specific) product versions in the country of importation is carried out, followed by their distribution within the country.

Figure 12.3 shows the complex spectrum of possible international logistics systems and their characteristics. Distinctions are made between variants such as direct export to the sales region, national organization of warehouses and distribution in the countries of importation, centralized logistics solutions for combined sales regions (e.g. continents, cultural regions etc) and production or assembly in the country of origin to make allowances for regional idiosyncrasies.[40]

[40] Cf. Giesa/Hagen (2003), p. 1 et seq.

12.4 Development of Logistics Services Offers

On the basis of business solutions in the industry, service, and trading sectors, different business models have been developed for logistics service providers. With increased outsourcing of logistical activities to service providers, new concepts are continually arising, which incorporate contract-logistical services along with transport and storage tasks (see Sect. 5.5.3). With regard to sector-specific business models, a distinction can be drawn between services focusing on consumer goods and industrial, contract-logistical services.[41] The characteristics of *consumer goods distribution* are:

- Staple goods (which can be placed on a pallet) in cardboard boxes
- Use of barcodes
- Not labeled for distribution to a particular receiver; instead, article-specific distribution
- Sell-by dates and production batches (which both are controlled)
- Distribution from production sites and central warehouses of the producers to the warehouses and stores of the trading sector
- Single-User-Systems for individual customers or Multi-Client-Warehouses for several customers from one or several sectors

Through *consumer goods contract logistics*, leading logistics service providers offer comprehensive distribution networks with corresponding storage facilities, transshipment, and cross-docking points. To operate distribution centers, warehouse management systems and integrated order processing systems are available. Other critical success factors include a regional or even national truck-distribution system, access to transport networks, high IT competence, qualified staff, and high-quality technical equipment.

The following success factors are prevalent in *industrial contract logistics* business models[42]:

- IT competence and capability of IT integration
- Willingness to offer additional non-logistical services
- Customer-specific and sector-specific know-how
- Innovation skills to improve business models of clients

The service portfolio in industrial contract logistics mainly includes non-logistical tasks, apart from the merely logistical ones. Examples of this are assembly and mounting services, packaging, operation of call centers for customer service, and quality inspections in production logistics. IT competence plays an essential role in this business model since suppliers, producers and logistical service providers need to be information-technologically connected according to sector-specific standards.[43]

[41] Cf. Tripp (2004), p. 12 et seq., Klaus (2007), p. 97 et seq.

[42] Cf. Tripp (2004), p. 24.

[43] Cf. Giesa and Hagen (2003), p. 43 et seq.

Logistics service providers internationalize their network structures in accordance with the requirements of the industry and trade sector, whereas there are differences between individual providers and their service offers. No service provider offers comprehensive, world-wide transport networks yet.[44] European networks, however, are well developed. Specialized, multi-modal service providers have emerged on the basis of (standardized) inter-modal network services in conjunction with (individual) logistics services. These so-called *integrators* act as both CEP service providers as well as road, air, sea, and rail freight transport providers, while offering a number of additional logistics services (see Sect. 5.5.2).[45]

Apart from contract-logistical service providers and integrators that offer *logistics from a single source*, niche providers may also assert themselves on the market. Niche providers specialize their logistics services in a specific industry segment (e.g. liquid raw materials transport for the food industry) or a special service (e.g. crane services). Business models for rendering standard logistics services, such as transport by making use of economies of scale, have been reviving for some time. IT-aided management (route planning, GPS tracking etc.) and a sufficiently large fleet of trucks can help position an economically attractive offer on the market.[46]

Figure 12.4 summarizes the logistics service offer. As a whole, it shows the service elements of a multi-modal full-range provider who specializes in many industries.

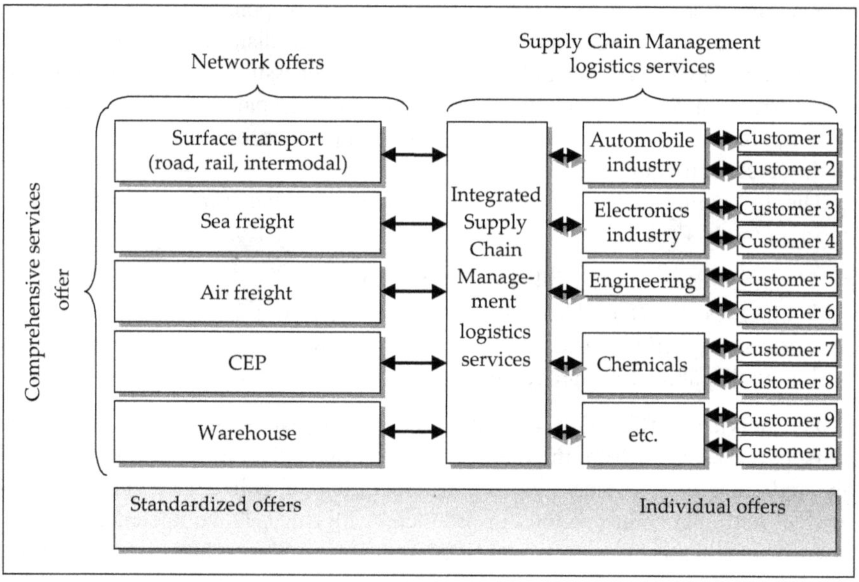

Fig. 12.4 Possible combinations of logistics services (Cf. Lange (2000), p. 199)

[44] Cf. Lieb and Lang (2003), p. 448.

[45] Cf. Lieb and Lang (2003), p. 454.

[46] Cf. Klaus and Kille (2008), p. 89 et seq.

Review Questions

1. What are logistical industry solutions?
2. Name business models for the distribution of consumer goods.
3. What are the logistical requirements for the mail-order business?
4. How does the Internet change business models and logistics systems?
5. What internationalization strategies and logistics strategies associated with them do you know?
6. How are Europe-wide and global distribution systems designed?
7. What requirements do business models have to meet in industrial contract logistics?
8. Which business models are supported by logistics service providers?
9. Explain the supply chain of the textile trade.
10. What are the logistical requirements of e-commerce?

Additional Literature

Buchholz, J., Clausen, U., & Vastag, A. (Ed.) (1998). *Handbuch der Verkehrslogistik*. Heidelberg: Springer.

David, A., & Stewart, R. (2010). *International logistics: The management of international trade operations*. Mason: Cengage Learning.

Frunzke, H., Boldt, O., & Garbisch, O. (2004). Management und Kundenbeziehungen in der Logistikdienstleistungsbranche. In H.-C. Pfohl (Ed.), *Netzkompetenz in Supply Chains. Grundlagen und Umsetzung*. Wiesbaden: Gabler.

Giesa, F., & Kopfer, H. (2000). Management logistischer Dienstleistungen der Kontraktlogistik. *In: Logistik Management*, Nr. 1, Nürnberg.

Klaus, P., & Kille, C. (Ed.) (2008). *Die TOP 100 der Logistik*. Hamburg: DVV.

Mangan, J., Lalwani, C., Butcher, T., & Javadpour, R. (2011). *Global logistics and supply chain management*. Chichester: Wiley.

Piontek, J. (1994). *Internationale Logistik*. Stuttgart: Kohlhammer.

Stroh, M. (2006). *A practical guide to transportation and logistics* (3rd ed.). Dumant: Logistics Network.

Tripp, C. (2003). Chancen und Risiken mittelständischer System-Stückgutkooperationen in Deutschland. Eine Untersuchung der Fraunhofer ATL Nürnberg im Auftrag der bayrischen Hypo-Vereinsbank AG München. Nürnberg: a study of the institute of Fraunhofer.

Tripp, C. (2004). *Kooperationen und Kontraktlogistik*. Nürnberg: Fraunhofer Institut für Integrierte Schaltungen.

Wolters, H. (1995). Modul- und Systembeschaffung in der Automobilindustrie – Gestaltung der Kooperation zwischen europäischen Hersteller- und Zulieferunternehmen, Wiesbaden: Deutscher Universitätsverlag.

Zimmermann, B. (2004). Kontraktlogistik als Zukunftsmarkt der Logistikdienstleistungswirtschaft? Mittelstandskongruenz und Entwicklung eines mittelstandsgerechten Vertriebsmodells, Nürnberg: University of Erlangen-Nürnberg.

Outline Solutions to Case Studies 13

13.1 Solution to Case Study 6-1

a

	Pallet shelving	Live storage shelving	Block storage
Base area of Euro pallet 1.2 x 0.8 m = 0.96 m²	Requirement for 6,000 pallets 5,760 m²	5,760 m²	5,760 m²
Number of storage levels possible with 13 m clearance height given	1m+0.5 m clearance 13 m : 1.5 m = 8.6 = 8 levels	= 8 levels	= 1 level
Space requirements	5,760 m² : 8 levels = 720 m²	5,760 m² : 8 levels = 720 m²	= 5,760 m²
Allowance for the level of storage space utilization	45 % Additional space for structures, aisles etc. needs to be taken into consideration = 1,600 m²	65 % = 1,108 m²	80 % = 7,200 m²
Alternative with the least space requirements		X	

b

Due to the broadrange of products and since items are stocked on and retrieved from opened pallets, pallet shelving is the most advisable option. This is also true since pallet shelving has only marginally higher space requirements than live storage shelving.

H. Gleissner and J.C. Femerling, *Logistics*, Springer Texts in Business and Economics, 285
DOI 10.1007/978-3-319-01769-3_13, © Springer International Publishing Switzerland 2013

13.2 Solution to Case Study 6-2

Time required per pallet: 311 m/2 m/s = 155.5 s

Net time required for transport = 155.5 s × 5,190 pallets = 807,045 s

Utilization time per forklift truck per week = 15 h × 5 days = 75 Std.

Conveying capacity utilization = 40 %

Net time available for pallet transport = 75 h × 40 % = 30 h

Net time = 30 h = 30 × 3,600 s = 108,000 s

Number of forklift trucks: 807,045 s/108,000 s = 7.5 i.e. approximately 8 forklift trucks are required

13.3 Solution to Case Study 6-6

(a) 2 × 3 storages spaces when minimum stock quantity is reached + 2 × 6 storages spaces for replenishment = 18

(b) 2 × average stock quantity ((3 + 9)/2 = 6) = 12

An average stock quantity is used for the calculation within a chaotic storage strategy since it is assumed that not all articles of a warehouse are re-ordered at the same time.

13.4 Solution to Case Study 8-2

In the case of hub-and spoke systems: 10 connections (with n = 10 depots plus 1 hub)

Point-to-point: 9 + 8 + ... + 2 + 1 = 45 connections (with n = 10 depots without hub)

or: [10 * (9)]/2 = 45 (with n = 10 depots without hub).

Investments in a hub must amortize within the operating life and the structure must be expandable, i.e. integration of additional depots must be possible.

13.5 Solution to Case Study 8-5

13.5.1 Selective Inventory

Decentralized storage of fast-moving goods at subsidiaries and customized delivery with slow-moving goods from the central warehouse. This results in a reduction of inventory within the distribution network while the number of regional warehouses stays the same and transport costs may increase.

13.5.2 Regional Warehouses

The subsidiaries in Germany and Spain are directly supplied by the suppliers while the independent vendors are supplied by the subsidiaries. Interim storage in the central warehouse is thus eliminated and the product range can be adjusted to the individual subsidiaries. A decrease of inventory within the distribution network is counterbalanced by an increase in transport costs.

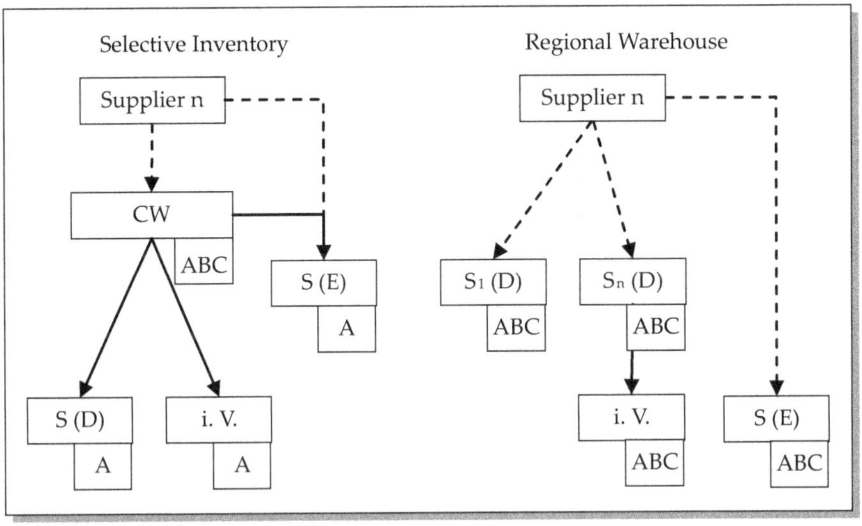

13.5.3 C Articles Management

The central warehouse only stores fast-moving goods. Slow moving goods are stored by the suppliers until they are requested by the subsidiaries to which they are then supplied directly. A significant reduction in inventory thus results for a large portion of C articles while transport costs only increase slightly.

13.5.4 Direct Supply

All suppliers directly supply the subsidiaries. Only small-sized trucks can be used to ensure their maximum capacity utilization. The average distance of transport increases in comparison to the use of a central warehouse.

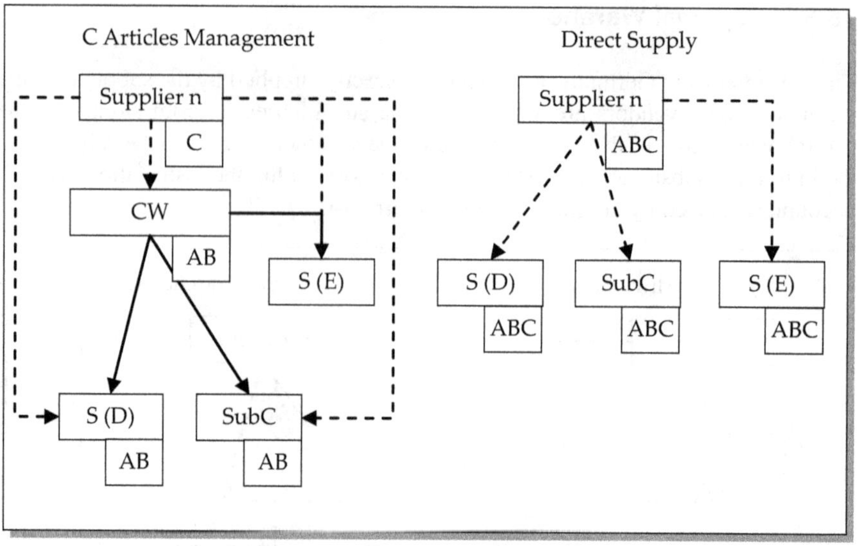

In order to assess these different options and to opt for the most favorable alternative, a use-value analysis can often be applied. To this end, specific assessment criteria are defined which are then given a certain amount of points. Furthermore, the criteria may be weighted by means of certain factors. The most suitable criteria for the problem at hand are:

- Market (Service level, Speed of service, subcontractors are supplied separately),
- Internal processes (Improved warehouse operations, capacity utilization reserves),
- Monetary aspects (transport costs, handling and warehousing costs, inventory costs).

Selective inventory or introducing a separate C article management system appears to be a reasonable solution here. Combining both principles should also be considered. Due to the high amount of direct supply to Spain, special consideration should be given to the location of the central warehouse.

13.6 Solution to Case Study 11-2

(a) Decrease in the operational logistics costs

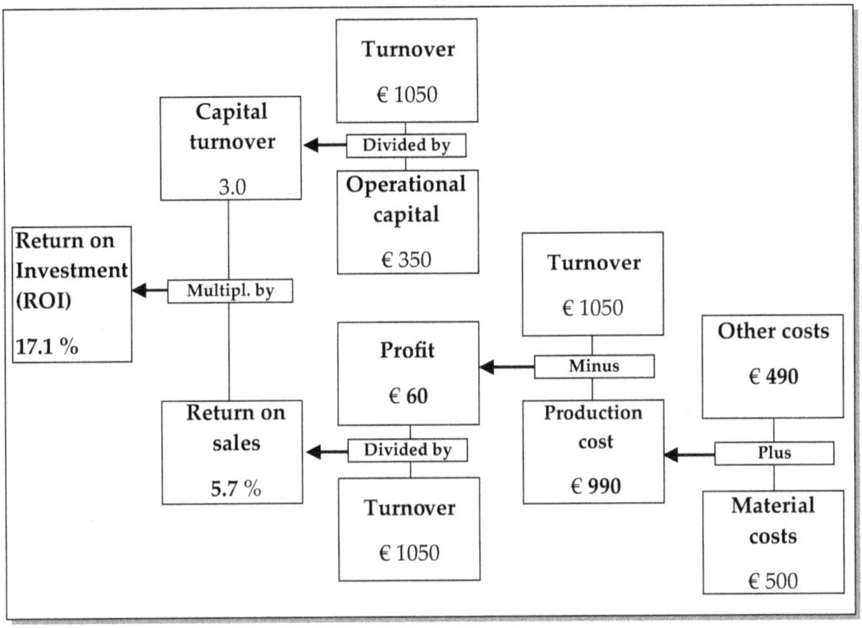

As can be seen, a relatively minimal decrease in logistics costs by 10 % – and thus a decrease of other costs by **€** 10 units – yields an improved ROI by 2.8 percentage points.

(b) Reduction of inventory

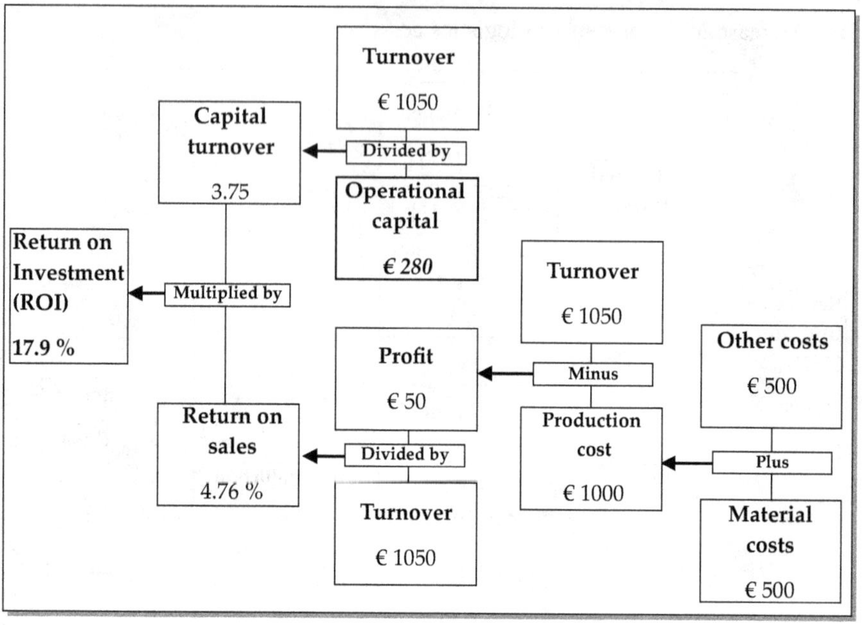

Reducing inventory and thus reducing operational costs by 20 % leads to an increase in ROI by 3.6 percentage points. This demonstrates that measures to decrease inventory can have a significant effect on the overall yields of the company.

(c) Increase in turnover

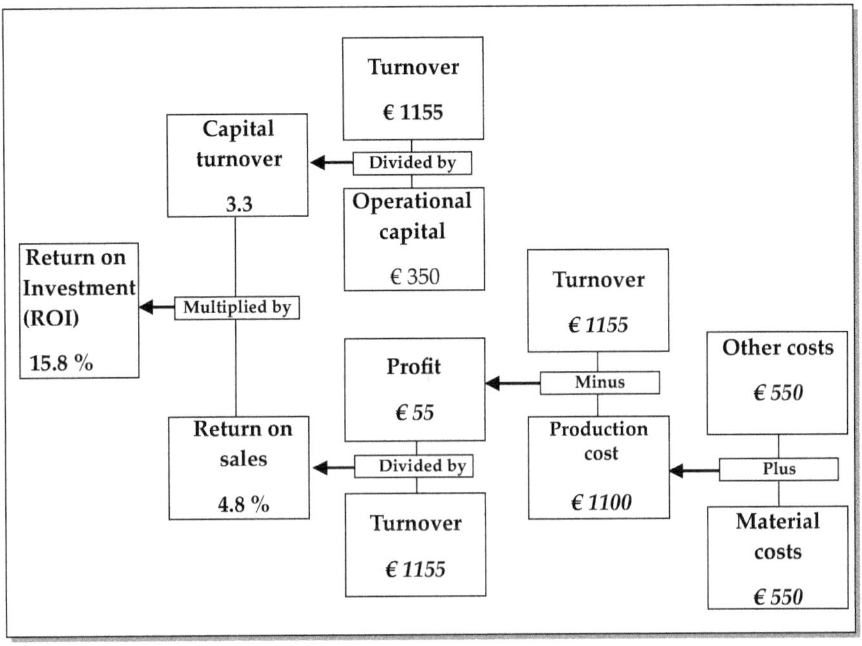

An increase in turnover by 10 % merely results in an increase in ROI by 1.4 percentage points. This shows that decreasing the logistics costs contributes more to the ROI than an increase in turnover by the same percentage does. Moreover, a decrease in logistics costs is usually more easily achieved within a company than an increase in turnover. Increases in turnover are primarily dependent on market conditions that are rather difficult to influence.

Bibliography

Aberle, G. (2009). *Transportwirtschaft: Einzelwirtschaftliche und gesamtwirtschaftliche Grundlagen*, München/Wien.

Aden, D. (2001). Neue Chancen für Logistik-Dienstleister durch die Globalisierung. In H. Baumgarten (Ed.). *Logistik im E-Zeitalter. Die Welt der globalen Logistiknetzwerke* (S. 89–98). Frankfurt.

Alicke, K. (2005). *Planung und Betrieb von Logistiknetzwerken: Unternehmensübergreifendes Supply Chain Management*. Berlin.

Arminger, G. (2004). Absatz- und Bestellprognosen im CPFR-Prozess für den Handel. In *Supply Chain Management* (S. 45–52), o. Jg., Nr. 3.

Arnold, D., Kuhn, A., Isermann, H., & Tempelmeier, H. (2008). *Handbuch Logistik*. Heidelberg: Springer.

Auffermann, C. (2007). Sinnvoller Schritt: Die Logistik für Vollsortimenter und Discounter unterscheidet sich grundsätzlich. In Log., o. Jg., Nr. 2, S. 24–27.

Baumgarten, H. (Ed.). (2001). *Logistik im E-Zeitalter. Die Welt der globalen Logistiknetzwerke*, Frankfurt.

Baumgarten, H., & Thoms, J. (2002). *Trends und Strategien in der Logistik. Supply Chains im Wandel*, Berlin.

Baumgarten, H., & Walter, S. (2000). *Trends und Strategien in der Logistik*. Berlin.

Baumgarten, H., et al. (Ed.). (2003). *Springer Experten System Logistik-Management, Band 1*. Berlin.

Bea, F., & Haas, J. (2009). *Strategisches Management* (5. Edition). Stuttgart.

Becker, H. P. (2008). *Investition und Finanzierung: Grundlagen der betrieblichen Finanzwirtschaft* (2. Updated Edition). Wiesbaden.

Becker, T. (2006). *Quo Vadis Zulieferpark? Neue Zusammenarbeitsmodelle in der automobilien Zulieferkette*. Dissertation, Universität Kassel, Hamburg.

Beckmann, K. (2007). *Logistik*. Rinteln.

Bender, W. F. (1997). In J. Bloech, & G. B. Ihde, (Ed.). *Vahlens Großes Logistik Lexikon* (S. 295–297). München.

Berg, C. C. (2003). Marketingkonzept für Güterverkehrszentren. In *Internationales Verkehrswesen 55* (S. 417–418). Jg., Heft 9.

Berliner Flughäfen. (2011). *Berlin Brandenburg Airport Willy Brandt – Imagebroschüre*. Berlin.

Berndt, T. (2001). *Eisenbahngüterverkehr*. Stuttgart u.a.

Berning, R. (2002). *Prozessmanagement und Logistik: Gestaltung der Wertschöpfung*. Berlin.

Bjelicic, B. (2005). *Unterlagen zur Vorlesung Verkehrsbetriebslehre I: Landverkehr und Schifffahrt*. Universität Mannheim.

Bjelicic, B., & Kostka, G. (2000). Moderne Finanzierungskonzepte für Logistikzentren. In R. Hossner (Ed.). *Jahrbuch der Logistik 2001* (S. 164–167). Mönchengladbach.

Blom, F., & Harlander, N. A. (2003). *Logistik-Management: Der Aufbau ganzheitlicher Logistikketten in Theorie und Praxis*. Renningen.

H. Gleissner and J.C. Femerling, *Logistics*, Springer Texts in Business and Economics,
DOI 10.1007/978-3-319-01769-3,

Bode, W., & Preuß, R. (2004). *Intralogistik in der Praxis: Komplettanbieter der Intralogistik.* Suhl.

Boeing (Ed.). (2006). *Product Review 2007 – Commercial Airplanes,- Freighters.* Chicago.

Börner-Kleindienst, M. (2006). Logistikimmobilien: Kategorien, Determinanten und Marktteilnehmer. In B. Falk, & T. Momme (Ed.). *Handbuch Gewerbe- und Spezialimmobilien* (S. 447–472). Köln.

Boysen, N. (2008). Über die Synchronisierung von Güterströmen in der Umschlaglogistik. In *Zeitschrift für Betriebswirtschaft 78* (S. 1285–1315). Jg. H. 12.

Boysen, N., & Tiedemann, S.-K. (2010). Design und Betrieb von Cross Docks: Planungsprobleme für einen effizienten Warenumschlag in lagerlosen Umschlagzentren. In WiSt, o. Jg. Heft 2, S. 54–60.

Brandes, D. (2003). Was weg ist, muss hin: Im Discount hat Einfachheit Vorrang. In H. Merkel, & B. Bjelicic (Ed.). *Logistik und Verkehrswirtschaft im Wandel: Unternehmensübergreifende Versorgungsnetzwerke verändern die Wirtschaft. Festschrift für Gösta B.* Ihde (S. 199–204), München.

Brecht, Gratkowski, v., Kuhlmann, & Spörker (1982). *Gewinnung und Transport von Erdöl und Erdgas, Verarbeitung von Erdöl (Raffinerietechnik).* München/Wien.

Bretzke, W.-R. (1999). Industrie- versus Handelslogistik: Der Kampf um die Systemführerschaft. In *Logistik Management 1* (S. 81–95). Jg. Nr. 2.

Bretzke, W.-R. (2007). Bindung an Logistikdienstleister – Chance oder Gefahr durch Abhängigkeit. In W. Stölzle, J. Weber, E. Hofmann, & C. M. Wallenburg (Ed.). *Handbuch Kontraktlogistik. Management komplexer Logistikdienstleistungen* (S. 167–179). Weinheim.

Brinkmann, B. (2005). *Seehäfen – Planung und Entwurf.* Berlin.

Buchholz, J., Clausen, U., & Vastag, A. (Ed.). (1998). *Handbuch der Verkehrslogistik.* Heidelberg.

BMVBS Bundesministerium für Verkehr-, Bau- und Stadtentwicklung. (Ed.). (2009). *Verkehr in Zahlen 2009/2010.* Hamburg.

BMVBS Bundesministerium für Verkehr-, Bau- und Stadtentwicklung. (Ed.). (2011). *Verkehr in Zahlen 2010/201.* Hamburg.

Caninenberg, W. (2004). Mit Software Software testen. In *Software in der Logistik* (S. 34–38). München.

Chopra, S., & Meindl, P. (2009). *Supply Chain Management.* Upper Saddle River.

Christopher, M., Lowson, B., & Peck, H. (2009). Fashion logistics and quick response. In von J. Fernie, & L. Sparks (Ed.). *Logistics and retail management* (3. Edition). London/ Philadelphia: Kogan Page.

Clausen, U., Hein, A.-P., & Deymann, S. (2007). Potenziale globaler Direktbelieferung für Textil-logistikketten aus China. In G. Aberle (Ed.). *Internationales Verkehrswesen, 59* (S. 25–29). Jg., Nr. 1+2.

Corsten, H. (Ed.). (1994). *Handbuch Produktionsmanagement.* Wiesbaden.

Corsten, D., & Gössinger, R. (2007). *Einführung in das Supply Chain Management.* München.

Dehler, M. (2001). *Entwicklungsstand der Logistik. Messung – Determinanten – Erfolgswirkungen.* Wiesbaden.

Delfmann, W. (1998). Hub-and-Spoke-Systeme. In P. Klaus, & W. Krieger (Ed.). *Gabler –Lexikon Logistik* (S. 177–178). Wiesbaden.

Delfmann, W., & Reihlen, M. (Ed.). (2003). *Controlling von Logistikprozessen.* Stuttgart.

Deutsche Bahn AG. (Ed.). (2008). *Geschäftsbericht 2008.* Berlin: Deutsche Bahn AG.

Deutsche Binnenreederei AG. (Ed.). (2007). *Transportlogistik Innovativ Europaweit (Geschäfts-broschüre).* Berlin.

Deutsche Post, AG. (Ed.). (2007). *Geschäftsbericht 2006.* Bonn.

Deutscher Speditions- und Logistikverband (DSLV). (Ed.). (2010). *Zahlen, Daten, Fakten aus Spedition und Logistik 2010.* Bonn 2011.

Deutsches Institut für Normung (DIN) – Normen in ihrer jeweils aktuellen Fassung.

Donier Consulting GmbH. (2006). *Erfolgsfaktoren und Bewertungsmöglichkeiten von PPP-Projekten für Verkehrsinfrastruktur: Eine internationale und verkehrsträgerübergreifende Analyse*. Berlin.

Eichler, B. (2003). *Beschaffungsmarketing und -logistik*. Berlin.

Eicke, H. v., & Femerling, J. C. (1991). *Modular Sourcing: Ein Konzept zur Neugestaltung der Beschaffungslogistik*. München.

Eisenhart-Rothe, F. von, & Jütte, S. (2003). Innovative Logistikfinanzierung rund um die supply chain. In e. V. Bundesvereinigung Logistik (BVL) (Ed.). *Finanzierung – eine neue Dimension der Logistik* (S. 151–170). Berlin.

Elsenbach, I. M. (2006). Tracking und Tracing – IT – Instrumente zum wertorientierten supply chain management. In: C. Mauch, & H. Wildemann (Ed.). *Handbuch IT-management* (1. Edition) (S. 437–463). München.

European Logistics Association (ELA). (Ed.). (2005). *Terminology in logistics – Terms and definitions*. Brüssel: European Logistics Association.

FedEx Corporation. (Ed.). (2006). *2006 Annual report*. Memphis.

Feinen, K. (2003). Leasing von Logistik-Immobilien. In Bundesvereinigung Logistik (Ed.). *Finanzierung – eine neue Dimension der Logistik* (S. 187–199). Berlin.

Feldkamp, W. (2001). Logistik goes "glocal" – Entwicklung und Trends aus der Sicht der DaimlerChrysler AG, Logistik PKW. In H.-C. Pfohl (Ed.). *Jahrhundert der Logistik: Customer Related-Glocal-E-Based* (S. 167–184). Berlin.

Femerling, J. C. (2003). B2C- und B2B-Logistik als Erfolgsfaktor der New Economy. In H. Merkel, & B. Bjelicic (Ed.). *Logistik und Verkehrswirtschaft im Wandel: Unternehmensübergreifende Versorgungsnetzwerke verändern die Wirtschaft. Festschrift für Gösta B.* Ihde (S. 199–204), München

Fiege, H., Nowak, H., & Mayer, C. (2007). Kontraktlogistik in der Hightech-Industrie. In W. Stölzle, J. Weber, E. Hofmann, & C. M. Wallenburg (Ed.). *Handbuch Kontraktlogistik. Management komplexer Logistikdienstleistungen* (S. 657–677). Weinheim.

Fischer, W., & Dittrich, L. (2003). *Materialfluss und Logistik. Optimierungspotenziale in Transport und Lagerwesen*. Berlin.

Fortmann, K.-M., & Kallweit, A. (2007). *Logistik*. (2. Edition). Stuttgart.

Franke, W., & Dangelmaier, W. (Ed.). (2006). *RFID – Leitfaden für die Logistik*. Wiesbaden.

Freichel, S. L. K. (2002). Distributionsmanagement – Gestaltung internationaler Logistiksysteme. In W. Stölzle, & K. Gareis (Ed.). *Integrative Management- und Logistikkonzepte* (S. 263–287). Wiesbaden.

Frese, E. (1987). *Grundlagen der Organisation. Die Organisationsstruktur der Unternehmung*. Wiesbaden.

Frese, E. (2005). *Grundlagen der Organisation. Entscheidungsorientiertes Konzept der Organisationsgestaltung* (9. Edition). Wiesbaden.

Friemut, U., Carl, H., Oster, M., & Seterhenn, J. (1997). Qualitätsregelkreise – Stellschraube für Dienstleister. In *Logistik heute, 1997* (S. 96–100). Nr. 10.

Frunzke, H., Boldt, O., & Garbisch, O. (2004). Management und Kundenbeziehungen in der Logistikdienstleistungsbranche. In H.-C. Pfohl *Netzkompetenz in Supply Chains. Grundlagen und Umsetzung* (S. 265–284). Wiesbaden.

Gabele, E., Dannenberg, J., & Kroll, M. (2001). *Immobilien-Leasing: Vertragsformen, Vor- und Nachteile, steuerliche Analyse* (4. Updated Edition). Wiesbaden.

Giesa, F., & Hagen, N. (2003). Logistiknetzwerke in der global verteilten Automobilproduktion. In H. Baumgarten, et al. (Ed.). *Springer Experten System Logistik-Management, Band 1* (S. 1–17). Berlin. Abschnitt 4/04/04.

Giesa, F., & Kopfer, H. (2000). Management logistischer Dienstleistungen der Kontraktlogistik. In *Logistik management* (S. 43–53). Nr. 1.

Gladen, W. (2011). *Performance measurement – Controlling mit Kennzahlen* (5. Edition). Wiesbaden.

Gleißner, H. (2000). *Logistikkooperationen zwischen Industrie und Handel*. Göttingen.

Gleißner, H. (2003). Supply chain controlling für die europaweite Endkundendistribution von Großgütern und Möbeln im Versandhandel. In W. Stölzle, & A. Otto (Ed.). *Supply chain controlling in Theorie und Praxis: Aktuelle Konzepte und Unternehmensbeispiele* (S. 197–214). Wiesbaden.

Gomm, M. (2008). *Supply Chain Finanzierung: Optimierung der Finanzflüsse in Wertschöpfungsketten*. Berlin.

Göpfert, I. (2005). *Logistik Führungskonzeption*. München.

Greschner, J. (1996). Lernfähigkeit von Unternehmen. In *Schriften zur Unternehmensplanung, Band 38*. Frankfurt a. M.

Grochla. (1983). *Erfolgsorientierte Materialwirtschaft durch Kennzahlen*. Baden-Baden.

Grunow, M. (2001). Management von Produktions- und Distributionsnetzwerken in der chemischen Industrie. In H.-J. Sebastian, & T. Grünert (Ed.). *Logistik Management, Supply Chain Management und E-Business* (S. 323–335). Stuttgart.

GS1 Germany GmbH. (Ed.). (o.J.) *Positionspapier zur Nutzung von EANCOM® und GS1XML für die elektronische Kommunikation*. Köln.

GS1 Germany GmbH. (Ed.). (2003). *Management-Informationsbroschüre RFID*. Köln.

GS1 Germany GmbH. (Ed.). (2006a). *EANCOM® – Der Motor für EDI*. Köln.

GS1 Germany GmbH. (Ed.). (2006b). *EAN Identsysteme in der Anwendung*. Köln.

GS1 Germany GmbH. (Ed.). (2009a). *Die NVE und das GS1-Transportetikett in der Anwendung*. Köln.

GS1 Germany GmbH. (Ed.). (2009b). *EDI/eCommerce Einführung in den elektronischen Datenaustausch*. Köln.

Gudehus, T. (2010). *Logistik: Grundlagen, Strategien, Anwendungen* (4. Edition). Berlin.

Günther, H. O., & Kim, K. H. (Ed.). (2005). *Container terminals and automated transport systems: Logistics control issues and quantitative decision support*. Berlin/Heidelberg: Springer.

Hahn, D., & Kaufmann, L. (2002). *Handbuch Industrielles Beschaffungsmanagement* (2. Edition). Wiesbaden.

Hardt, A., & Karsch, C. (2007). Kontraktlogistik in der Chemieindustrie. In W. Stölzle, J. Weber, E. Hofmann, & C. M. Wallenburg (Ed.). *Handbuch Kontraktlogistik. Management komplexer Logistikdienstleistungen* (S. 639–656). Weinheim.

Heiserich, O.-E., Helbig, K., & Ullmann, W. (2011). *Logistik: Eine paxisorientierte Einführung* (4. vollständig überarbeitete und erweiterte Edition). Wiesbaden.

Hertel, J., Zentes, J., & Schramm-Klein, H. (2005). *Supply-Chain-Management und Warenwirtschaftssysteme im Handel*. Berlin/Heidelberg/New York.

Herter, M. (2000). *Strategisches Management der Logistik: Konzeptioneller Bezugsrahmen und empirische Analyse in deutschen und amerikanischen Industrieunternehmen*. Berlin.

Hofmann, E. (2005). Supply chain finance: Some conceptual insights. In: R. Lasch, & C. G. Janker (Ed.). *Logistik-management – Innovative Logistikkonzepte* (S. 203–214). Wiesbaden.

Hofmann, E., & Elbert, R. (2004). Collaboratives Cash flow Management. Financial Supply Chain Management als Herausforderung der Netzkompetenz. In H.-C. Pfohl (Ed.). *Netzkompetenz in Supply Chains – Grundlagen und Umsetzung* (S. 94–117). Wiesbaden.

Horvath & Partner. (Ed.). (2007). *Balanced Scorecard umsetzen*. Stuttgart.

Horvath & Partners. (2009). *Der Weg zu einem wirkungsvollen Controllingsystem* (7. Edition). München.

Ihde, G. B. (2001). *Transport, Verkehr, Logistik: Gesamtwirtschaftliche Aspekte und einzelwirtschaftliche Handhabung* (3. Edition). München.

Ihme, J. (2006). *Logistik im Fahrzeugbau*. Wien/Köln.

Isermann, H. (1994). *Logistik. Beschaffung, Produktion, Distribution*. München.

Jahns, C. (2003). *Balanced Scorecard – Arbeitsbuch*. St. Gallen.

Jansen, R. (1989). Alle Forderungen Berücksichtigen: Kostenerfassung und Leistungsverrechnung in der Verpackungsplanung. In *Beschaffung aktuell* (S. 78–81).

Janz, O. (2003). *Integriertes Transportnetzmanagement: Angebots- und nachfrageorientierte Planung und Steuerung komplexer Transportnetze*. Köln.

Jirasek, J. (1977). *Das Unternehmen – ein kybernetisches System: Eine Einführung für die Wirtschaftspraxis*. Berlin.

Jünemann, R. (1989). *Materialfluß und Logistik: Systemtechnische Grundlagen mit Praxisbeispielen*. Berlin.

Jünemann, R., & Schmidt, T. (1999). *Materialflusssysteme: Systemtechnische Grundlagen* (2. Edition). Berlin.

Kaplan, R., & Norton, D. (1997). *Balanced Scorecard – Strategien erfolgreich umsetzen*. Stuttgart.

Karrer, M., & Petzold, D. (2004). Performance Management bei Logistikdienstleistern- Steuerung von Logistikprozessen mit der Balanced Scorecard. In C. Schneider (Ed.). *Controlling für Logistikdienstleister, Konzepte, Instrumente, Anwendungsbeispiele, trends* (S. 91–106). Hamburg.

Keller, A. (2011). *Breitbandkabel und Zugangsnetze: Technische Grundlagen und Standards* (2. völlig neu bearbeitete Edition). Heidelberg.

Kernler, H. (2007). *Logistiknetze: Mit supply chain management erfolgreich kooperieren*. Heidelberg.

Kiesel, J. (1997). Produktions-controlling. In A.-W. Scheer (Ed.). *Rechnungswesen und EDV*. Heidelberg.

Klaas, T. (2002). *Logistik-organisation. Ein konfigurationstheoretischer Ansatz zur logistikorientierten Organisationsgestaltung*. Wiesbaden.

Klaas, T. (2005). Jenseits des Organigramms: Grundsätzliche Überlegungen zur ganzheitlichen Gestaltung der supply chain. In H. Kopfer, & C. Bierwirth (Ed.). *Logistik management, 7* (S. 8–20). Jg., Nr. 3.

Klauenberg, J. (2009). *Modellierung und Typisierung von Distributionszentren in Seehäfen*. Dissertation, Universität Rostock, Hamburg.

Klaus, P. (2002). *Die Dritte Bedeutung der Logistik – Beiträge zur Evolution Logistischen Denkens*. Hamburg.

Klaus, P. (2007). Markt für Kontraktlogistik – Volumen und Entwicklungen in Europa. In W. Stölzle, J. Weber, E. Hofmann, & C. M. Wallenburg (Ed.). *Handbuch Kontraktlogistik. Management komplexer Logistikdienstleistungen* (S. 89–112). Weinheim.

Klaus, P., & Kille, C. (2008). *Die Top 100 der Logistik* (5. Edition). Hamburg.

Klepzig, J. (2010). *Working-Capitel und Cash Flow: Finanzströme durch Prozessmanagement optimieren* (2. Edition). Wiesbaden.

Kloth, H. (1999). *Neue Konzepte der Distributionslogistik von Handelsunternehmen*. Dissertation, Duisburg.

Koch, B. (2006). *Entwicklungsplanung für Flughafengesellschaften: Entscheidungsunterstützung durch die integrierte Marktsprognose- und Finanzmodellierung*. Köln.

Koepke, R., & Frahne, K. (2010). *Der Vermietungsmarkt für Lager-/Logistikflächen, Logistikimmobilien-Report JonesLangLaSalle*. Deutschland.

Kopfer, H., & Bierwirth, C. (Ed.). (1999). *Logistik management: Intelligente I + K Technologien*. Berlin/Heidelberg.

Korf, W. (Ed.). (2005a). *Lorenz – Leitfaden für Spediteure und Logistiker in Ausbildung und Beruf, Bd. 1* (20. Edition). Hamburg.

Korf, W. (Ed.). (2005b). *Lorenz 2 – Leitfaden für Spediteure und Logistiker in Ausbildung und Beruf, Bd. 2* (16. Edition). Hamburg.

Kotzab, H. (1997). *Neue Konzepte der Distributionslogistik von Handelsunternehmen*. Wien.

Kuhn, A., & Hellingrath, B. (2002). *Supply chain management- Optimierte Zusammenarbeit in der Wertschöpfungskette* (2. Edition). Berlin.

Kummer, S. (2010). *Einführung in die Verkehrswirtschaft*. Wien.

Küpper, H.-U. (2008). *Controlling – Konzeption, Aufgaben und Instrumente* (5. Edition). Stuttgart.

Lammers, S., & Neubauer, M. (2005). Integrale Logistik-Konzepte in Europa am Beispiel der Firma Philips Consumer Electronics GmbH. In C. Neiberger, & H. Bertram (Ed.). *Waren um die Welt bewegen: Strategien und Standorte im Management globaler Warenketten*. (S. 47–54). Mannheim.

Lange, U. (2000). *Supply chain management und Netzwerkmanagement aus der strategischen Sicht des Logistikdienstleisters am Beispiel eines multimodalen Vollsortimenters*. Düsseldorf.

Lasch, R., & Lemke, A. (2002). *Defizite der Logistiksysteme von E-Commerce-Anbietern – Lösungsansätze für ein besseres Logistikmanagement, Ergebnisse eine Studie unter online-Shops, Schlussfolgerungen und Empfehlungen*. Dresden.

Laurent, M. (1996). *Vertikale Kooperationen zwischen Industrie und Handel*. Frankfurt.

Lay, G. (2007). *Betreibermodelle für Investitionsgüter: Verbreitung, Chancen, Auswirkungen und Erfolgsfaktoren*. Karlsruhe.

Lieb, T. (1997). Seehafenhinterlandverkehr. In J. Bloech, & G. B. Ihde (Ed.). *Vahlens Großes Logistik Lexikon* (S. 931–932). München.

Lieb, T. C., & Lange, U. (2003). Strategien und Organisationsstrukturen global integrierter Logistikdienstleister. In H. Merkel, & B. Bjelicic (Ed.). *Logistik und Verkehrswirtschaft im Wandel: Unternehmensübergreifende Versorgungsnetzwerke verändern die Wirtschaft. Festschrift für Gösta B*. Ihde (S. 446–459), München.

Lufthansa Cargo AG. (Ed.). (2005). *Cargo handling manual*. Frankfurt.

Mahler, K. (2008). *Logistikimmobilien: Optimierung des Investment- und Logistikerfolges*. Köln.

Mangold, K. (2001). Finanzierungs- und Leasingaktivitäten industrienaher Dienstleister. In R.-E. Breuer (Ed.). *Handbuch Finanzierung* (3. Edition) (S. 227–248). Wiesbaden.

Marbacher, A. (2001). *Demand and supply chain management*. Bern et al. 2001

Maurer, P. (2003) *Luftverkehrsmanagement: Basiswissen* (3. überarb. und erweiterte Edition). München/Wien.

Meeder, S. (2003). Flughafenwettbewerb und Flughafenstrategien. In H. Merkel, & B. Bjelicic (Ed.). *Logistik und Verkehrswirtschaft im Wandel: Unternehmensübergreifende Versorgungsnetzwerke verändern die Wirtschaft, Festschrift für Gösta B*. Ihde (S. 487–505), München.

Mensen, H. (2007). *Planung, Anlage und Betrieb von Flughäfen*. Berlin.

Merkel, H., & Bjelicic, B. (Ed.). (2003). *Logistik und Verkehrswirtschaft im Wandel: Unternehmensübergreifende Versorgungsnetzwerke verändern die Wirtschaft. Festschrift für Gösta B*. Ihde, München.

Merkel, H., & Heymans, J. (2003). Geschäftsmodelle im stationären Handel. In H. Merkel, & B. Bjelicic (Ed.). *Logistik und Verkehrswirtschaft im Wandel: Unternehmensübergreifende Versorgungsnetzwerke verändern die Wirtschaft. Festschrift für Gösta B* (S. 161–183). München: Ihde

Metro MGL GmbH. (2002). (Ed.). *Ganzheitliche Handelslogistik-Innovation in der Supply Chain durch einen "Internal 4 PL"*. Düsseldorf.

Meyer, C. (2011). *Betriebswirtschaftliche Kennzahlen und Kennzahlensysteme* (6. Edition). München.

Ministerium für Wirtschaft – NRW Landesinitiative Logistik. (Ed.). (2006). *Mittelstand und Energie des Landes NRW*. Düsseldorf.

Morgenstern, O. (1955). Note of the formulation of the theory of logistics. In *Naval Research Logistics Quarterly, Vol. 2, Issue 3*. Princeton.

Müller, A. (2005). *Strategisches Management mit der Balanced Scorecard*. Stuttgart et al.

Müller, J. & Ermel, R. (2007). Tendermanagement für die Kontraktlogistik. In W. Stölzle, J. Weber, E. Hofmann, & C. M. Wallenburg (Ed.). *Handbuch Kontraktlogistik. Management komplexer Logistikdienstleistungen* (S. 293–308). Weinheim.

Müller-Daupert, B., Gomm, F., Becker, J., & Zwissler, T. (2009). *Logistik Outsorcing. Ausschreibung – Vergabe – Controlling* (2. Edition). München.

Muncke, G., Dziomba, M., & Walter, M. (2008). Standort- und Marktanalysen in der Immobilienwirtschaft. In K.-W. Schulte, & S. Bone-Winkel *Handbuch Immobilien-Projektentwicklung*. Köln.

Neher, A. (2005). Internationale Logistikstrategien von Industrieunternehmen im Wandel. In C. Neiberger, & H. Bertram (Ed.). *Waren um die Welt bewegen: Strategien und Standorte im Management globaler Warenketten* (S. 33–45). Mannheim.

Nehm, A., & Schryver, C. (2007). Alternativen der Logistikimmobilien-Bereitstellung. In B. Bohlmann, & T. Krupp *Strategisches Management für Logistikdienstleister: Grundlagen und Praxisberichte* (S. 230–254). Hamburg.

Nehm, A., Veres-Homm, U., & Kille, C. (2009). *Logistikimmobilien in Deutschland: Markt und Standorte, Fraunhofer-Arbeitsgruppe für Technologien der Logistik-Dienstleistungswirtschaft (ATL)*. Nürnberg.

Nothardt, F., Schmitter, F., & Trede, T. (2007). Logistik in der Bekleidungsindustrie – quo vadis? In W. Stölzle, J. Weber, E. Hofmann, & C. M. Wallenburg (Ed.). *Handbuch Kontraktlogistik. Management komplexer Logistikdienstleistungen*. (S. 679–700). Weinheim.

Nuhn, H. (2005). Internationalisierung von Seehäfen Vom Cityport und Gateway zum Interface globaler Transportketten. In C. Neiberger, & H. Bertram (Ed.). *Waren um die Welt bewegen: Strategien und Standorte im Management globaler Warenketten* (S. 109–124). Mannheim.

Nuhn, H., & Hesse, M. (2006). *Verkehrsgeographie*. Paderborn.

Ockens, K. (2003). Investieren in Logistik-Immobilien. In Bundesvereinigung Logistik (Ed.). *Finanzierung – eine neue Dimension der Logistik* (S. 201–215). Berlin.

Österreichische Wasserstraßen-Gesellschaft mbH. (Ed.). (2007). *Manual of Danube navigation*, Wien.

Peridon, L., Steiner, M., & Rathgeber, A. W. (2009). *Finanzwirtschaft der Unternehmung* (14. Edition). Wiesbaden.

Peters, S. (2003). Dokumenten- und Informationslogistik für Finanzdienstleister – Rhenus Office Systems realisiert ganzheitliche Lösungen aus einer Hand. In Bundesvereinigung Logistik (Ed.). *Banklogistik – Lösungsansätze für interne und externe Dienstleister* (S. 105–119). Berlin.

Pfaff, D., Skiera, B., & Weiss, J. (2004). *Financial supply chain management*. Bonn: Galileo Press.

Pfohl, H.-C. (1972). *Marketing-Logistik*. Mainz: Distribution-Verlag.

Pfohl, H.-C. (1990). *Logistiksysteme*. Berlin/Heidelberg/New York.

Pfohl, H.-C. (2000). *Logistiksysteme: betriebswirtschaftliche Grundlagen* (6. neubearbeite und Updated Edition). Berlin.

Pfohl, H.-C. (2004a). *Logistikmanagement* (2. Edition). Berlin.

Pfohl, H.-C. (2004b). *Netzkompetenz in Supply Chains. Grundlagen und Umsetzung*. Wiesbaden.

Pfohl, H.-C. (2010). *Logistiksysteme: Betriebswirtschaftliche Grundlagen* (8. Edition). Berlin.

Pfohl, H.-C., Hofmann, E., & Elbert, R. (2003). Bedeutung der finanziellen Supply Chain in der unternehmerischen Praxis. In Bundesvereinigung Logistik (BVL) e. V. (Ed.). *Finanzierung – eine neue Dimension der Logistik* (S. 65–105). Berlin.

Pintsch, S. (2004). Gesunde Logistiklösungen für den Healthcare-Markt. In H. Baumgarten, I.-L. Darkow, & H. Zadek (Ed.). *Supply Chain Steuerung und Services: Logistik-Dienstleister managen globale Netzwerke – Best Practices* (S. 251–259). Berlin.

Piontek, J. (1994). *Internationale Logistik*. Stuttgart.

Pompl, W., Schuckert, M., & Möller, C. (2006). *Luftverkehr: Eine ökonomische und politische Einführung*. Berlin.

Porter, M. E. (1985). *Competitive advantage*. New York: Free Press.

Preißner, A. (2011). *Balanced Scorecard anwenden Kennzahlengestütze Unternehmenssteuerung*. München.

Railion Deutschland AG. (Ed.). (2007). *Güterwagenkatalog (CD-ROM)*. Mainz.

Reichmann, T. (2006). *Controlling mit Kennzahlen und Management-Tools – Die systemgestützte Controlling-Konzeption* (7. Edition). München.

Rödler, E., Rödler, R., & Müller, S. (2003). *Balanced Scorecard und MIS – Leitfaden zur Implementierung*, Bonn.

Rühl, M. (2008). Logistik in Westeuropa. In I. Göpfert, & D. Braun (Ed.). *Internationale Logistik in und zwischen unterschiedlichen Weltregionen* (S. 25–66). Wiesbaden.

Rümenapp, T. (2002). *Strategische Konfigurationen von Logistikunternehmen, Ansätze zur konsistenten Ausrichtung in den Dimensionen Strategie, Struktur und Umwelt*. Wiesbaden.

Sarbach, P. (2006). *Das Konzept des Financial Supply Chain Management: Entwicklungsstand, Auswirkungen und Nutzen bei Schweizer Grossunternehmen*. Bern: Arbeitsbericht Nr. 175 des Instituts für Wirtschaftsinformatik der Universität Bern.

Scheer, A.-W. (Ed.). (2008). *Rechnungswesen und EDV*. Heidelberg.

Scheer, A.-W., & Borowski, R. (1999). Supply Chain Management: Die Antwort auf neue Logistikanforderungen. In H. Kopfer, & Chr. Bierwirth (Ed.). *Logistik Management: Intelligente I + K Technologien* (S. 3–14). Berlin/Heidelberg et al.

Schenk, M., Richter, K., Jumar, U., & Herrmann, A. (2011). Galileo-Testfeld Sachsen-Anhalt – Anwen dungsschwerpunkte Verkehr und Logistik. In Th. Wimmer, & H. Wöhner (Ed.). *Intelligent wachsen, Kongressband 27. Deutscher Logistik-Kongress* (S. 565–600). Hamburg.

Schmid, H. (2001). Die schnelle Fabrik, Planung und Realisierung einer Fabrik für Computersysteme und deren Einbindung in eine logistisches Netzwerk. In H.-C. Pfohl (Ed.). *Jahrhundert der Logistik: Customer-Related-Glocal-E-Based* (S. 141–165). Berlin.

Schneider, C. (Ed.). (2004). *Controlling für Logistikdienstleister, Konzepte, Instrumente, Anwendungsbeispiele, Trends*. Hamburg.

Schorb, U., Halsband, E., & Anders, F. (2007). Kontraktlogistik in der Automobilindustrie. In W. Stölzle, J. Weber, E. Hofmann, & C. M. Wallenburg (Ed.). *Handbuch Kontraktlogistik. Management komplexer Logistikdienstleistungen* (S. 619–637). Weinheim.

Schubert, W. (Ed.). (2000). *Verkehrslogistik: Technik und Wirtschaft*. München.

Schulte, G. (1996). *Material- und Logistikmanagement* (1. Edition). München.

Schulte, G. (2001). *Material- und Logistikmanagement* (2. Edition). München.

Schulte, C. (2009). *Logistik: Wege zur Optimierung des Material- und Informationsflusses*. München.

Schütte, R., Vering, O., & Wiese, J. (2000). *Erfolgreiche Geschäftsprozesse durch standardisierte Warenwirtschaftssysteme*. Berlin/Heidelberg.

Sebastian H.-J., & Grünert, T. (Ed.). (2001). *Logistik Management, Supply Chain Management und E-Business*. Stuttgart.

Seitz, R. (2004). *Entwicklung eines Referenzmodells zur Erstellung eines Bewertungssystems für die Planung von Rechenzentren*. Diss, Universität Leipzig.

Sommer, K. (2003). Fahrzeugdistribution in der Automobilindustrie. In H. Merkel, & B. Bjelicic (Ed.). *Logistik und Verkehrswirtschaft im Wandel: Unternehmensübergreifende Versorgungsnetzwerke verändern die Wirtschaft* (S. 239–246). München: Festschrift für G. B. Ihde.

Spengler, T., Voß, S., & Kopfer, H. (Ed.). (2004). *Logistik Management: Prozesse, Systeme, Ausbildung*. Heidelberg.

Stadtler, H., & Kilger, Ch. (2002). *Supply chain management and advanced planning*. Berlin: Springer.

Stenzel, J. (2003). Finanzierung als Dienstleistungskomponente. In Bundesvereinigung Logistik (Ed.). *Finanzierung – eine neue Dimension der Logistik* (S. 139–150). Berlin.

Stieglitz, A. (1999). *Die Reorganisation handelslogistischer Versorgungsketten*. München.

Stölzle, W. (1999a). *Industrial relationships*. München.

Stölzle, W. (1999b). Dimensionen der Gestaltung des Organisationsproblems. In J. Weber, & H. Baumgarten (Ed.). *Handbuch Logistik: Management von Material- und Warenflußprozessen* (S. 571–589). Stuttgart.

Stölzle, W., & Gareis, K. (2002). Konzepte der Beschaffungslogistik – Anforderungen und Gestaltungsalternativen. In D. Hahn/L. Kaufmann (Eds.), *Handbuch Industrielles Beschaffungsmanagement* (S. 402–423). Wiesbaden.

Stölzle, W., & Otto, A. (Ed.). (2003). *Supply Chain Controlling in Theorie und Praxis: Aktuelle Konzepte und Unternehmensbeispiele*. Wiesbaden.

Stölzle, W., Heusler, K. F., & Karrer, M. (2004). *Erfolgsfaktor Bestandsmanagement: Konzept, Anwendungen, Perspektiven*. Duisburg.

Stölzle, W., Weber, J., Wallenburg, C. M., & Hofmann, E. (2007). *Handbuch Kontraktlogistik: Management komplexer Logistikdienstleistungen* (S. 35–54). Weinheim.

Straube, F. (2004). *E-Logistics – Ganzheitliches Logistikmanagement*. Berlin.

Straube, F., Dangelmaier, W., Günthner, W. A., & Pfohl, H.-C. (2005). In Bundesvereinigung Logistik e. V. (Ed.), *BVL Trends und Strategien in der Logistik – Ein Blick auf die Agenda des Logistik-Managements 2010*. Hamburg.

Sydow, J., & Möllering, G. (2004). *Produktion in Netzwerken: Make, buy & cooperate*. München: Vahlen.

ten Hompel, M., & Schmidt, T. (2007). *Warehouse management*. Berlin: Springer.

ten Hompel, M., et al. (2007). *Software in der Logistik*. München.

TNT N. V. (Ed.). (2007). *2006 Annual report*. Amsterdam.

Tripp, C. (2003). *Chancen und Risiken mittelständischer System-Stückgutkooperationen in Deutschland. Eine Untersuchung der Fraunhofer ATL Nürnberg im Auftrag der bayrischen Hypo-Vereinsbank AG München*. Nürnberg.

Tripp, C. (2004). *Kooperationen und Kontraktlogistik*. Nürnberg.

United Parcel Service, Inc. (Ed.). (2007). *UPS annual report 2006*. Atlanta.

Vahrenkamp, R. (2005). *Logistik: Management und Strategien* (5. fully updated and extended edition). München.

Vahrenkamp, R. (2007). *Logistik: Management und Strategien*. München.

Vahrenkamp, R. (2008). *Produktionsmanagement*. München/Wien.

Verein Deutscher Ingenieure (VDI) e. V. – VDI-code with respect to the up-to-date version.

Vis, I. F. A., & Harika, I. (2005). Comparison of vehicle types at an automated container terminal. In H.-O. Günther, & K. H. Kim (Ed.). *Container terminals and automated transport systems: Logistics control issues and quantitative decision support*. (S. 38–63). Berlin/Heidelberg.

Weber, J. (2001). *Logistikcontrolling*. Stuttgart: Springer.

Weber, J. (2002a). *Logistikkostenrechnung – Kosten-, Leistungs- und Erlösinformation zur erfolgsorientierten Steuerung der Logistik*. Heidelberg.

Weber, J. (2002b). *Logistik- und Supply Chain Controlling* (5. Edition). Stuttgart.

Weber, J., & Schäffer, U. (2000). *Balance Scorecard & Controlling, Implementierung – Nutzen für Manager und Controller – Erfahrungen in deutschen Unternehmen*. Wiesbaden.

Weber, J., & Schäffer, U. (2011). *Einführung in das Controlling* (13.Edition). Stuttgart.

Weid, H. (1995). Wettbewerbsvorteile durch Elektronic Data Interchange (EDI). In H. Baumgarten, & G. B. Ihde (Ed.). *Schriftenreihe der Bundesvereinigung Logistik*. Bremen.

Wildemann, H. (2002). *Distributionslogistik*. München.

Wildemann, H. (2007). Entscheidungsprozesse beim Outsourcing komplexer logistischer Aufgaben. In W. Stölzle, J. Weber, E. Hofmann, & C. M. Wallenburg (Ed.). *Handbuch Kontraktlogistik. Management komplexer Logistikdienstleistungen* (S. 133–149). Weinheim.

Wildemann, H. (2010). *Logistik Prozeßmanagement*. München.

Witten, P., & Karies, R. (2003). Logistik im Distanzhandel. In H. Merkel, & B. Bjelicic (Ed.). *Logistik und Verkehrswirtschaft im Wandel: Unternehmensübergreifende Versorgungsnetzwerke verändern die Wirtschaft* (S. 185–197). München: Festschrift für Gösta B. Ihde.

Wolters, H. (1995). *Modul- und Systembeschaffung in der Automobilindustrie – Gestaltung der Kooperation zwischen europäischen Hersteller- und Zulieferunternehmen*. Wiesbaden.

Zentes, J. (2004). Marketing-Effektivität vs. Logistik-Effizienz: Theoretische Überlegungen und empirische Befunde. In T. Spengler, S. Voß, & H. Kopfer (Ed.). *Logistik Management: Prozesse, Systeme, Ausbildung* (S. 255–270). Heidelberg.

Zimmermann, B. (2004). *Kontraktlogistik als Zukunftsmarkt der Logistikdienstleistungswirtschaft? Mittelstandskongruenz und Entwicklung eines mittelstandsgerechten Vertriebsmodells*. Nürnberg.

Index

H. Gleissner and J.C. Femerling, *Logistics*, Springer Texts in Business and Economics, DOI 10.1007/978-3-319-01769-3,

CPSIA information can be obtained
at www.ICGtesting.com
Printed in the USA
LVHW081002130422
716098LV00003B/31

9 783319 017686